ANTIOXIDANTS IN SYSTEMS OF VARYING COMPLEXITY

Chemical, Biochemical, and Biological Aspects

ANTIOXIDANTS IN SYSTEMS OF VARYING COMPLEXITY

Chemical, Biochemical, and Biological Aspects

Edited by
Lyudmila N. Shishkina, DSc
Alexander N. Goloshchapov, PhD
Larissa I. Weisfeld, PhD

Reviewers and Advisory Board Members
Gennady E. Zaikov, DsC
Anatoly I. Opalko, PhD

APPLE
ACADEMIC
PRESS

Apple Academic Press Inc.	Apple Academic Press Inc.
3333 Mistwell Crescent	1265 Goldenrod Circle NE
Oakville, ON L6L 0A2	Palm Bay, Florida 32905
Canada	USA

Library and Archives Canada Cataloguing in Publication

Title: Antioxidants in systems of varying complexity : chemical, biochemical, and biological aspects / edited by Lyudmila N. Shishkina, DSc, Alexander N. Goloshchapov, PhD, Larissa I. Weisfeld, PhD.

Names: Shishkina, Lyudmila N., editor. | Goloshchapov, Alexander N., editor. | Weisfeld, Larissa I., editor.

Description: Includes bibliographical references and index.

Identifiers: Canadiana (print) 20190143185 | Canadiana (ebook) 20190143274 | ISBN 9781771888509 (hardcover) | ISBN 9780429325168 (ebook)

Subjects: LCSH: Antioxidants.

Classification: LCC TP159.A5 A58 2019 | DDC 547/.23—dc23

CIP data on file with US Library of Congress

Apple Academic Press also publishes its books in a variety of electronic formats. Some content that appears in print may not be available in electronic format. For information about Apple Academic Press products, visit our website at **www.appleacademicpress.com** and the CRC Press website at **www.crcpress.com**

About the Editors

Lyudmila N. Shishkina, DSc

Professor, Biophysics, and Head, Department of Physicochemical Problems in Radiobiology and Ecology, Emanuel Institute of Biochemical Physics, Russian Academy of Sciences, Moscow, Russia

Lyudmila N. Shishkina, DSc in Chemistry, is Professor in Biophysics and Head of the Department of Physicochemical Problems in Radiobiology and Ecology at the Emanuel Institute of Biochemical Physics of the Russian Academy of Sciences, Moscow, Russia, and a member of the Scientific Counsel of the Russian Academy of Sciences on Radiobiology. She is a laureate of the medal "In Memory of Academician N. M. Emanuel" for Achievements in Chemical and Biochemical Physics. In the last five years, she has given nine invited lectures and 17 oral presentations at international conferences. She has published 225 research papers in several leading Russian and international journals as well as several textbooks and monographs. She is a chapter author and co-editor of the book *Chemistry and Technology of Plant Substances: Chemical and Biochemical Aspects*, published by Apple Academic Press.

Alexander N. Goloshchapov, PhD

Deputy Director, Emanuel Institute Biochemical Chemistry; Head of the Laboratory of Physical and Chemical Bases of Biological Systems, Russian Academy of Sciences, Moscow, Russia

Alexander N. Goloshchapov, PhD in Biology, is Deputy Director of the Emanuel Institute of Biochemical Chemistry and Head of the Laboratory of Physical and Chemical Bases of Biological Systems of the Russian Academy of Sciences, Moscow, Russia. He is a specialist in the fields of physical and chemical problems of cells and biological membranes. He is a member of the International EPR Society. As part of a creative team, he takes part in Russian and international programs and has participated in the creation of the patent "Preventative Preparation, Contributing to Increase Life Expectancy." He participated in Russian and international scientific conferences and has more than 150 scientific publications to his credit. His scientific interests are connected to biochemical physics, chemical kinetics, natural and synthetic antioxidants, and neuro-degenerative diseases of man. He is the reviewer and advisory board member with Apple Academic Press for the book *Chemistry and Technology of Plant Substances: Chemical and Biochemical Aspects.*

Larissa I. Weisfeld, PhD

Chief Specialist, Emanuel Institute of Biochemical Physics, Russian Academy of Sciences, Moscow, Russia

Larissa I. Weisfeld, PhD, is a Chief Specialist at the Emanuel Institute of Biochemical Physics of the Russian Academy of Sciences, Moscow, Russia; and a member of the All-Russia Vavilov Society of Geneticists and Breeders. She is also the author more then 300 articles published in scientific and conference proceedings and holds seven patents for inventions. She is the co-author of four new winter wheat cultivars registered in the State Register of the Russian Federation. Her main field of interest concerns basic problems of chemical mutagenesis, mutational selection, and the mechanism of action of p-aminobenzoic acid. She has worked as a scientific editor in the publishing house Nauka ("Sciences" in Russian) (Moscow) and with the journals *Genetics* and *Ontogenesis.* She is the author of several book and has co-edited several books with Apple Academic Press, including *Ecological Consequences of Increasing Crop Productivity: Plant Breeding and Biotic Diversity; Biological Systems, Biodiversity, and Stability of Plant Communities; Temperate Crop Science and Breeding: Ecological and Genetic Study; Heavy Metals and Other Pollutants in the Environment: Biological Aspects; Chemistry and Technology of Plant Substances: Chemical and Biochemical Aspects;* and *Temperate Horticulture for Sustainable Development and Environment: Ecological Aspects.*

Contents

Contributors

Yuliya V. Afanasyeva
All-Russia Horticultural Institute of Vegetable of Breeding, Agrotechnology, and Nursery,
Laboratory of Field Cultures, Junior Researcher, 4, Zagoryevskaya St., Moscow 115598, Russia,
Tel.: +79269360238, E-mail: yuliya_afanaseva_90@bk.ru

Olga M. Alekseeva
PhD, Emanuel Institute of Biochemical Physics of Russian Academy of Sciences,
Department of Physical and Chemical Basis of Regulation of Biological Systems, Senior Scientist,
4, Kosygin St., Moscow 119334, Russia, Tel.: +79163690530, E-mail: olgavek@yandex.ru

Sarra A. Bekuzarova
DSc Professor, Gorsky State Agrarian University, Chair of Plant Cultivation, Honored the Inventor of
the Russian Federation, Kirov St., 37, Vladikavkaz, Republic of North Ossetia–Alania 362040, Russia,
Tel.: +79618259796, E-mail: bekos37@mail.ru

Nina A. Bome
DSc Professor, Tyumen State University, Institute of Biology, Head of the Department of Botany,
Biotechnology and Landscape Architecture, Volodarsky St., 6, Tyumen 625003, Russia,
Tel.: +79129236177, E-mail: bomena@mail.ru

Elena B. Burlakova
DSc Professor, Emanuel Institute of Biochemical Physics of Russian Academy of Sciences,
Head of Laboratory of Physicochemical Basis of Regulation of Biological Systems, Kosygin St.,
4, Moscow, 119334, Russia

Irina I. Chesnokova
PhD, The A.O. Kovalevsky Institute of Marine Biological Research of Russian Academy of Sciences,
Laboratory of Ecotoxicology, Researcher, Nahimov Av., 2, 299011, Sevastopol, Russian Federation,
Tel.: +79788307440, E-mail: mtk.fam@mail.ru

Olga V. Domanskaya
Tyumen State University, International Institute of Cryology and Cryosophy, Junior Researcher,
Volodarsky St., 6, Tyumen 602003, Russia, Tel.: +79068757661, E-mail: olga-nv@bk.ru

Vladimir O. Domanskii
Tyumen Industrial University, Junior Researcher, Volodarsky St., 38, Tyumen 625000, Russia,
Tel.: +79292002988, E-mail: vdomanskiy@gmail.com

Nina S. Domnina
PhD, Saint Petersburg State University, Institute of Chemistry, Department of Macromolecular
Compound Chemistry, Associate Professor, Saint Petersburg State University,
26 University Avenue, Petergof 198504, Russia, Tel.: +78124286840, +79215718492,
E-mails: ninadomnina@mail.ru, n.domnina@spbu.ru

Alexander E. Dontsov
DSc, Emanuel Institute of Biochemical Physics of Russian Academy of Sciences,
Department of Physico-Chemical Base of Perception, Leading Scientist, 4, Kosygin St.,
Moscow 119334, Russia, Tel.: +74959397422, E-mail: adontsovnick@yahoo.com

Tugan A. Dulaev
Gorsky State Agrarian University, Post-Graduate Student, Plant Growing Department,
Republic of North Ossetia-Alania, 37 Kirov St., Vladikavkaz 362040, Russia,
Tel.: +79888347434, E-mail: tugand@mail.ru

Tatyana V. Gavruseva
PhD, The A.O. Kovalevsky Institute of Marine Biological Research of Russian Academy of Sciences,
Laboratory of Ecotoxicology, Senior Scientist, Nahimov Av., 2, 299011, Sevastopol,
Russian Federation, Tel.: +79247910461, E-mail: gavrt2004@mail.ru

Alexander N. Goloshchapov
PhD, Emanuel Institute of Biochemical Physics of Russian Academy of Sciences, Laboratory of
Physicochemical Basis of Regulation of Biological Systems, Head of Laboratory, Deputy Director,
4, Kosygin St., Moscow 119334, Russia, Tel.: +74959380561, E-mail: golan@sky.chph.ras.ru

Natalia E. Ionova
PhD, Assistant Professor, Kazan Federal University, Department of Plant Physiology, 18,
Kremlyovskaya St., Kazan 420008, Russia, Tel.: +79047634718, E-mail: alekta-meg@list.ru

Natalya V. Khrustova
PhD, Emanuel Institute of Biochemical Physics of Russian Academy of Sciences,
Laboratory of Physicochemical Problems in Radiobiology and Ecology, Researcher, 4,
Kosygin St., Moscow 119334, Russia, Tel.: +74959397838, E-mail: khrnata@gmail.com

Yury A. Kim
DSc, Professor, Institute of Cell Biophysics of Russian Academy of Sciences,
Laboratory of Cell Cultures and Cell Engineering, Leading Scientist, 3, Institutskaya St.,
Pushchino, Moscow Region 142292, Russia, E-mail: yuk01@rambler.ru

Natalia N. Kolokolova
PhD, Associate Professor, Tyumen State University, Institute of Biology, Department of Botany,
Biotechnology and Landscape Architecture, Senior Lecturer, Volodarsky St.,
6, Tyumen 625003, Russia, Tel.: +791292339944, E-mail: campanella2004@mail.ru

Nona L. Komissarova
PhD, Emanuel Institute of Biochemical Physics of Russian Academy of Sciences,
Department Chemistry of Antioxidants, Senior Scientist, Kosygin St., 4, Moscow 119334, Russia,
Tel.: +74959397353, E-mail: komissarova@polymer.chph.ras.ru

Tatyana B. Kovyrshina
The A.O. Kovalevsky Institute of Marine Biological Research of Russian Academy of Sciences,
Laboratory of Ecotoxicology, Researcher, Nahimov Av., 2, 299011, Sevastopol, Russian Federation,
Tel.: +79788307440, E-mail: mtk.fam@mail.ru

Mikhail V. Kozlov
PhD, Emanuel Institute of Biochemical Physics of Russian Academy of Sciences, Researcher, 4,
Kosygin St., Moscow 119334, Russia, Tel.: +74959397186, E-mail: wer-swamp@yandex.ru

Anna V. Krementsova
PhD, Emanuel Institute of Biochemical Physics of Russian Academy of Sciences,
Laboratory of Physic-Chemical Base of Biological Systems Regulation, Senior Scientist,
4, Kosygin St., Moscow 119334, Russia, Tel.: +79169107689, E-mail: akrementsova@mail.ru

Anastasiya V. Malkova
PhD, Emanuel Institute of Biochemical Physics of Russian Academy of Sciences,
Department Chemistry of Antioxidants, Junior Researcher, Kosygin St., 4, Moscow 119334, Russia

Lidiya I. Mazaletskaya
PhD, Emanuel Institute of Biochemical Physics of Russian Academy of Sciences, Moscow,
Laboratory of Photo and Chemiluminescent Processes, Leading Scientist, 4, Kosygin St.,
Moscow 119334, Russia, Tel.: +74959397171, +74959397385, E-mail: lim@sky.chph.ras.ru

Svetlana M. Motyleva
PhD, All-Russia Horticultural Institute of Vegetable of Breeding, Agrotechnology, and Nursery,
Leading Scientist, Head of the Laboratory of Physiology and Biochemistry, 4, Zagoryevskaya St.,
Moscow 115598, Russia, Tel.: +79102052710, E-mail: motyleva_svetlana@mail.ru

Mikhail A. Ostrovsky
DSc, Emanuel Institute of Biochemical Physics of Russian Academy of Sciences,
Head of Department Physico-Chemical Base of Perception, Kosygin Str., 4, Moscow 119334,
Russia, Tel.: +74991357073, E-mail: ostrovsky3535@mail.ru

Margarita N. Ovsyannikova
PhD, Emanuel Institute of Biochemical Physics of Russian Academy of Sciences,
Department Chemistry of Antioxidants, Senior Scientist, Kosygin St., 4, Moscow 119334, Russia

Larisa S. Pogodina
PhD, Biological Department of Lomonosov Moscow State University, Senior Scientist,
Lenin Hills, 1, Page 12, Moscow 119234, Russia, Tel.: +74959394567, E-mail: lpogod@mail.ru

Natalia V. Polyakova
Tyumen State University, International Institute of Cryology and Cryosophy, Assistant Researcher,
Volodarsky St., 6, Tyumen 602003, Russia, Tel.: +79129256873, E-mail: aleksus.2010@mail.ru

Dmitry A. Postnikov
DSc, Professor, Russian State Agrarian University, Moscow Timiryazev Agricultural Academy,
Professor of the Department of Ecology, Timiryazevskaya St., 49, Moscow 127550, Russia,
Tel.: +79629404545, E-mail: dpostnikov@rambler.ru

Irina I. Rudneva
DSc, Professor, the A.O. Kovalevsky Institute of Marine Biological Research of Russian Academy of
Sciences, Leading Scientist, Head of the Laboratory Ecotoxicology, Nahimov Av. 2, 299011,
Sevastopol, Russian Federation, Tel.: +79787491704, E-mail: svg-41@mail.ru

Liudmila V. Semenyak
DSc, VNIRO Russian Federal Institute of Fisheries and Oceanography, Laboratory of
Consolidated Forecast, Leading Scientist, 17, Verhnyaya Krasnoselskaya St., Moscow 107140, Russia,
Tel.: +79853922486, E-mail: lvsemen@yandex.ru

Natalia B. Sereznikova
Emanuel Institute of Biochemical Physics of Russian Academy of Sciences, Junior Researcher,
Kosygin Str., 4, Moscow 119334, Russia | Biological Department of Lomonosov
Moscow State University, Educational Master, Lenin Hills, 1, Page 12, Moscow 119234, Russia,
Tel.: +74959394567, E-mail: natalia.serj@yandex.ru

Valentin G. Shaida
Researcher, The A.O. Kovalevsky Institute of Marine Biological Research of Russian Academy of
Sciences, Researcher of the Laboratory Ecotoxicology, Nahimov Av., 2, 299011, Sevastopol,
Russian Federation, Tel.: +79787491708, E-mail: svg-41@mail.ru

Nataliya I. Sheludchenko
Emanuel Institute of Biochemical Physics of Russian Academy of Sciences,
Laboratory of Physicochemical Problems in Radiobiology and Ecology, Researcher,
4, Kosygin St., Moscow 119334, Russia, Tel.: +74959397385, E-mail: nish48.mail.ru

Lyudmila N. Shishkina

DSc, Emanuel Institute of Biochemical Physics of Russian Academy of Sciences,
Head of Department Laboratory of Physicochemical Problems in Radiobiology and Ecology,
4, Kosygin St., Moscow 119334, Russia, Tel.: +74991374101, Fax: +74959397186,
E-mail: shishkina@sky.chph.ras.ru

Elena V. Shtamm

DSc, Emanuel Institute of Biochemical Physics, Russian Academy of Science, 4, Kosygin St.,
Moscow 119334, Russia; VNIRO Russian Federal Research Institute of Fisheries and Oceanography,
Laboratory of Physico-Chemical Problems of Radiobiology and Ecology, Leading Scientist, 17,
Utter Krasnoselskaya St., Moscow 107140, Russia, Tel.: +74991476090, E-mail: lvsemen@yandex.ru

Viacheslav O. Shvydkiy

PhD, Emanuel Institute of Biochemical Physics, Russian Academy of Science,
Laboratory of Physico-Chemical Problems of Radiobiology and Ecology, Senior Scientist,
4, Kosygin St., Moscow 119334, Russia, Tel.:. +79161177244, E-mail: slavuta58@gmail.com

Yurii I. Skurlatov

DSc, Professor, Semenov Institute of Chemical Physics of Russian Academy of Sciences,
Laboratory of Chemical Safety Problems, Head of Department, 4–1, Kosygin St., Moscow 119334,
Russia, E-mail: yskurlatov@gmail.com

Suluchan K. Temirbekova

DSc, All-Russian Research Institute of Phytopathology, Head of the Laboratory for Selection for
Resistance to Abiotic and Biotic Stress Factors, Leading Scientist, Odintsovo Regom,
Bolshie Vyazemy, 5, Institute St., Moscow Region 143050, Russia, Tel.: +79162249618,
E-mail: sul20@yandex.ru

Julia A. Treshchenkova

PhD, Emanuel Institute of Biochemical Physics of Russian Academy of Sciences,
Laboratory of Physicochemical Basis of Regulation of Biological Systems, Senior Scientist,
Kosygin St., 4, Moscow, 119334, Russia, Tel.: +74959397390, E-mail: tresch@sky.chph.ras.ru

Violetta B. Voleva

PhD, Emanuel Institute of Biochemical Physics of Russian Academy of Sciences,
Head of Department Chemistry of Antioxidants, Kosygin St., 4, Moscow 119334, Russia,
Tel.: +74959397286, E-mail: komissarova@polymer.chph.ras.ru

Larissa I. Weisfeld

PhD, Chief Specialist, Emanuel Institute of Biochemical Physics of Russian Academy of Sciences,
4, Kosygin St., Moscow 119334, Russia, Tel.: +79162278685, E-mail: liv11@yandex.ru

Elena A. Yagolnik

PhD, Tulsky State University, Department of Biology, Associate Professor, 92, Prospect Lenin,
Tula 300012, Russia, Tel.: +79202726083, E-mail: yea_88@mail.ru

Pavel P. Zak

DSc, Professor, Emanuel Institute of Biochemical Physics of Russian Academy of Sciences,
Department of Physico-Chemical Base of Perception, Leading Scientist, Kosygin Str., 4,
Moscow 119334, Russia, Tel.: +79169494893, E-mail: pavelzak@mail.ru

Irina V. Zhigacheva

DSc, Emanuel Institute of Biochemical Physics of Russian Academy of Sciences,
Laboratory of Physical and Chemical Basis of Regulation of Biological Systems, Leading Scientist,
4, Kosygin St., Moscow, 119334, Russia, Tel.: +74959397409, E-mail: zhigacheva@mail.ru

Abbreviations

A	Angstrom (angstrom)
AAI	acute alcohol intoxication
ABA	antibacterial activity
ADP	adenine dinucleotide phosphate
Ag	Argentum
AHH	acute hypobaric hypoxia
AO	antioxidants
AOA	antioxidant activity
APA	antiperoxide activity
ArH	aromatic hydrocarbons
ATP	adenosine triphosphate
BAS	biologically active substances
BLAST	basic local alignment search tool
Bp	basal processes
BSA	bovine serum albumin
CAT	catalase
Cd	cadmium
CGS	centimeter-gram-second
CHR	choroid
CL	chemiluminescence
cm	centimeter
CO	carbon monoxide
CO_2	carbon dioxide molecule
CoQ	coenzyme Q
Cu	copper
D	dextran
DASM	differential adiabatic scanning microcalorimeter
DBI	double bond index
DMPC	dimyristoylphosphatidylcholine
DMSO	dimethyl sulfoxide
DNA	deoxyribonucleic acid
DOPA	dihydroxyphenylalanine
DPPH	2,2-diphenyl-1-picrylhydrazyl
$DPPH-SO_3Na$	2,2-diphenyl-1-picrylhydrazyl sulfonic acid

DSC	differential scanning calorimetry
DZN	diazinon
E	endothelium
EDTA	ethylenediaminetetraacetic acid
EPR	electron paramagnetic resonance
ER	endoplasmic reticulum
ERA	environmental risk assessment
ESR	electron spin resonance
ETS 123	European Convention for the Protection of Vertebrate Animals Used for Experimental and other Scientific Purposes
F	fenestra
FA	fatty acids
FA	fenozan-acid
FCR	Folin–Ciocalteu reagent
FDP	fructose-1,6-diphosphate
FeS	iron sulfide
FIA	fluid-injection apparatus
FIT	fluorescein isothiocyanate
g	gram
GOST	State Standards of the Russian Federation
GPx	glutathione peroxidase
GR	glutathione reductase
GSH	glutathione
H	hydrogen atom
H^+	hydrogen ion
H_2O	water molecule
H_2O_2	hydrogen peroxide
Hb	hemoglobin
HEPES	4-(2-hydroxyethyl)-1-piperazineethanesulfonic acid
HES	hydroxyethyl starch
Hg	mercury
HMAO	hybrid macromolecular antioxidants
$HMAO_D$	hybrid macromolecular antioxidants on the base of Dextran
$HMAO_{HES}$	hybrid macromolecular antioxidants on the base of hydroxyethyl starch
$HMAO_{PEG}$	hybrid macromolecular antioxidants on the base of polyethylene glycol
$HMAO_{PVA}$	hybrid macromolecular antioxidants on the base of polyvinyl alcohol
$HO\cdot$	hydroxyl radical

HP	hindered phenols
HP+SD	high pressure and shear deformation
HPLC	high performance liquid chromatography
HS⁻	hydrosulfide ion
IAA	indole-3-acetic acid
IBCP RAS	Institute of Biochemical Physics of Russian Academy of Sciences
IPM	interphotoreceptor matrix
K	Kelvin
K	potassium
KCL	potassium chloride
kDa	kilodalton
kg/ha	kilogram per hectare
l	litre
LCST	low critical solubility temperature
LDG	lactate dehydrogenase
LED	light emitting diodes
LG	lipofuscin granules
LH	lipids
LPO	lipid peroxidation
LWS	long wavelength sensitive
lx	Lux
M (or mol)	mole
M	mitochondria
M	molecular mass
MAAs	mycosporine-like amino acids
MB	myeloid body
MG	melanin granules
mg/L	milligram per liter
$MgCl_2$	magnesium chloride
ml	milliliter
mm Hg	millimeter of mercury
mm/sec	millimeters per second
MMAs	mycosporine-like amino acids
MRS-agar	Man, Rogosa, and Sharpe
MTs	metallothioneins
MWS	middle wavelength sensitive
Na_2S	sodium sulfide
NaCl	sodium chloride
NAD	nicotinamide adenine dinucleotide

NADH	dihydronicotinamide adenine dinucleotide
$NADP^+$	nicotinamide adenine dinucleotide phosphate
NADPH	dihydro nicotine amide adenine dinucleotide phosphate
NCBI	National Center for Biotechnology Information
nm	nanometer
$nMoO_3$	molybdenum oxide (VI)
NMR	nuclear magnetic resonance
NO•	nitric oxide
OD	oil droplets
P	phosphorus
PABA	*p*-aminobenzoic acid
PC	phosphatidylcholine
PCR	polymerase chain reaction
PE	phosphatidylethanolamine
PEG	polietilenglicol
PGRs	plant growth regulators
Ph	phagosome
PHEN	potassium phenosan
PL	phospholipids
POS	phagocytosis of photoreceptor outer segments
PUFA	polyunsaturated fatty acids
PVA	polyvinyl alcohol
Q	quercetin
RAS	Russian Academy of Russia of Sciences
RCIM	Russian Collection of Industrial Microorganisms
RCR	respiratory control rates
RNA	ribonucleic acid
ROS	reactive oxygen species
RPE	retinal pigment epithelium
RSA	radical-scavenging activity
RSH	reduced sulfur compounds
RSR	respiratory control rate
SD	shear deformations
SM	sphingomyelin
SO_2	sulfur oxide
SO_4^{2-}	sulfate ion
SOD	superoxide dismutase
SOPR	standard operating procedures of the researcher
SQ	semiquinone
SWS	short wavelength sensitive

TBA	thiobarbituric acid
TBARS	thiobarbituric acid reactive substances
t-Bu	*tert*-butyl
TLC	thin-layer chromatography
TMAO	trimethylamine oxide
TMRE	tetramethylrhodamine ethyl ester
TO•	phenoxyl radical
TP	α-tocopherol
Tris	tris(hydroxymethyl)aminomethane $(HOCH_2)$ $3CNH_2$
TUNEL	terminal deoxynucleotidyl transferase dUTP nick end labeling
ULD	ultra-low doses
UV	ultraviolet radiation
WD	water deficiency
WSFG	water-soluble fraction of gasoline
X-ray	X-radiation (Roentgen radiation)
Zn	zinc

Symbols

%PL	share of phospholipids in the total lipid composition
$(AP^*)^{2-}$	aminophthalate anion
λ_{max}	wavelength maximum
$(SQ)_2Cu$	bis-semiquinolate
$(SQ)_3Cr$	tris-semiquinolate
$(SQ)_3M$	tris-semiquinolate of the metal
$[AO]_j$	concentration of the antioxidant in the reaction system
$\sum EOPL$	sum of the more easily oxidizable fractions of phospholipids
$\sum POPL$	sum of the more poorly oxidizable fractions of phospholipids
μl	microliter
$^1O_2$	singlet oxygen
2,4-DTBP	2,4-di-*tert*-butylphenol
2,6-DTBP	2,6-di-*tert*-butylphenol
3,6-Cat	3,6-di-*tert*-butyl catechol
3,6-Q	3,6-di-*tert*-butyl-ortho-benzoquinone
3,6-QH$^+$	protonated quinone
3,6-SQ	3,6-semiquinone
A2E	pyridinium bis-retinoid
ADC	digital format by a converter
ANPH	sodium anphen
AP^{2-}	aminophthalate anion
arb.un.	arbitrary unit
B	basal protrusions of plasmalemma
band 4,1 and 4,2	proteins of bands 4,1 and 4,2
BF	4-bromo-(or acetoxy-)methyl-2,6-di-*tert*-butylphenol
BI	basal infoldings of plasmalemma
Br	Bruch's membrane
C	cones mitochondria
CBA/C57Bl	hybrid mice
CH_2O	carbohydrate
CH_4	methane
cM	changed forms of mitochondria (ring-shaped and dumbbell-like)

Da	unified atomic mass unit or Dalton
Fe^{2+}	divalent ion of iron
ha	hectare in metric system
HMAO-I	first series of HMAO
$h\nu$	photon, quantum of light
InH	inhibitor of the radical reactions, antioxidant
J/cm^2	joule/square centimeter
J/K	heat capacity
K/min	grades of Kelvin at minute
k_7	the rate constant of the interaction of antioxidants with peroxyl radicals
kDa $[H_2O_2]$	rate of hydrogen peroxide decomposition by catalase mechanism
KH_2PO_4	potassium dihydrogen phosphate
KK	4-hydroxy-3,5-di-*tert*-butylcinnamic acid
Km	Michaelis constant
Kp	constant of interaction with superoxide radicals
LOOH	hydroperoxides of lipids
$M \pm m$	average arithmetic mean $\pm$ its root-mean-square error
$M \times sec^{-1}$	the initiation rate of the autooxidation reaction (moles per second)
M^+	cation of metal
M^+/M^{2+}	variable valency metal in the reduced (M^+) and oxidized (M^{2+}) form
M^{2+}	divalent cation of metal
M^{3+}	trivalent cation of metal
mol	unit for measuring the amount of a substance in the International System Units (SI)
MT-Me	MT with metal ions
MTs-Me	complexes of MT with metal ions
O_2	molecular oxygen (dioxygen)
O_2^-	superoxide radical
$O_2^{\cdot-}$ or O_2^-	superoxide anion radical
°C	grades of Celsius
pH	index of acid reaction of the aquatic environment
Phenosan	β-(4-hydroxy-3,5-di-*tert*-butylphenyl) propionic acid
pK_a	negative logarithm of the acid-base equilibrium constant
ROH	organic molecule containing an alcohol group OH
$S°$	elemental sulfur

$S_2O_3^=$	sulfite ion
$S_n^=$	polysulfide containing in sulfur atoms
t	ton (1000 kilograms)
T	transepithelial channel
t/ha	ton per hectare
Tmax (°C)	temperature of maximum of phase transition
units mg^{-1}	amount of enzyme associated with 1 mg of protein
V_3	rates of oxidation of substrates in the presence of ADP
V_4	rates of oxidation in a state of rest (rates of oxidation of the substrate with the exhaustion of ADP)
W/cm^2	watt/centimeter^2
W_0	the initiation rate of radicals
W_{cat}	rate of the catalase decomposition of H_2O_2
W_{per}	rate of decomposition of hydrogen peroxide
α	dilution factor of the sample when it is mixed with the reaction system
α-TF	α-tocopherol
α-tocopherol	vitamin E
β-carotene	beta-carotene
β-spectrin	cytoskeleton protein
γ	substitution degree of the polymer chain by HP fragments
ΔAOA	difference of values of the antioxidant activity between the experimental and control samples
ΔCp	change of relative heat capacity
ε_{DPPH}	molar extinction coefficient of DPPH
λ	wavelength
λ_{em}	wavelength of emission
λ_{ex}	wavelength of excitation
μm	Micrometer
τ	the induction period
τ_{RH}	the induction period of the thermal autooxidation of methyl oleate
e^-	electron
OH	hydroxyl radical
$C_6H_5CH(OOH)CH_3$	ethylbenzene α-hydroperoxide

Foreword

Antioxidants reduce the damaging effects of free radicals on cell structures. Free radicals, having unpaired electrons, are very unstable and actively interact with other molecules. In a living organism, free radicals are an integral part of normal metabolism. At the same time, a number of external factors can promote the additional generation of free radicals, such as smoking, environmental pollution, radiation, drugs, pesticides, industrial solvents, and ozone. These circumstances, in fact, determine the multidirectional nature of the action of antioxidants on living systems. In other words, under different conditions, the same antioxidant can have effects of the opposite direction or have no such effects. Thus, studies of the fine mechanisms of action for different classes of antioxidants on living systems seem important and relevant. This book contains chapters devoted to such a detailed and comprehensive study of the effects of synthetic and natural antioxidants. A number of chapters in the book are devoted to the description of the effects of specific antioxidants and specimens of natural antioxidants. Below are specific examples.

The first chapter of the book (*V.B. Vol'eva, N. L. Komissarova, M. N. Ovsyannikova, N. S. Domnina, and A. V. Malkova*) introduces us to experiments on the chemistry of antioxidants such as hindered phenols (HP) carried out at the N. M. Emanuel Institute of Biochemical Physics of Russian Academy of Sciences. The chapter focuses on hybrid antioxidants combining in one molecule a redox-active phenolic core and a structural element (charged onium anchor, lipophilic long chain alkyl radical-float, hydrophilic biological and synthetic polymers) that provides biocompatibility, transport properties, the ability to interact with cellular structures, increased antioxidant activity. Particular attention in this chapter is given to the unique example of a diatomic HP with symmetrically shielded hydroxyl groups, the 3,6-di-*tert*-butyl catechol (3,6-Cat) and its redox conjugated derivatives, semiquinone, and *o*-quinone.

The second chapter (*L.N. Shishkina, M. V. Kozlov, L. I. Mazaletskaya, N. V. Khrustova, and N. I. Sheludchenko*) describes the application of the KINS computer program to analyze the kinetics of oxidation of plant cell components in standard model systems. This allowed the authors to identify the contribution of the main compounds to the inhibitory efficiency of plant components, depending on the stage of the oxidation process.

In the fifth chapter (*O. M. Alekseeva, A. V. Krementsova, A. N. Goloch-shapov, and Y. A. Kim*), the features of the antioxidant-melafen are investigated. Melafen is a plant growth regulator. It is shown that its effects are associated with structural changes in membrane proteins, actin, and erythrocyte spectrin. These effects are especially pronounced in the induced hemolysis of red blood cells. The same biochemical model was used for the study of a synthetic antioxidant in Chapter 3–the potassium salt of phenosan (*O. M. Alekseeva, Yu. A. Kim, E. A. Yagolnik*).

Antioxidants as adaptogens and plant growth regulators are considered in Chapter 4 (*I. V. Zhigacheva*). Individual effects of the synthetic antioxidant potassium salt of phenosan, the exposure of gamma radiation at a dose of 15 cGy and their combined effect on the activity of aldolase and lactate dehydrogenase in the cytoplasm, microsomes and synaptosomes of the mice brain at different duration after the cessation of action, as well as the structural state of the membranes and the composition of the mice, brain lipids have been studied in Chapter 9 (*J.A. Treshenkova, A. N. Goloshchapov, L. N. Shishkina, and E. B. Burlakova*).

In Chapter 13 (*S. A. Bekuzarova, L. I. Weisfeld, and T. A. Dulaev*), a method of using nitrogen-fixing cultures by treating seeds with a solution of sugar syrup and *p*-aminobenzoic acid (PABA) is proposed. PABA is a physiologically active natural compound, refers to group B vitamins, has antioxidant properties, and is necessary for the growth of many microorganisms, in particular, bacteria.

The study of the physicochemical parameters of the antioxidant activity of the leaves of the oil-bearing safflower crop, carried out in Chapter 11 (*S. K. Temirbekova, Y. V. Afanasyeva, D. A. Postnikov, S. M. Motyleva, and N. E. Ionova*), showed significant differences in their values, depending on the phase of plant growth, the climate conditions and the region of cultivation.

In addition to studying the mechanisms of action of individual antioxidants, this book also presents a chapter on the study of enzymes of oxidative stress. Chapter 12 (*O. V. Domanskaya, N. A. Bome, N. N. Kolokolova, V. O. Domanskii, N. V. Polyakova*) deals with the study of the relationship between the temperature of cultivation of bacteria of the genus Bacillus isolated from the Arctic permafrost and the activity of enzymes of oxidative stress. Catalysis shows the increase in catalase activity at low positive temperatures, which indicates the participation of antioxidant protection enzymes in adaptation to cold stress conditions.

I would like to draw attention to the work (Chapter 10: *I. I. Rudneva, I. I. Chesnokova, T. B. Kovyrshina, T. V. Gavruseva, and V. G. Shaida*) on a comparative study of the diversity of low molecular weight antioxidants in aquatic organisms and the reaction of the antioxidant status of various species to environmental changes. The authors have established a link between the global and local harmful factors of the marine environment and the role of antioxidants.

Interesting data are presented in Chapter 6 (*E. V. Stamm, V. O. Shvydkiy, Y. I. Skurlatov, and L. V. Semenyak*) about the role of the disbalance in the oxidation-reduction processes for the formation of the toxic properties of the aquatic environment.

And of course, special attention should be given to a review (*A. E. Dontsov and M. A. Ostrovsky*) of own and literary data on the structure, biochemistry, and function of the screening pigments of the eyes of vertebrates and invertebrates and humans–melanosomes and ommochromes. The authors of Chapter 7 made an important conclusion that these screening pigments have a pronounced antioxidant activity.

It is undoubtedly of interest in terms of a model system for assessing the potential of antioxidant action and a review article (Chapter 8: *P. P. Zak, A. E. Dontsov, L. S. Pogodina, N. B. Sereznikova, M. A. Ostrovsky*) on the morphofunctional study of retinal pigment epithelium and choroid on Japanese quail as accelerated aging model.

In conclusion, I want to note that reading this book will provide a lot of interesting scientific information on the diversity of antioxidant effects on living systems.

—Ilya N. Kurochkin, DSc
Professor
Director of Emanuel Institute of Biochemical Physics of
Russian Academy of Sciences

Introduction

Oxidation of organic substances by molecular oxygen is a chain process involving active free radicals (peroxide and alkoxyl ones), as well as reactive oxygen species (superoxide anion radical, hydroxyl radical, singlet oxygen). Oxidation processes often lead to negative consequences both in chemical technologies, where hydrocarbons, alcohols, and acids undergo transformations, and in the case of biological processes during the transformation of lipids, proteins, and nucleic acids. Substances inhibiting oxidation are called antioxidants. The mechanism for their action lies in breaking the reaction circuits. Antioxidant molecules interact with active radicals to form inactive products. Oxidation is also slowed down in the presence of the substances that destroy hydroperoxides. In this case, the rate of formation of free radicals decreases.

The theory of oxidative chain processes and the chemistry of antioxidants were originally developed for chemical processes: in the gas phase–in the early XX century, later for liquid-phase and solid-phase processes. The history of science owes academician N.M. Emanuel, who applied the methods of chemical kinetics to the study of the inhibitory effect of antioxidants in pathological conditions of the organism, aging, oncology, and radiation-induced lesions, thus making the revolutionary breakthrough of the antioxidant doctrine in biology. As a result, a new direction–biochemical physics–was included in the science and the largest scientific school of physical and chemical biology was created. Having become one of the keystones of life sciences, the antioxidant doctrine is currently celebrating another triumph. N.M. Emanuel's pupils are actively searching and researching the mechanism for the action of new antioxidants both in Russia and abroad. The proposed collective monograph introduces the reader to the latest achievements in the study of chemistry and biological activity of antioxidants gained by scientific groups of well-known chemists and biophysicists.

The monograph presents a large array of research findings of new synthetic antioxidants. The search for effective antioxidants among the substituted spatially hindered phenols has shown that not only the antioxidant activity, but also the biocompatibility and interaction of phenols with cellular structures changes depending on the substituents in the molecule, their nature, and physical and chemical properties. Screening of new compounds has made it possible to identify promising hybrid structures–potential bioantioxidants.

Another set of studies concerns a synthetic antioxidant–potassium salt of phenosan. The effect of this antioxidant on the enzymatic status and structural state of the mice brain membranes is considered. The biochemical model reflecting the most important details of the antioxidant action of phenosan potassium salt has been created.

A significant part of the monograph is devoted to the results of the study of natural antioxidants and the effect of antioxidants on plants. The action of antioxidants as adaptogens and plant growth regulators is shown. A computerized method for the analysis of the oxidation kinetics of plant cell components, which allows one to identify the key regularities of the antioxidant activity of plant components, has been developed. The antioxidant effect of safflower leaves has been found to depend on the phase of plant growth. Detailed consideration is given to the antioxidant properties of *para*-aminobenzoic acid.

The greatest interest is traditionally associated with the effect of antioxidants on animals and humans. The monograph presents the results of studying the effect of natural antioxidant melafen on the structural characteristics of membrane proteins in red blood cells. Also shown is an increase in the catalytic activity of bacterial cells at low temperatures and subsequent action of antioxidant protective enzymes in the adaptive mechanisms in cold stress conditions.

A comparative study of the effect of antioxidants on aquatic organisms and the response of the antioxidant activity of various organisms to the changes in the environment has been carried out. Thus, a fundamental relationship between the influence of adverse environmental factors and the role of antioxidant protection is established. The importance of the antioxidant status of organisms in the formation of toxic properties in the aquatic environment is shown.

The original results of research into the action of antioxidants in the process of photoreception are presented. The model systems of the antioxidant action based on the retinal pigment epithelium have been developed. Data on the most important inhibitory properties of visual pigments and their role in protection against the changes associated with aging are presented.

To sum up, we can say with confidence that today the biochemical physics of antioxidants is steadily developing and constantly adding breakthrough results to the data bank, thus confirming its status of one of the main life sciences. The monograph eloquently demonstrates the importance of the studies related to natural antioxidants, the need for synthetic research in the field of new compounds, the development of practical recommendations to prevent pathological conditions of the human body and improve the quality of life.

—**Anton V. Lobanov**
DSc of Chemical Sciences, N. N. Semenov Institute of Chemical Physics,
Russian Academy of Sciences,
Professor of Moscow State Pedagogical University.

PART I

Synthesis and Physicochemical Properties of Antioxidants in the Chemical and Biochemical Model Systems

CHAPTER 1

Structure Modification of Bioantioxidants Based on Hindered Phenols

VIOLETTA B. VOLEVA,[1] NONA L. KOMISSAROVA,[1]
MARGARITA N. OVSYANNIKOVA,[1] NINA S. DOMNINA,[2] and
ANASTASIYA V. MALKOVA[1]

[1]*Emanuel Institute of Biochemical Physics of Russian Academy of Sciences, Kosygin St., 4, Moscow 119334, Russia,*
E-mail: komissarova@polymer.chph.ras.ru

[2]*Institute of Chemistry of Saint Petersburg State University, 26 University Avenue, Petergof 198504, Russia*

ABSTRACT

This work is a brief review of research conducted at the Emanuel Institute of Biochemical Physics of Russian Academy of Sciences (IBCP) reflecting modern trends in the chemistry of hindered phenols (HP).

The greatest attention is paid to hybrid antioxidants based on HP combining in one molecule variously functional fragments: a redox-active phenolic core and a structural element that provides biocompatibility, transport properties, the ability to interact with cellular structures, increased antioxidant activity, etc.

Hybrid antioxidants are represented by two classes of antioxidants: "float" compounds containing a charged onium anchor and lipophilic long chain alkyl radical-float, as well as hybrids of HP with hydrophilic biological and synthetic polymers (dextrans, hydroxyethyl starch (HES), polyvinyl alcohol (PVA), and oligomeric polyethylene glycols). A comprehensive study was conducted, including the development of synthesis methods and the evaluation of antioxidant activity in model reactions with free radicals, diphenylpicrylhydrazyl, and its water-soluble sulfonate salt derivative. Water shell of the macromolecule plays an important role in the activity of hybrid antioxidants. This activity depends on the location of the redox active phenolic core inside or outside the shell determined by the length of the spacer separating the phenolic core from the polymer chain. Hybrid antioxidants with a different number of HP fragments grafted to the polymer were synthesized and studied in organic, water-organic, and aqueous media.

A special section is devoted to the unique example of a diatomic HP with symmetrically shielded hydroxyl groups, the 3,6-di-*tert*-butyl catechol (3,6-Cat) and its redox conjugated derivatives, semiquinone, and *o*-quinone.

This triad demonstrates the properties of a universal antioxidant capable of interacting with various active oxygen forms, including singlet oxygen, as well as the unusual reactivity associated with the ease of electronic transfer in the triad, chelating effect of neighboring hydroxyl/carbonyl groups, and sensitivity to the action of pressure and shear deformation.

The triad of 3,6-Cat – semiquinone – *o*-quinone also demonstrates a cascade mechanism of the antioxidant activity, in which the total activity is provided not only by the original compound, but also by the products of its conversions *in situ*. Examples of such a mechanism were also found in the study of antibacterial activity (ABA) of HP. A correlation between antibacterial and antiradical activity was revealed. A special part of the review deals with examples of syntheses based on HP related to the regioprotective function of *tert*-butyl groups.

1.1   INTRODUCTION

Hindered phenols (HP), being antioxidants of a wide spectrum of action, are used in solving applied problems and fundamental research. The most practical significance in the recent years has been acquired by β-(4-hydroxy-3,5-di-*tert*-butylphenyl) propionic acid (phenosan-acid, FA) and its methyl ester (methylox, phenosan-1) obtained by alkylation of 2,6-di-*tert*-butylphenol (2,6-DTBP) with methyl acrylate. A kinetic study of this process carried out at the IBCP [1–4] with the development of a new catalytic system made it possible to achieve a quantitative yield of the product with a high degree of purity. Phenoaan-1 gives rise to two rows of antioxidants, for technical and for medico-biological purposes (Figure 1.1). Antioxidants for technical purposes, inhibitors of thermooxidative degradation of various materials, were synthesized by transesterification with various alcohols (ethylene glycol, diethylene glycol, thiodiethylene glycol, pentaerythrol, etc.).

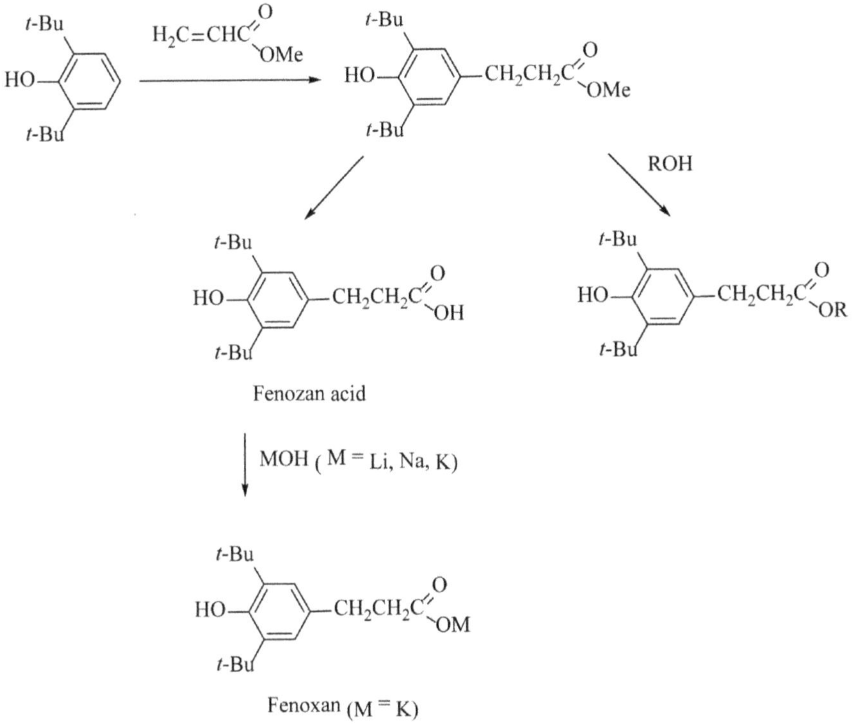

FIGURE 1.1   Synthesis of phenosan-1 and antioxidants based on it.

FA and its alkali salts are used as bioantioxidants. In particular, they are convenient objects for the study of the ultra-low dose's effect [5, 6]. However, the modern approach to creating bioantioxidants requires more specialization, the combination of antioxidant properties with the ability to targeted delivery and structural interactions with the protected area of biosystems. This is achieved in the hybrid molecules, fragments of which provide the necessary polyfunctionality.

1.2   HYBRID MACROMOLECULAR ANTIOXIDANTS (HMAO)

An example of this type of hybrids is the "float" HP, containing in their composition a charged onium group (anchor) with a lipophilic long-chain alkyl substituent (float). A set of float HP was synthesized by the quaternization of the choline ester of FA by alkyl halides with different chain length (Figure 1.2).

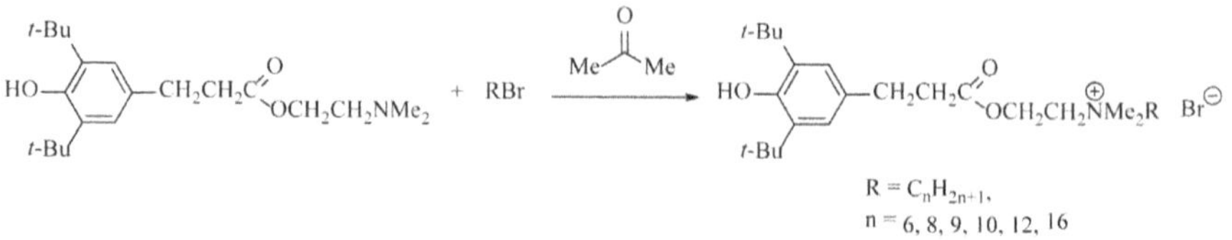

$R = C_nH_{2n+1}$,
$n = 6, 8, 9, 10, 12, 16$

FIGURE 1.2 Quaternization of the choline ester of phenosan acid by alkyl halides.

The process is carried out in acetone, where the resulting ammonium salts have low solubility and precipitate during the reaction shifting the equilibrium towards the product.

Such a structure of the antioxidant allows interacting with a charged lipid bilayer of cell membranes to maintain the normal level of lipid peroxidation (LPO). Float antioxidants were extensively studied at IBCP in various biological models [7].

In recent years, we created and investigated a special class of hybrid macromolecular antioxidants (HMAO) based on hydrophilic biological and synthetic polymers of medical purpose with chemically bounded HP fragments [8]. This approach to the design of biologically active substances (BAS) can be called biomimetic. Many molecules of natural compounds are hybrids of low and high molecular weight substances with different types of bonds – covalent, ionic, coordinative.

The attachment of active fragments with certain chemical or biological properties to the carrier polymers provided a promising way in the development of new medical compounds [9]. This approach provides considerable advantages: proper solubility, increased stability of a bioactive component, reduced toxicity, and durable effect. Biopolymers (dextran, starch, chitosan, etc.) and water-soluble synthetic polymers (polyvinylpyrrolidone, polyacrylic acid, polyvinyl alcohol (PVA), polyethylene glycol (PEG), etc.) were usually used as a polymer base.

A comprehensive study of HMAO is carried out including the development of synthetic methods, structural identification, physicochemical characteristics of HMAO and their dependence on the nature of the environment, testing on biological models, and determining the most promising ways of practical use. The possibility of such a study is provided by a variety of structural modifications of HMAO: nature and molecular mass (M) of the base polymer, a number of grafted fragments of HP, type of HP-polymer covalent bond, nature, and length of the spacer inserted between the HP core and the polymer chain. Dextran (D) with M 6,000, 10,000, 18,000, 40,000, 70,000, 200,000, hydroxyethyl starch (HES), and PVA were chosen for the synthesis of HMAO as a macromolecular base. These polymers are widely used in biology and medicine. Some of them are the basis of blood plasma volume expanders (Figure 1.3).

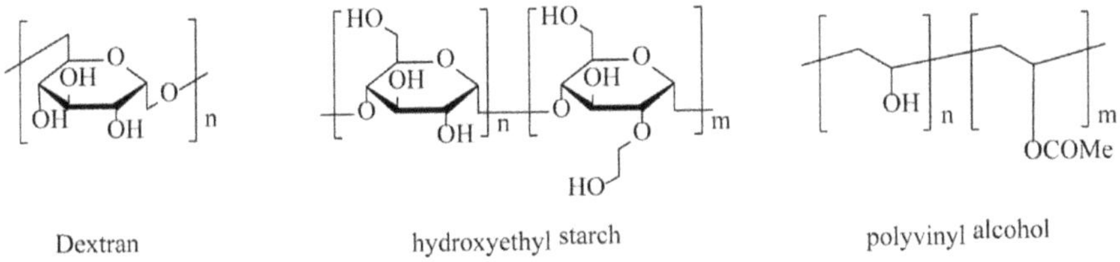

Dextran   hydroxyethyl starch   polyvinyl alcohol

FIGURE 1.3 Hydrophilic polymers used in the synthesis of hybrid macromolecular antioxidants.

Functionalized derivatives of HP capable of covalent bonds formation with hydroxyl groups of polymers were used for the chemical modification of polymers. These are β-(4-hydroxy-3,5-di-*tert*-butylphenyl)propionic acid (phenosan acid, FA), 4-bromo-(or0 acetoxy-)methyl-2,6-di-*tert*-butylphenol (BF), 4-hydroxy-3,5-di-*tert*-butylcinnamic acid (KK), 4-(β-tosyloxy) ethoxymethyl-2,6-di-*tert*-butylphenol (Ts-2F), 4-(β-tosyloxy)butoxymethyl-2,6-di-*tert*-butylphenol (Ts-4F).

Several HMAO series with different structural parameters were synthesized. The first series of HMAO (HMAO-I) is represented by hybrid compounds with ester linkages between polymer and HP. This series was obtained by the condensation of the polymer with FA used as condensing agent dicyclohexylcarbodiimide and catalyst dimethyl-aminopyridine. We received some hybrids based on D (HMAO$_D$-I), HES (HMAO$_{HES}$-I), and PVA (HMAO$_{PVA}$-I) with different substitution degree of the polymer chain by HP fragments (γ, mol%) by varying the ratios of reagents and their concentrations, and the esterification time (Figure 1.4). Dextran with different M was used for the synthesis of HMAO$_D$-I. The substitution degree of the polymer by HP fragments was determined by UV spectrophotometer (aromatic chromophore 270 nm) and NMR that gave similar results.

The solubility of HMAO in water depends on the nature of polymer and γ value. The maximum substitution degree corresponding to a complete dissolution of HMAO in water (border of water solubility) was 10 mol%. With further increase in substitution degree, HMAO started to dissolve in water-organic mixtures (water-ethanol, water-dioxane), and at maximum substitution degree (dextran with M 40000, γ=47.7 mol%), polymeric product started to dissolve in dioxane, chloroform, and benzene. All HMAO dissolved very well in DMSO.

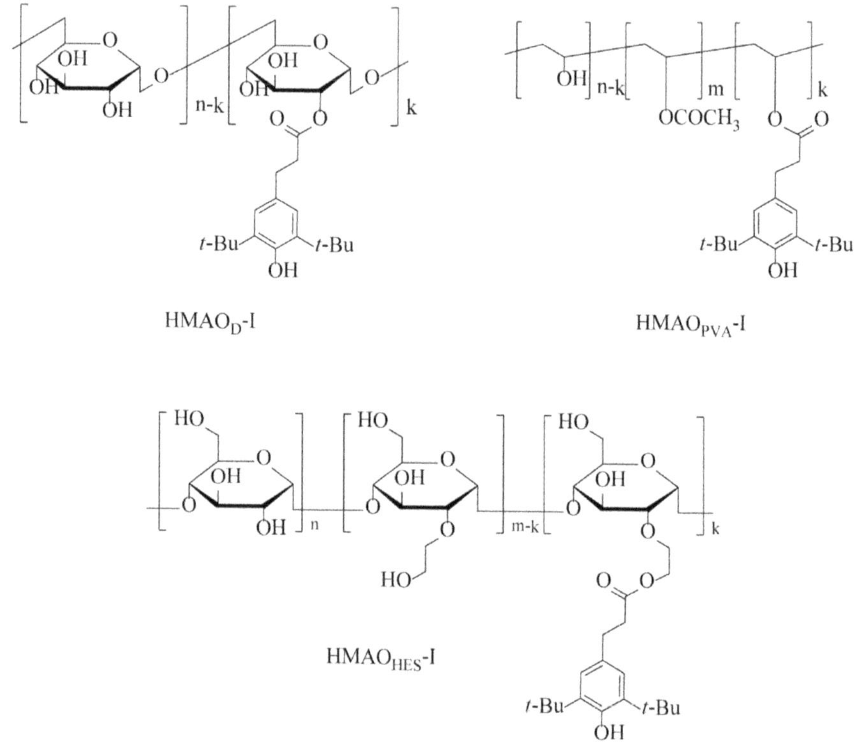

HMAO$_D$-I   HMAO$_{PVA}$-I

HMAO$_{HES}$-I

FIGURE 1.4 Examples of hybrid macromolecular antioxidants with ester linkages between the polymer and phenosan acid (the first series).

The second series of macromolecular antioxidant with simple ether linkages HP-polymer (HMAO-II) was obtained by the interaction of OH-groups of the polymer and 4-bromo- (or acetoxy)methyl-2,6-di-*tert*-butylphenol. The products contented different amounts of chemically bounded fragments of HP and demonstrated different solubility in water (Figure 1.5).

HMAO$_D$-III and HMAO$_D$-IV were prepared using 4-bromomethyl-2,6-di-*tert*-butylphenol and ethylene and butylene glycols, respectively. Synthesis of this HMAO series involved the isolation of intermediate toluenesulfonates derivatives (Ts-2F) and (Ts-4F) followed by the reaction with the polymer (Figure 1.6).

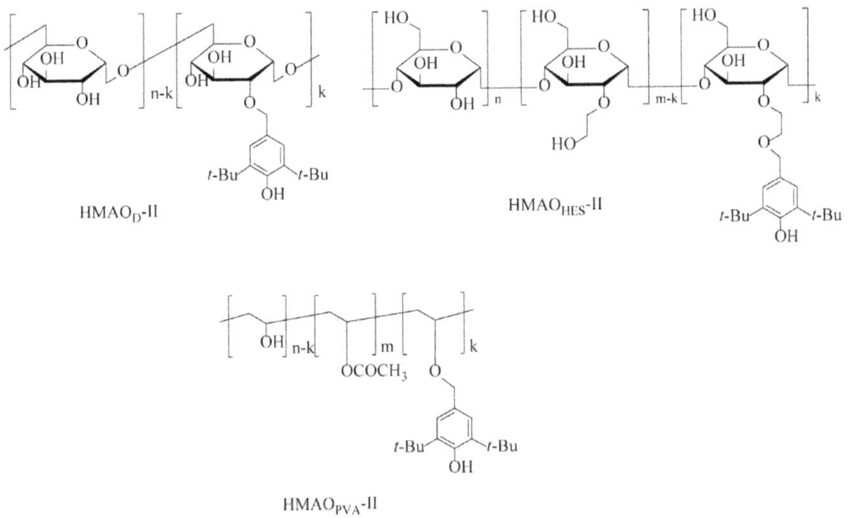

FIGURE 1.5 Examples of hybrid macromolecular antioxidants with ether linkages between the polymer and hindered phenols (the second series).

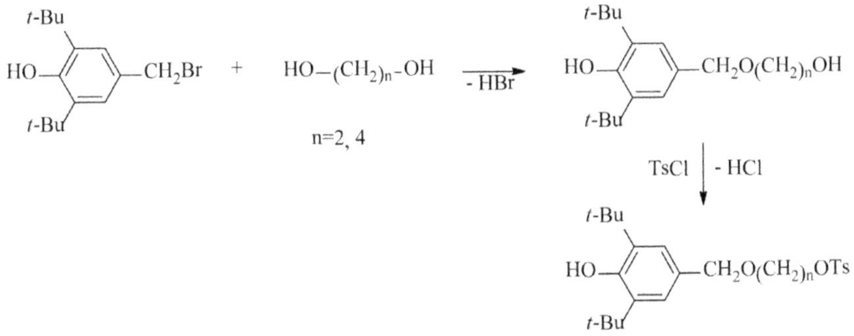

FIGURE 1.6 Synthesis of 4-(β-tosyloxy)ethoxymethyl-2,6-di-*tert*-butylphenol ($n = 2$) and 4-(β-tosyloxy)butoxymethyl-2,6-di-*tert*-butylphenol ($n = 4$).

This approach to the HMAO synthesis allowed varying the length of the spacer. Dextran with different M was used (Figure 1.7).

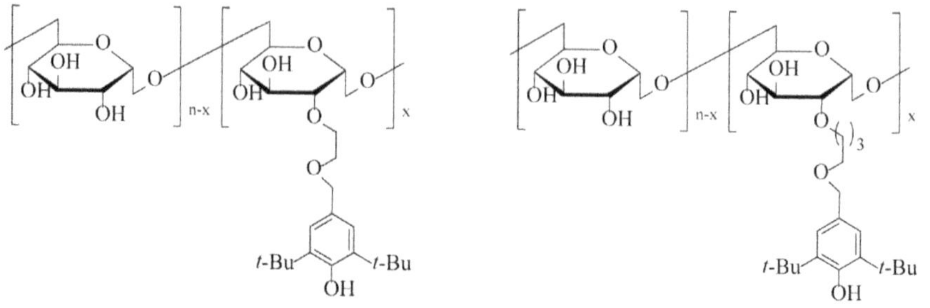

HMAO$_D$-III HMAO$_D$-IV

FIGURE 1.7 Hybrid macromolecular antioxidants on the base of dextran the third and fourth series with different length of the spacer.

To evaluate the radical-scavenging activity (RSA) of the compounds obtained, a kinetic study of their reactions with 2,2-diphenyl-1-picrylhydrazyl (DPPH) and the sodium salt of 2,2-diphenyl-1-picrylhydrazyl sulfonic acid (DPPH-SO_3Na) was carried out. Earlier, the correlation of RSA with antioxidant activity was demonstrated for the interaction of DPPH with low molecular weight HP [9] (Figure 1.8).

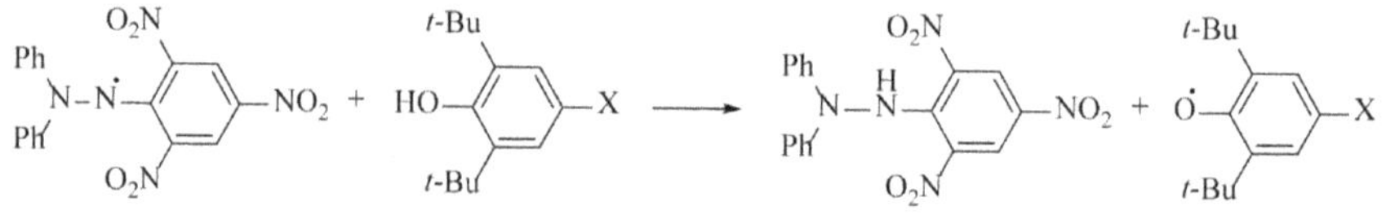

DPPH DPPH-H

FIGURE 1.8 Interaction of 2,2-diphenyl-1-picrylhydrazyl with low molecular weight hindered phenols.

The interaction of macromolecular antioxidants with DPPH was carried out at 20°C under the condition of a pseudo-first order reaction. The rate constants were measured both in organic solvents (benzene, chloroform, and dioxane) and in dioxane-water mixtures with different component ratios (Table 1.1). It was found that the reaction rate constants for low molecular weight HP in chloroform and benzene exceed the constants for HMAO, whereas in the case of dioxane and in dioxane-water mixtures, an inverse relationship was observed.

TABLE 1.1 Rate Constants of the Reactions of Antioxidants With 2,2-Diphenyl-1-Picrylhydrazyl (a) and Sodium Salt of 2,2-Diphenyl-1-Picrylhydrazyl sulfonic Acid (b) in Various Solvents

Antioxidant	K (l/mol·s)				
	Chloroform (a)	Benzene (a)	Dioxane (a)	Dioxane – water 1:1, (b)	Water (b)
Fenozan-acid (FA)	0.188±0.009	0.142±0.008	0.056±0.004	1.2±0.05	
Phenoxan				3.0±0.2	35.7±1.2
HMAO$_D$-I, $\gamma =$ 47.7%, M 40000	0.153±0.008	0.083±0.005	0.135±0.007		
HMAO$_D$-I, $\gamma = 10$%, M 40000				3.5±0.2	1100±50
HMAO$_D$-II, M 40000					270±13
HMAO$_D$-III, M 40000					356±18
HMAO$_D$-IV, M 40000					28.5±1.5
HMAO$_{HES}$-I					544±18
HMAO$_{HES}$-II					226±10
HMAO$_{PVA}$-I					95±5
HMAO$_{PVA}$-II					70±3

(*Note:* a – 2,2-diphenyl-1-picrylhydrazyl (DPPH); b – 2,2-diphenyl-1-picrylhydrazyl sulfonic acid (DPPH-SO$_3$Na); γ – substitution degree of the polymer chain by HP fragments; K is the reaction rate constant).

The rate constants increase for both low molecular weight HP (phenoxan) and for HMAO$_D$-I as the water content in the mixed dioxane-water medium increases. So does the difference of the rate constants (Figure 1.9).

Thus, the medium strongly affects the rate of antioxidant reactions with DPPH. The HMAO are much more sensitive to changes in medium composition compared with low molecular ones. So, HMAO$_D$-I has the highest rate constants of interaction with DPPH in the mixed solvent with the maximum water content.

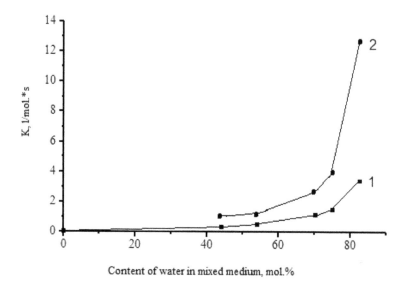

FIGURE 1.9 Dependence of the rate constants of 2,2-diphenyl-1-picrylhydrazyl reactions with phenosan (1) and with hybrid macromolecular antioxidants on the base of dextran (the first series) ($\gamma = 10.0\%$, $M = 40\ 000$) (2) on the composition of the solvent.

A water-soluble analog of DPPH: sodium salt of DPPH-SO$_3$Na, was used for studying RSA in pure water for all the antioxidants. This allowed covering the whole range of solvents: dioxane, dioxane-water mixtures, water (Figure 1.10).

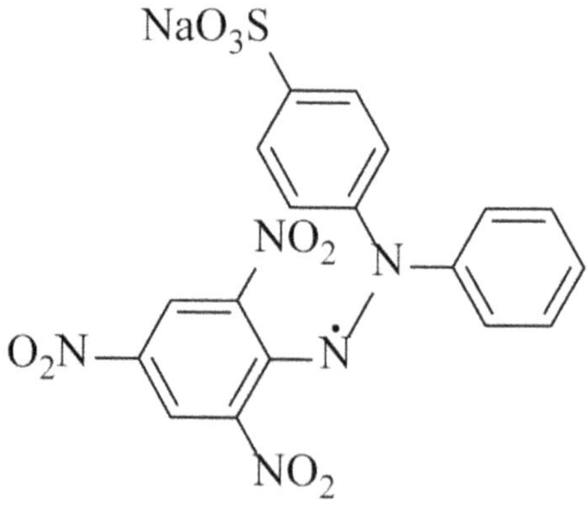

FIGURE 1.10 The water-soluble sodium salt of 2,2-diphenyl-1-picrylhydrazyl sulfonic acid (DPPH-SO$_3$Na).

DPPH-SO$_3$Na in a mixed solvent is less active than DPPH, regardless of the nature of an antioxidant. The mixture of dextran and low M HP (phenosan) in reactions with DPPH and DPPH-SO$_3$Na have the same rate constants as phenosan in the absence of polymer, which indicates the correctness of DPPH-SO$_3$Na use in kinetic studies [10, 11].

The obtained data (Table 1.1) indicate that in pure water, the activity of HMAO strongly exceeds the activity of a low molecular weight antioxidant phenosan. The difference in reaction rate constants for HMAO$_D$ is determined by the nature of the linkages and its size. The maximum value of the constant is fixed for HMAO$_D$-I with an ester bond. As the length of the linkages of the same nature increases, a decrease in the rate constant, for example, for HMAO$_D$-III, is observed. For HMAO$_D$-IV, in which the HP fragments are maximally removed from the dextran polymer chain, the value of the constant approaches the value of the constant for phenoxan.

From presented results (Table 1.1) it follows that in water and in water-organic media the higher RSA of HMAO in comparison with the low molecular weight analog is due to a hybrid structure, namely to the presence of a hydrophilic chain in the HMAO.

It is known that the nature of the medium, as a rule, has little effect on the rate of radical reactions. However, the data in Figure 1.11, which depict

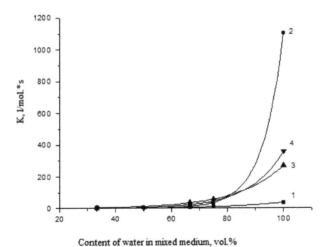

Content of water in mixed medium, vol.%

FIGURE 1.11 Dependence of the rate constants for the interaction of sodium salt of 2,2-diphenyl-1-picrylhydrazyl sulfonic acid with phenoxan (1) and with different hybrid macromolecular antioxidants (2–4) on the composition of mixed solvents: 2 – hybrid macromolecular antioxidants on the base of dextran (the first series) (γ = 9.1%); 3 – hybrid macromolecular antioxidants on the base of dextran (the second series) (γ = 9.1%); 4 – hybrid macromolecular antioxidants on the base of dextran (the third series) (γ = 9.1%). (M = 40,000).

the dependence of HMAO rate constants on the composition of solvents, indicate the presence of such an effect: when the water content decreases, the constants decrease also, and for an equal water-dioxane ratio the constants for all the antioxidants have similar values.

The evident dependence on the composition of the medium contradicts with the generally accepted radical mechanism of hydrogen atomtransfer from the HP molecule to DPPH-SO$_3$Na [12] (Figure 1.12).

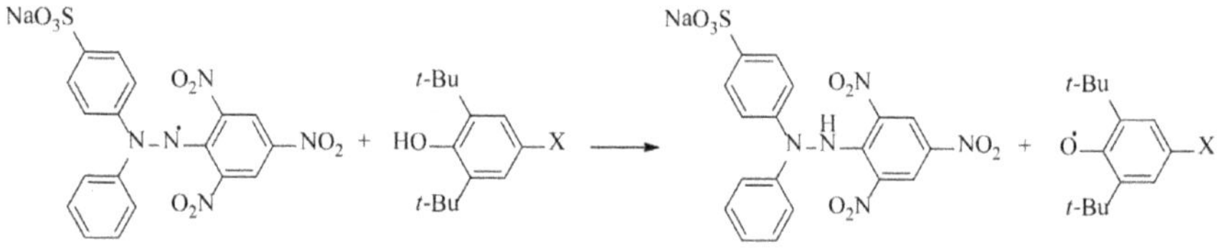

FIGURE 1.12   Radical mechanism of the interaction of hindered phenols and sodium salt of 2,2-diphenyl-1-picrylhydrazyl sulfonic acid.

It is reasonable to assume that the ion-radical mechanism with the formation of a cation-radical intermediate occurs in aqueous media (Figure 1.13).

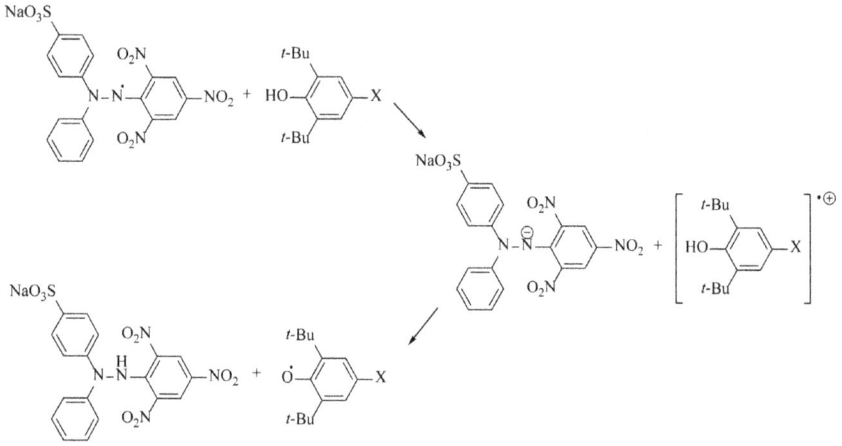

FIGURE 1.13   Ion-radical mechanism of the interaction of hindered phenols and sodium salt of 2,2-diphenyl-1-picrylhydrazyl sulfonic acid in aqueous media.

To confirm this assumption the Grünwald-Winstein kinetic criterion [13] was used: $lgK/K_0 = m \cdot Y$, where Y is a measure of the ionizing capacity of the solvent; m is a factor that reflects the sensitivity of a reaction to a change of Y; K is the reaction rate constant; K_0 is the reaction rate constant in a standard solvent (80% ethanol). Kinetic study of phenosan reaction with DPPH-SO$_3$Na in dioxane-water mixtures with a known ionizing ability revealed the linear character of this dependence on the parameter Y, which agrees with the ion-radical mechanism. The ionizing capacity of the solvate shell of dextran ($Y = 6.71$) was calculated for the samples of HMAO$_D$-I, and this value exceeded the value $Y = 3.4$ for pure water [11]. This can be explained by the change in the structure of water in the solvate shell of dextran, leading to reduction of energy expenses for the destruction of hydrogen bonds during the solvation of the radical-cation.

The activating effect of the hydrated shell on the interaction of HMAO with DPPH-SO$_3$Na is manifested in the extreme dependence of the rate constants on the M of dextran (Figure 1.14).

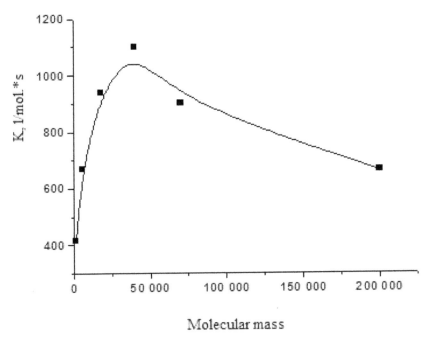

Molecular mass

FIGURE 1.14 Dependence of the reaction rate constants for the interaction of hybrid macromolecular antioxidants on the base of dextran (the first series) ($\gamma = 9.1\%$) with the sodium salt of 2,2-diphenyl-1-picrylhydrazyl sulfonic acid in the water on the molecular mass of dextran.

It is known that the volume of the hydrated shell of hydrophilic polymers grows with increasing of M [14]. Consequently, the rate of the redox reaction should also increase with increasing of M values. This actually occurs up to a certain M value, after which steric obstacles begin to influence, leading to a decrease in the reaction rate constants.

Hydrophilic polymers: dextran, HES, and PVA, which have been used to obtain HMAO, are widely used in medical practice as components of blood plasma volume expanders. Their solutions serve as effective remedies for keeping the blood pressure on an appropriate level and at the same time provide a positive effect on blood viscosity. Nevertheless, in acute blood loss, these solutions fail to prevent the secondary effects, such as oxidative stress. The release of free oxygen species after oxygen deficiency leads to lipid peroxide oxidation and, thus, to the cell membrane destruction [15]. Introduction of HMAO to volume expander composition can be one of the ways to solve this problem. Proper medico-biological experiments were conducted in the V.A. Almazov Institute of Cardiology (St. Petersburg) on laboratory animals (rats). The most active $HMAO_D$ and $HMAO_{HES}$ were studied as such additives [16–18].

The infusion of blood volume expander after the blood loss allowed retaining acceptable blood pressure level for an hour, but did not help to stabilize oxidative status. Blood volume expansion with HMAO additive ($2 \cdot 10^{-6}$ mol/l, taking into account the γ value), made possible not only to sustain but also to keep control of the oxidative damage. A special series of experiments showed that the survival rate of animals increased up to 52% compared to the control group (8%) if HMAO was introduced into blood flow before a massive blood loss. Thus, HMAO with chemically bounded fragments of HP possesses high antioxidant activity and membrane stabilizing properties and provide prolonged action of antioxidants. The possibility of varying the structural parameters makes it possible to adapt HMAO for solving various problems of biology, medicine, and agriculture.

Thus, a new class of phenolic antioxidants based on hydrophilic polymers with chemically bounded HP fragments (HMAO) was created. Structural factors of high RSA of HMAO in aqueous media were studied. The kinetic method showed that the mechanism of redox-processes in water involves the intermediate formation of ion-radical particles. The important role of supramolecular structures of HMAO in aqueous solutions was emphasized, as well as the location of the redox-active phenolic core inside or outside the hydrated shell. Biocompatibility and RSA of HMAO were demonstrated on living biological models.

Recently, the HMAO series has been expanded due to using PEG as a base polymer (*M* from 3,400 to 22,000) for the synthesis of HMAO [19–21]. A number of PEG derivatives (HMAO$_{PEG}$) have been synthesized by using carboxy-substituted HP. The molecular structure of the HMAO$_{PEG}$ represents a PEG polymer containing antioxidants fragments as end-groups that differ in activity and hydrophobicity. Water solutions of HMAO$_{PEG}$ are thermosensitive and may be characterized by the lowest critical solubility temperature (LCST) in the range 20–90°C. It was established that the LCST value of HMAO$_{PEG}$ depends on the antioxidant structure (Table 1.2).

TABLE 1.2 The Low Critical Solubility Temperature (LCST) Values for Hybrid Macromolecular Antioxidants on the Base of Polyethylenglicols With Different Hindered Phenols

Hindered phenols	M×10^{-3}	LCST, °C
β-(4-hydroxy-3,5-di-*tert*-butylphenyl)propionic acid	3.4	22
	3.9	38
	6.8	64
	21.6	>100
β-(4-hydroxy-2-methyl-6-*tert*-butylphenyl)propionic acid	3.4	43
	3.9	54
	6.8	75
	21.6	>100
4-hydroxy-3,5-di-*tert*-butylcinnamic acid	3.9	49
	6.8	76
	21.6	>100

It was established that the LCST value of HMAO$_{PEG}$ depends on the antioxidant structure (Table 1.2). The hydrophobicity of HMAO$_{PEG}$ changes in a following set: β-(4-hydroxy-3,5-di-*tert*-butylphenyl) propionic acid > 4-hydroxy-3,5-di-*tert*-butylcinnamic acid >β-(4-hydroxy-2-methyl-6-*tert*-butylphenyl) propionic acid. The highest level of RSA was detected for aqueous solutions of HMAO with fragments of 4-hydroxy-3,5-di-*tert*-butylcinnamic acid (value *K* was more than 10^5 mol/l s).

The change in LCST revealed that HMAO$_{PEG}$ with fragments of β-(4-hydroxy-3,5-di-*tert*-butylphenyl) propionic acid transformed into HMAO$_{PEG}$ with fragments of 4-hydroxy-3,5-di-*tert*-butylcinnamic acid due to autooxidation under the action of atmospheric oxygen and moisture (Figure 1.15).

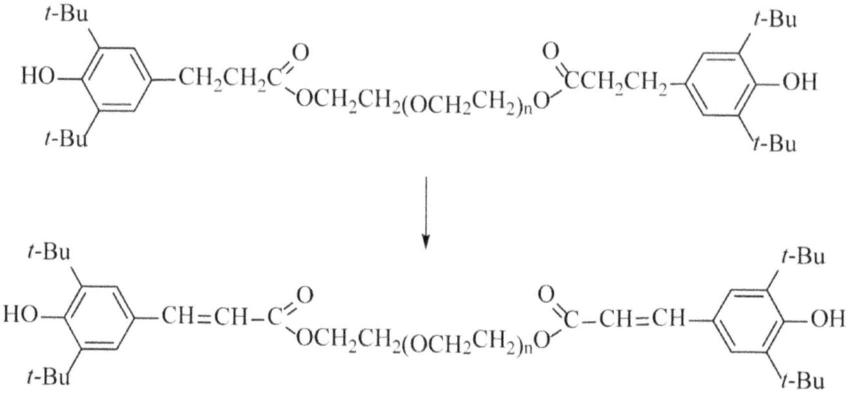

FIGURE 1.15 Auto-oxidation of hybrid macromolecular antioxidants on the base of polyethyleneglycols under the action of atmospheric oxygen and moisture.

This example of HP auto-oxidation in a hydrophilic medium with the formation of a new antioxidant is not a single one. In particular, similar redox-transformations were observed in the study of antibacterial activity (ABA) of HP.

1.3 ANTIBACTERIAL ACTIVITY (ABA) AS A BIOLOGICAL TEST FOR THE ANTIRADICAL ACTIVITY OF STERICALLY HINDERED PHENOLS (HP)

As a result of testing a large number of HP, the compounds with significant ABA were detected, which may be associated with normalization of peroxidation processes of the lipid cell membrane unbalanced in the infected organism. But to date, evidence has been obtained that the antioxidant can influence the ultrastructural organization of the infectious agent that causes a disruption of almost all metabolic processes in the cell, including a membrane-oxygen complex, mesosoma, partitions dividing [22]. However, of particular interest are compounds whose activity can be related to structures generated directly in a bioprocess. Such compounds include "half-screened" mono-*ortho*-substituted phenols. During the bioprocess, they are able to undergo *ortho*-hydroxylation with the formation of diatomic phenols (catechols), which determines the level of ABA comparable with ABA of known catechols (Table 1.3).

TABLE 1.3   Antibacterial Activity (ABA) of Alkyl-Substituted Phenols and 3,5-di-*tert*-butyl Catechol

Test-culture	Substituted phenols					
	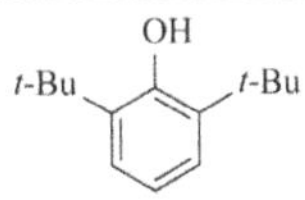 2,6-di-*t*-Bu phenol	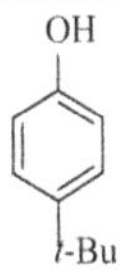 4-*t*-Bu phenol	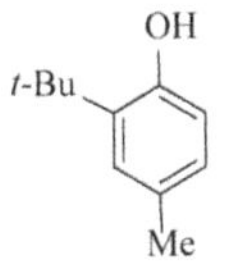 2-*t*-Bu-4-Me phenol	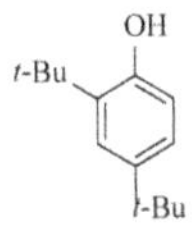 2,4-di-*t*-Bu phenol	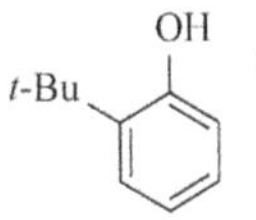 2-*t*-Bu phenol	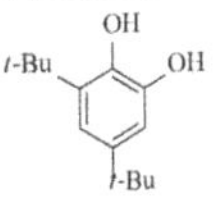 3,5-di-*t*-Bu catechol
	Area of sterile zone, mm²					
Escherichia coli 1157	39	46	193	185	147	179
Staphylococcus albus	50	79	168	572	218	218

Experimental confirmation of the effect of *in situ* stimulated ABA was obtained with of 2,4-di-*tert*-butylphenol, transforming in 3,5-di-*tert*-butyl-catechol via hydroxylation (Figure 1.16).

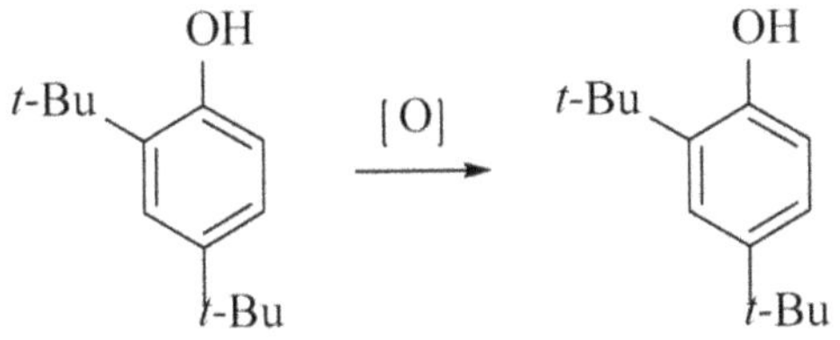

FIGURE 1.16   *Ortho*-hydroxylation of 2,4-di-*tert*-butylphenol.

ABA was determined as the result of the direct action of the substance to a living bacterial cell according to the previously described procedure [23]. As test cultures, *Escherichia coli* 1157 and *Staphylococcus albus* were used on an agar nutrient lawn. The ABA data were presented as the area of the sterile zone formed around the application point of the test substance. The presence of chemical compounds (both initial and formed in the bioprocess) was determined by analyzing the acetone extract from the substrate taken from the sterile zone. The extracts were analyzed by thin-layer chromatography (TLC) on Silufol-UV 254 plates with reference.

It has been shown that 3,5-di-*tert*-butylcatechol is formed only in the case of microorganisms participation in the process. In the control experiment on medium without sowing microorganisms, the original phenol does not undergo such a transformation.

An example of a stepped mechanism of ABA was also found in a series of formyl substituted HP, where compounds with an open OH group did not exhibit appreciable activity, and an increase of ABA was observed in acylated derivatives (Table 1.4).

TABLE 1.4   Antibacterial Activity (ABA) of Formyl-Substituted Phenols

Test-culture	Formyl-substituted phenols		
	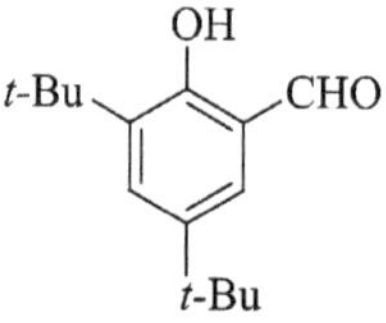	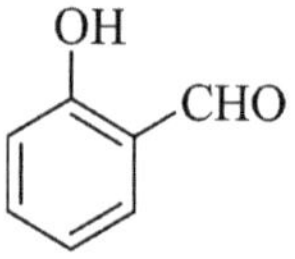	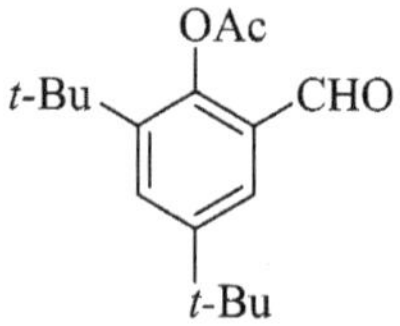
	Area of sterile zone, mm²		
Escherichia coli 1157	42	38	84
Staphylococcus albus	55	28	73

This effect may be explained by the difference in the oxidation ability of the formyl to the carboxyl group. Compounds with an open hydroxyl are stable to oxidation, while acylated derivatives are readily oxidized to the corresponding acids. Open hydroxyl being an intramolecular antioxidant suppresses the radical process of oxidation to the carboxyl group (Figure 1.17).

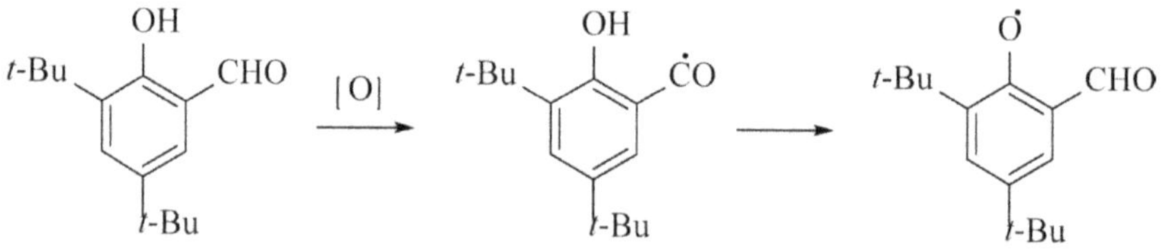

FIGURE 1.17   Auto-inhibition of the oxidation of 3,5-di-*tert*-butylsalicyl aldehyde.

In the acylated derivatives, the antioxidant function of the OH group is blocked, so characteristic for aromatic aldehydes oxidation to acids is observed (Figure 1.18). The possibility of such oxidation *in situ* determines ABA for formyl-substituted HP. The actual antibacterial agent is not the original formylphenol, but the product of its redox transformation [24].

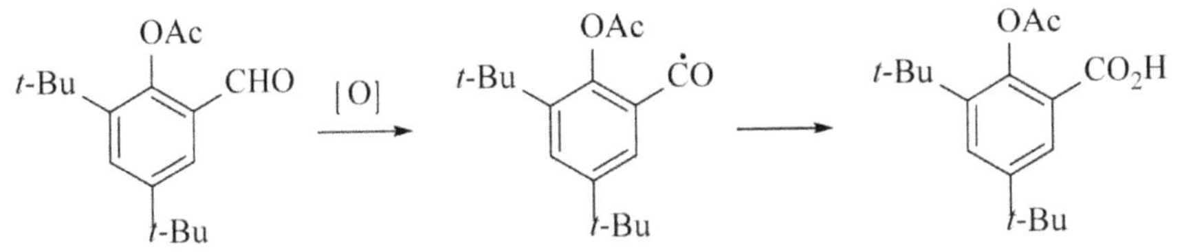

FIGURE 1.18   Oxidation of acylated 3,5-di-*tert*-butylsalicylic aldehyde.

Correlation of biological activity with the structure of products of oxidative and solvolitic transformations of test compounds allows to identify structural effectors of activity and use them for prognostic assessments. Thus, in the stepwise transformations of di-*tert*-butylcatechols in hydrophilic media, hydroxyquinones are formed (Figure 1.19).

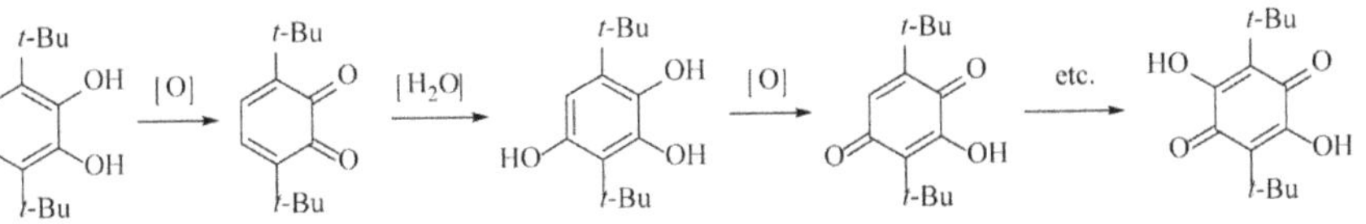

FIGURE 1.19   The stepwise transformations of 3,6-di-*tert*-butylcatechol in hydrophilic media.

The assumption of their possible ABA has been experimentally confirmed [25].

1.4   REDOXCOUPLE 3,6-DI-*TERT*-BUTYLCATECHOL – 3,6-DI-*TERT*-BUTYL-*O*-BENZOQUINONE AS A UNIVERSAL ANTIOXIDANT

The best possibility of studying the cascade (stepwise) mechanism of activity has provided the use of symmetrically hindered 3,6-di-*tert*-butyl catechol (3,6-Cat), first described in 1972 [26]. Earlier attempts to directly synthesize it by alkylation with alcohols and olefins using acidic catalysts, as well as catalysts for the selective *ortho*-alkylation of phenols (alcoholates and phenolates of aluminum and similar compounds) led to the formation of the isomeric 3,5-di-*tert*-butyl catechol (3,5-Cat). The 3,6-Cat claimed in Ref. [27] was the result of erroneous identification.

Selective *ortho*-alkylation with the formation of 3,6-Cat was carried out under the action of isobutylene in xylene at 100–150°C and 16 atm pressure in the presence of titanium bis-catecholate (Figure 1.20).

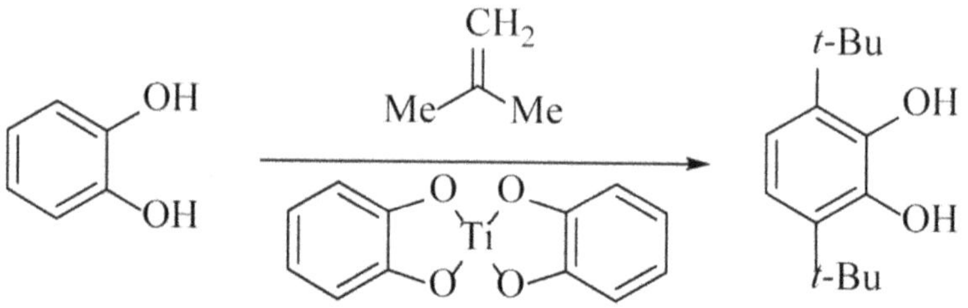

FIGURE 1.20   Selective *ortho*-alkylation of catechol.

The product was characterized by the ^{1}H NMR method, which unambiguously established the position of the *tert*-butyl groups in the aromatic ring. The spectrum contains singlets at 6.59 ppm (ArOH); 5.11 ppm (OH) and 1.31 ppm (*t*-Bu) in the ratio 1:1:9. The NMR spectrum of the 3,5-Cat synthesized according to [27] contains three pairs of singlets: 6.75 and 6.58 ppm (ArH); 5.34 and 4.93 ppm (OH); 1.33 and 1.19 ppm (*t*-Bu).

An excellent opportunity to identify *tert*-butylated catechols is given by TLC on Silufol UV-254 in the mixture hexane-ether (6:1). The entire set of catechols presenting in the alkylate: isomeric 3,5-Cat and 3,6-Cat, mono 3- and 4-*tert*-butyl catechols and the starting catechol – differ not only in R_f, but also in the color of the spots on the chromatograms that develops when plates are exposed in the air [28].

3,6-Cat was studied as a new antioxidant in various model systems [29]. It was shown that an important role in its activity plays the intermediate member of the redox triad (semiquinone, SQ) (Figure 1.21).

However, further investigations showed that the total antioxidant activity of 3,6-Cat is determined by all the members of the redox-triad, including 3,6-di-*tert*-butyl-*o*-benzoquinone (3,6-Q). This was evidenced by the results of 3,6-Q photooxygenolysis. 3,6-Q undergoes decarbonylation to 2,5-di-*tert*-butylcyclopentadienone under irradiation with visible light ($\lambda > 380$ nm) in benzene or hexane (Figure 1.22).

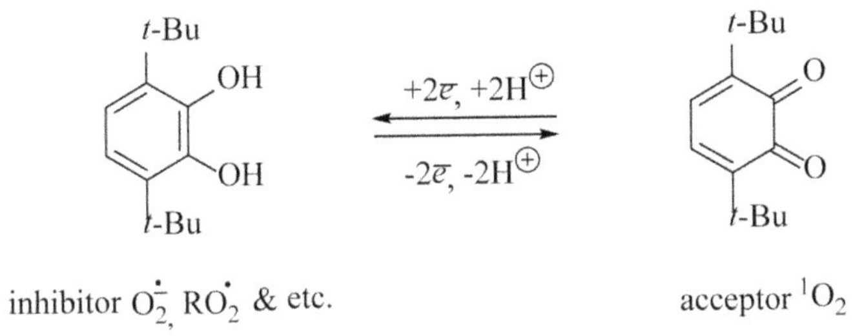

inhibitor $O_2^{\cdot-}$, $RO_2^{\cdot}$ & etc.   acceptor $^1O_2$

FIGURE 1.21 Redox triad 3,6-di-*tert*-butyl catechol– semiquinone –3,6-di-*tert*-butyl-*o*-benzoquinone.

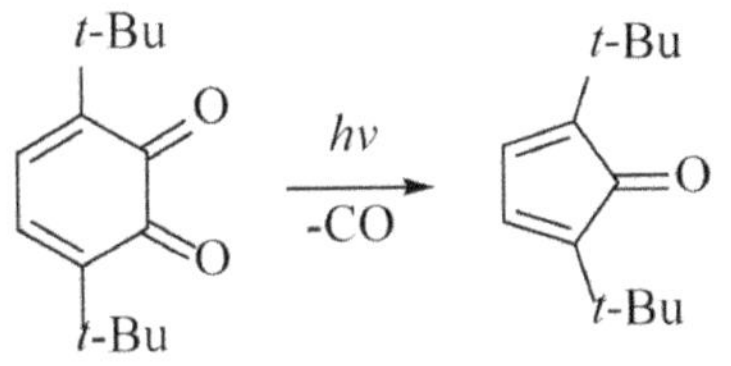

FIGURE 1.22 Photo-decarbonylation of 3,6-di-*tert*-butyl-*o*-benzoquinone.

The formation of 1,2-dipivalilethylene via cyclic peroxide adduct of dienone with oxygen is observed when photolysis is conducted in the presence of oxygen [30] (Figure 1.23).

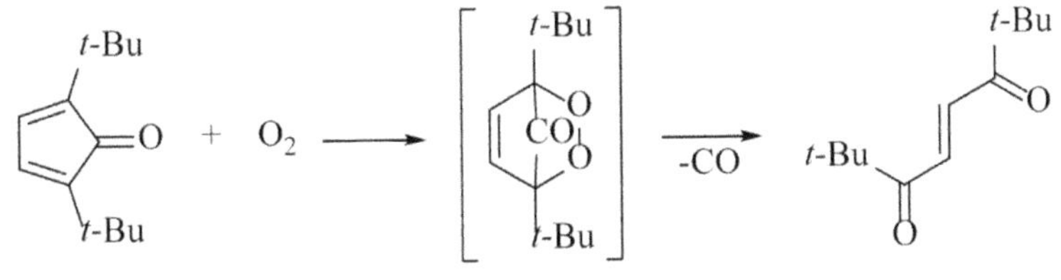

FIGURE 1.23 Photolysis of 2,5-di-*tert*-butylcyclopentadienone in the presence of oxygen.

This conversion is a complete analog of the reaction of 2,5-diphenylcyclopentadienone with singlet oxygen, which is used as one of the chemical tests for singlet oxygen (Figure 1.24).

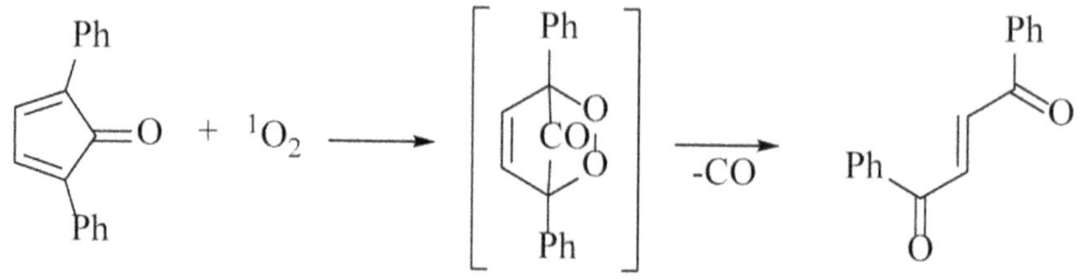

FIGURE 1.24 Chemical test for singlet oxygen.

The participation of the oxygen inhibiting decarbonylation changes the direction of the photoprocess. Photooxygenolysis becomes the main process. Its products are depivalylethylene (minor product) and di-*tert*-butylmuconic anhydride, predominant in the reaction mixture, formed by the interaction of $^1O_2$ with the original quinone (Figure 1.25).

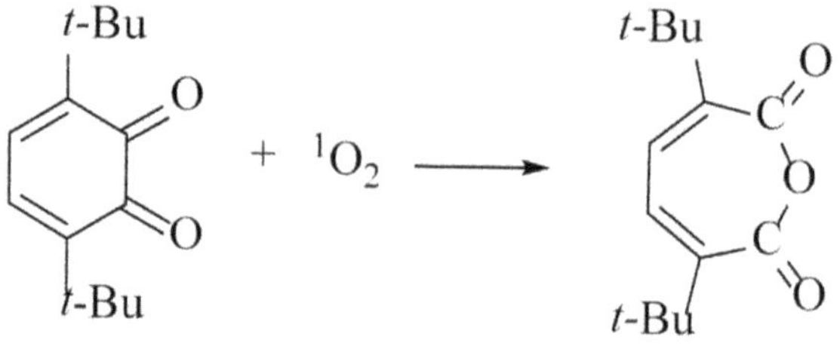

FIGURE 1.25 Photooxygenolysis of 3,6-di-*tert*-butyl-*o*-benzoquinone.

Thus, 3,6-Q indifferent to $^3O_2$ but capable of $^1O_2$ acceptance may be used as a chemical $^1O_2$ indicator along with cyclopentadienone. The ability of the 3,6-Q to interact with $^1O_2$ combined with the high antioxidant activity of the redox-conjugated 3,6-Cat can characterize the couple 3,6-Cat – 3,6-Q as the universal antioxidant (Figure 1.26).

Similar transformations of natural substituted catechols and quinones can occur under the action of atmospheric oxygen and sunlight.

The triad catechol – semiquinone – quinone can be complemented by cyclic ethers of the 3,6-Cat with the benzodioxolane structure formed as a result of 3,6-Cat condensation with carbonyl compounds (Figure 1.27).

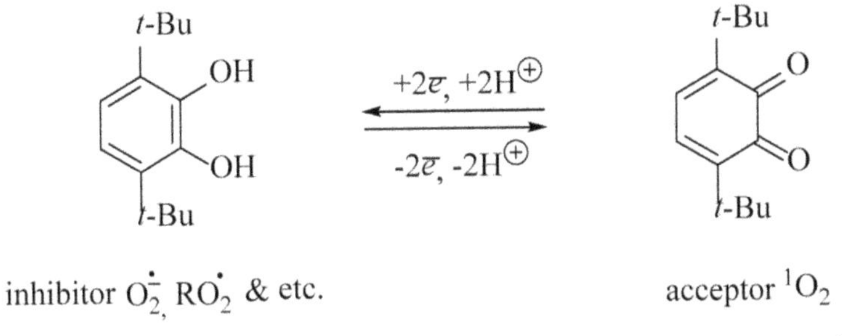

FIGURE 1.26 The couple 3,6-di-*tert*-butyl catechol – 3,6-di-*tert*-butyl-*o*-benzoquinone as the universal antioxidant.

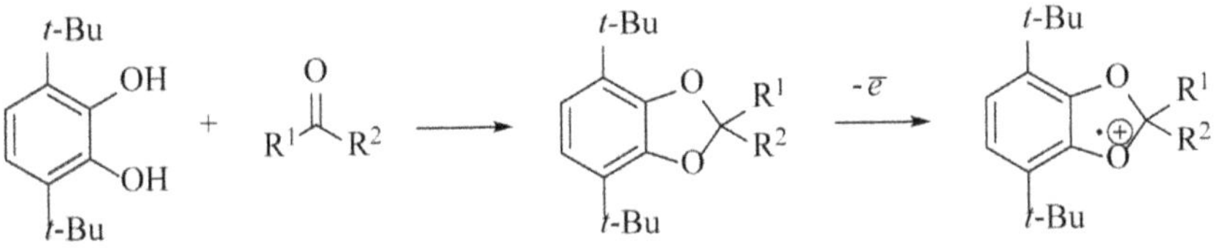

FIGURE 1.27 Benzodioxolanes formation from 3,6-di-*tert*-butyl-catechol and their oxidation to cation-radicals.

Benzodioxolanes without freehydroxyl groups is capable of one-electron oxidation with the formation of cation-radicals that are sufficiently stable for ESR observation.

The spectral parameters of a series of cation-radicals based on benzo-dioxolanes with various substituents R^1 and R^2 ($R^1 = R^2 = Me$; $R^1 = Me$, $R^2 = Et$) were studied. It was established that heterocyclic cation-radicals were formed as intermediates in the oxidation of monocarboxylic esters of catechol that undergo degenerated isomerization with the migration of the acyl group [31] (Figure 1.28).

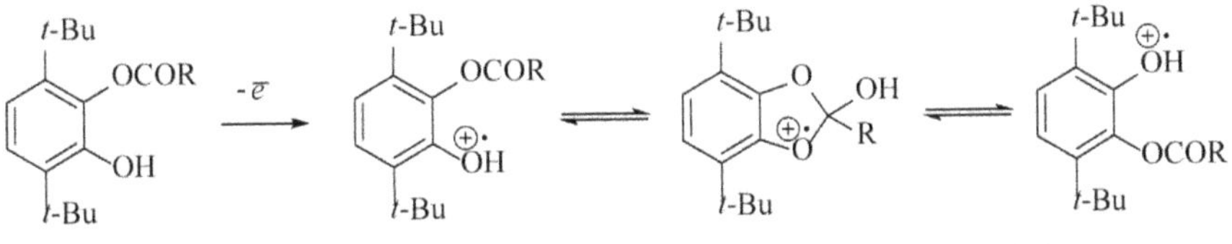

FIGURE 1.28 Oxidation of monocarbonic esters of 3,6-di-*tert*-buthylcatehol.

The ability to donate an electron in combination with the proton acceptor properties of benzodioxolanes makes it possible to use them as inhibitors of the thermooxidative destruction of polymers with the elimination of hydrogen halides (polyvinyl chloride and similar polyhalogenolefins) [32].

The presence of *tert*-butyl groups in HP molecules provides a steric and positional protection, which is used in synthesis for carrying out regioselective transformations. Nevertheless, HP have the sensitivity to deformation and, under conditions of extreme effects of high pressure and shear deformation (HP+SD), exhibit unusual reactivity and undergo transformations that are unrealizable or even forbidden under normal conditions. The most useful for such studies was 3,6-Q.

Thus, 3,6-Q, being stable compound, under the influence of HP+SD on Bridgeman steel anvils undergoes Diels-Alder dimerization. It was found that in addition to the dimer of 3,6-Q, a complex with iron appears, presumably preceding the dimer (Figure 1.29).

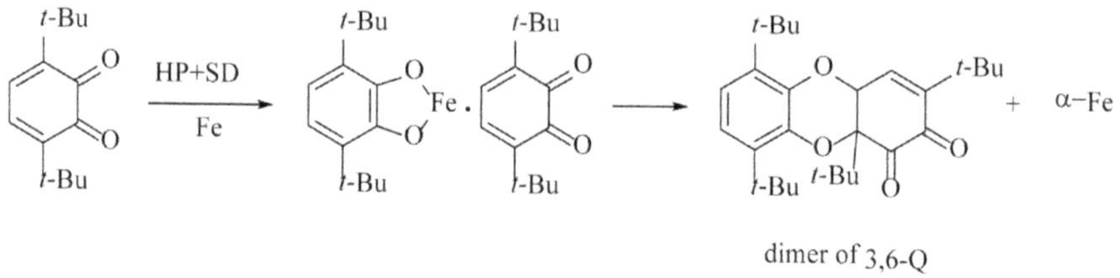

dimer of 3,6-Q

FIGURE 1.29   Conversion of 3,6-di-*tert*-butyl-*o*-benzoquinone under the action of high pressure and shear deformation.

On the anvils with chrome-covered working surfaces, the formation of chromium tris-semiquinolate (SQ)$_3$Cr is observed (Figure 1.30).

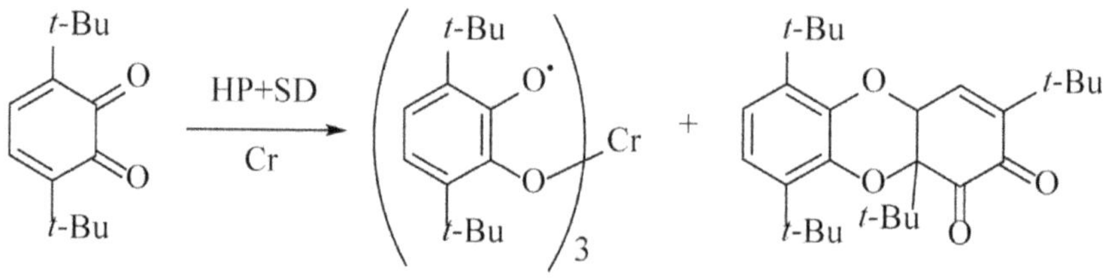

FIGURE 1.30   Conversion of 3,6-di-*tert*-butyl-*o*-benzoquinone on chromium-covered Bridgeman anvils.

The targeted synthesis of chromium tris-semiquinolate was affected by the action of HP+SD on the binary mixture of 3,6-Q with chromium powder [33]. The complex was isolated and identified by mass spectrometry and electron absorption spectroscopy; the electron distribution between the metal and ligands was studied by the ESR method [34].

In addition to chromium, copper, zinc, and aluminum were used to cover the working surfaces of the anvils. Under normal conditions, these metals do not enter into direct interaction with 3,6-Q. In the case of copper, this is due to the unfavorable ratio of redox potentials of the reagents. Zinc and aluminum are effective reducing agents, but they do not exist in the form of singly charged ions and are therefore incapable of one electron donating. In the redox reactions of 3,6-Q, the one-electron transfer plays the determinative role, so zinc and aluminum are indifferent to 3,6-Q in ordinary conditions.

Under HP+SD, all three metals form metal complexes when interacting with 3,6-Q. The complex with copper was identified as bis-semiquinolate–$(SQ)_2Cu$ [35]; Triradical $(SQ)_3Al$ registered as a result of the interaction with aluminum. In the case of zinc, there were two types of metal complexes: zinc bis-semiquinolate $(SQ)_2Zn$ and ion-radical salt $(SQZnCat)_2$ (Figure 1.31).

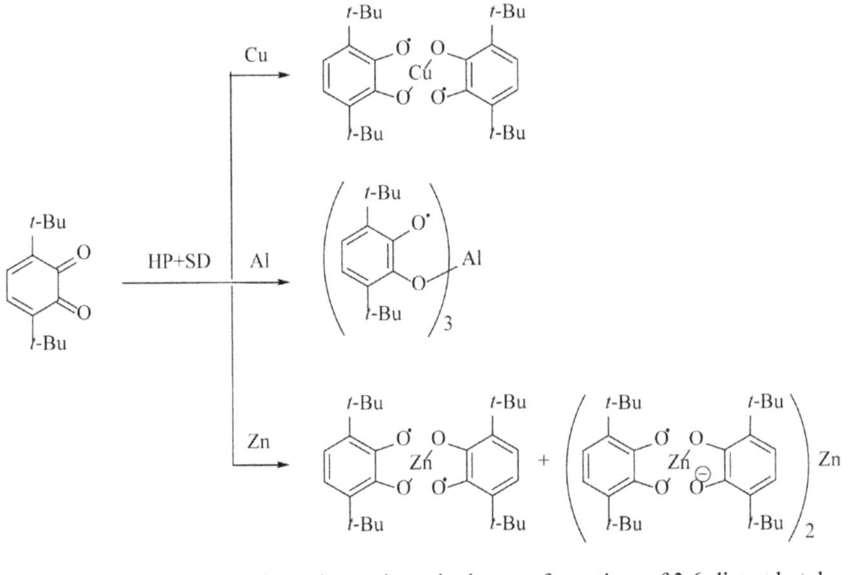

FIGURE 1.31 Formation of metal complexes in the transformations of 3,6-di-*tert*-butyl-*o*-benzoquinone under the action of high pressure and shear deformation (HP+SD).

It is obvious that under the action of HP+SD, an increase in the degree of donor-acceptor interactions occurs with a change in the redox characteristics of substances. The possibility of cooperative interactions and the implementation of multimolecular processes are increased.

The overwhelming majority of organic compounds are capable of coordination reaction with metals, so the possibility of forced by HP+SD metal participation should be taken into account when carrying out solid-phase organochemical conversions using equipment that allows direct contact of the substance with the metal.

It is important to note that the presence of oxide films on metals is not hampered the formation of metal complexes of 3,6-Q under the action of HP+SD so metals can be successfully replaced by oxides. Thus, the synthesis of $(SQ)_3Cr$ was affected by the action of HP+SD on a binary mixture of 3,6-Q with Cr_2O_3. Chromium oxide is one of the compounds with the strongest metal-oxygen bond and is not subjected to deoxygenation under the action of HP+SD [36]. So it is natural to assume that the observed displacement of oxygen from Cr_2O_3 by the quinone ligand includes the stage of formation of the coordination compound 3,6-Q and Cr_2O_3 with an increased degree of oxidation of the metal and subsequent intracomplex oxidation of oxygen ligands.

Similarly, other metal compounds, such as halides and carboxylates, can react. One of the interesting aspects of the conversion of these compounds is the possibility of solid-phase generation of radical particles. Thus, in the reaction of 3,6-Q with $CrCl_3$, the presence of 4-chloro-substituted 3,6-Q among the products indicates the formation of chlorine radicals (Figure 1.32).

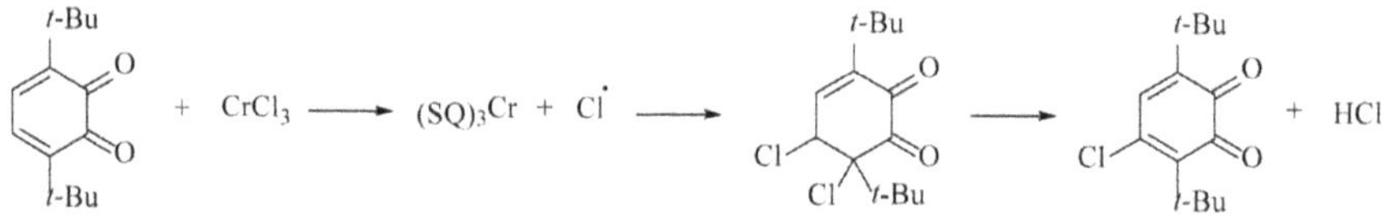

FIGURE 1.32   Interaction of 3,6-di-*tert*-butyl-*o*-benzoquinone with $CrCl_3$ under the action of high pressure and shear deformation.

The formation of $(SQ)_3Cr$ under the action of HP+SD also occurs when 3,6-Q interacts with the compounds where the metal is in the maximum oxidation state, for example, with Na_2CrO_4. The individual chromates of alkali metals are stable under conditions of HP+SD. Therefore, we have to admit the formation of 3,6-Q-Cr^{VI} complex and the subsequent oxidation of oxygen ligands in this complex to explain the observed transformation. Intermediate stages of this process may include the formation of peroxide groups at the metal atom (Figure 1.33).

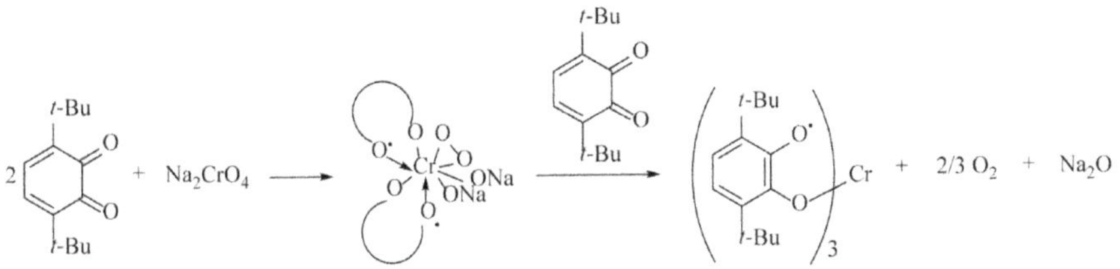

FIGURE 1.33   Interaction of 3,6-di-*tert*-butyl-*o*-benzoquinone with sodium chromate under the action of high pressure and shear deformation.

In addition to Na_2CrO_4, we used $NaVO_3$, Na_2WO_4, and Na_2MoO_4 in solid-phase interaction with 3,6-Q. In all cases, three-ligand complexes of metals are formed, which are identical to those synthesized from 3,6-Q and metal powders.

We confirmed the possibility of the formation of hyper-valent coordination compounds with the participation of 3,6-Q as an extra ligand in a parallel study of liquid-phase transformations of 3,6-Q on the example of its interaction with SiF_4. It turned out that the 3,6-Q – SiF_4 system exhibits the properties of a strong oxidant, which, in particular, is capable of acetone dehydrogenation [37]. Since individual 3,6-Q and SiF_4 are not strong oxidants, the hyper-valent complex 3,6-Q – SiF_4 should correspond to the observed redox process. Thus, it is possible to construct strong oxidants based on 3,6-Q and compounds of elements having a high affinity to oxygen and being in the maximum oxidation state. Protonated quinone (3,6-QH$^+$) should be considered as the simplest example of such a system. The participation of 3,6-QH$^+$ is explained by the reduction of 3,6-Q under the action of HCl [38] (Figure 1.34).

A similar process leading to a reduction to 3,6-Cat occurs under the conditions of HP+SD during the interaction of 3,6-Q with H_2O. This illustrates the changes in acid-base and redox ratios under the action of HP+SD (Figure 1.35).

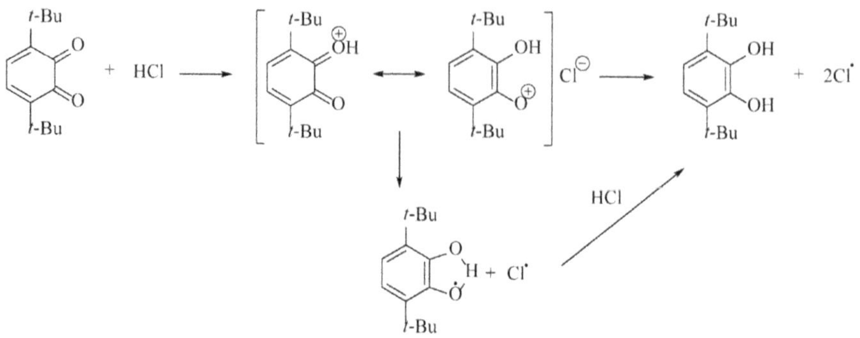

FIGURE 1.34 Reduction of 3,6-di-*tert*-butyl-*o*-benzoquinone (3,6-Q) by hydrogen chloride.

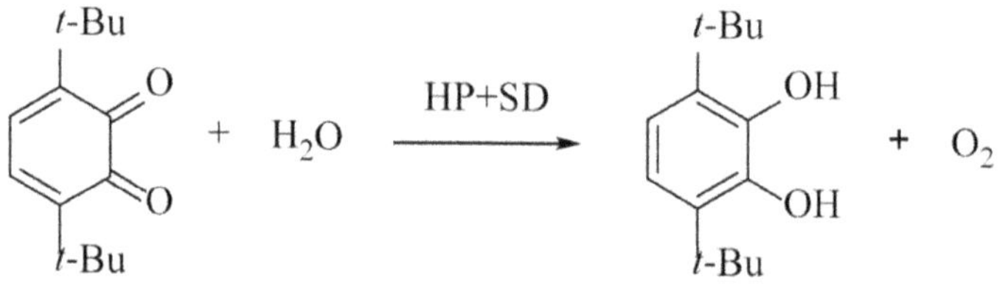

FIGURE 1.35 Reduction of 3,6-di-*tert*-butyl-*o*-benzoquinone by water under the action of high pressure and shear deformation.

1.5 PECULIARITIES OF TRANSFORMATIONS OF *TERT*-BUTYL-SUBSTITUTED PHENOLS

The above examples show that the presence of *tert*-butyl groups in the molecules of HP significantly affects the reactivity. The selectivity of processes increases; transformations that are not characteristic of unsubstituted analogs become possible. The study of specific reactions of HP is of undoubted practical interest, as it opens new synthetic possibilities, and also provides the information necessary for establishing mechanisms of activity of HP used as bioantioxidants, drugs, and plant protection products.

Thus, the conversion of 2,4-di-*tert*-butylphenol (2,4-DTBP) by means of bromination and oxidation is a convenient way for preparing hydroxy- and dihydroxy-dibenzofurans and related compounds. The variant bromination – oxidation includes the synthesis of 6-bromo-2,4-DTBP and its oxidation with manganese triacetate in proton media (acetic acid, alcohols) [39]. The main stages of the process are shown in the diagram (Figure 1.36).

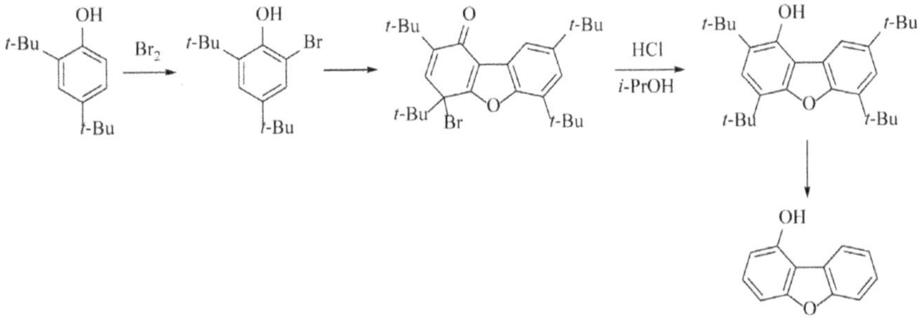

FIGURE 1.36 Synthesis of hydroxy-dibenzofurans from 2,4-di-*tert*-butylphenol via bromination – oxidation.

The composition of the oxidation products, the yield of the key quinobromide and the selectivity of the process depend on the nature of the solvent and additives. The most selective is oxidation in isopropyl alcohol, where a high yield of quinobromide is achieved, and conversion of bromophenol to hydroxydibenzofuran is carried out without separation of the stages. In pure isopropyl alcohol, the reduction proceeds slowly, but it sharply accelerates in the presence of HCl. Tetra-*tert*-butyl dibenzofuran is formed in the case where the oxidation of the original phenol to bisphenol precedes the bromination, [40] (Figure 1.37).

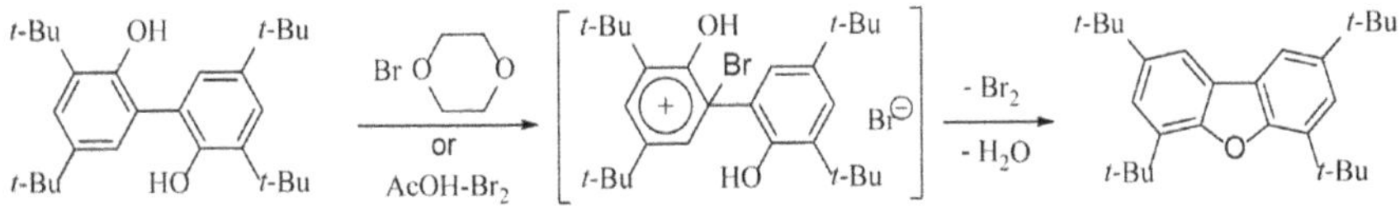

FIGURE 1.37 Formation of dibenzofuran as a result of the bromination of bisphenol.

Heterocyclization under the action of brominating agents occurs both in solutions (acetic acid – bromine) and in the solid phase in the reaction of bisphenol with dioxane dibromide [41]. The process is carried out through the formation and dehydrodegalogenation of the intermediate σ-complex having the structure of protonated *o*-quinobromide. The participation of halogen in this process can be called phantom bromination, as the introduction of bromine into the final product (dibenzofuran) does not occur. Unsubstituted *o*-bisphenol undergoes bromination under identical conditions; dibenzofuran is not detected (Figure 1.38).

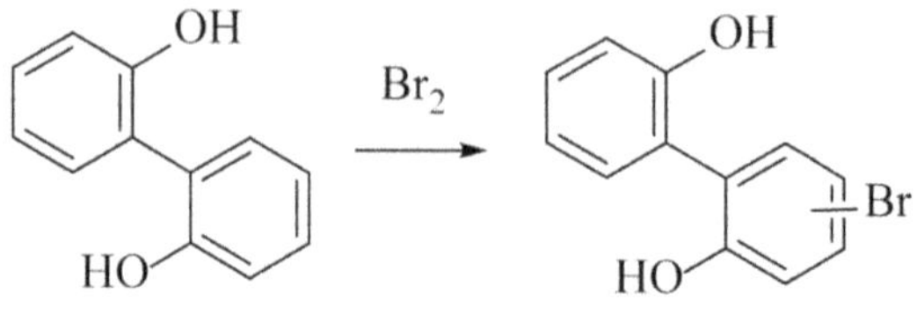

FIGURE 1.38 Electrophilic bromination of unsubstituted *o*-bisphenol.

We wanted to use 2,4-DTBP in a well-known Duff reaction with urotropine in the presence of H_3BO_3 in ethylene glycol to synthesize 2,4-di-*tert*-butyl-6-formylphenol. However, in this case, Duff reaction occurred unusually, and instead of the expected aldehyde, a benzoxazine derivative was formed [42] (Figure 1.39).

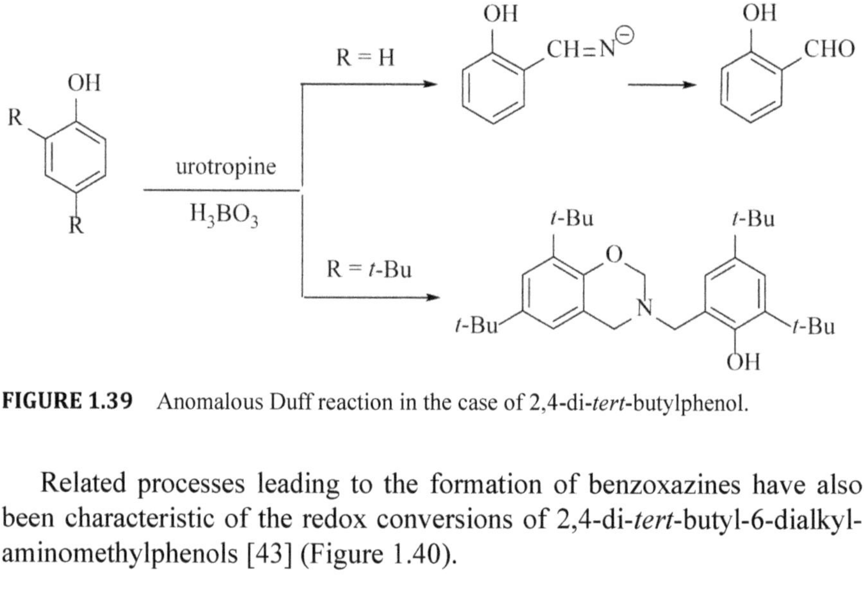

FIGURE 1.39 Anomalous Duff reaction in the case of 2,4-di-*tert*-butylphenol.

Related processes leading to the formation of benzoxazines have also been characteristic of the redox conversions of 2,4-di-*tert*-butyl-6-dialkyl-aminomethylphenols [43] (Figure 1.40).

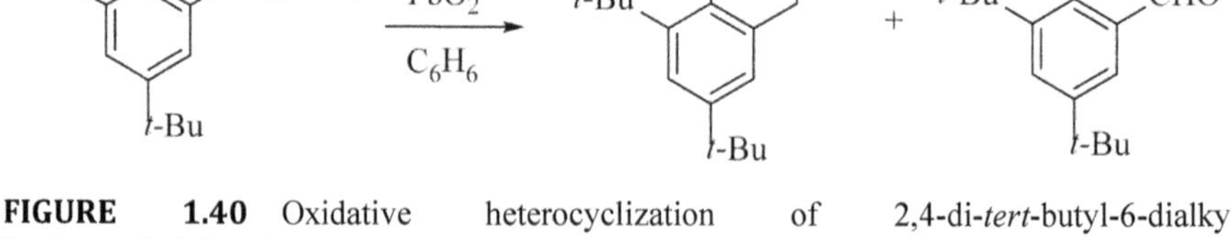

FIGURE   1.40 Oxidative   heterocyclization   of   2,4-di-*tert*-butyl-6-dialky laminomethylphenols.

Spontaneous dehydroheterocyclization with the formation of quaternized derivatives of the corresponding benzoxazines was observed in the quaternization of 6-dialkylaminomethyl-2,4-DTBP under the action of RBr in the presence of atmospheric oxygen. These processes were investigated by the NMR method [44] (Figure 1.41).

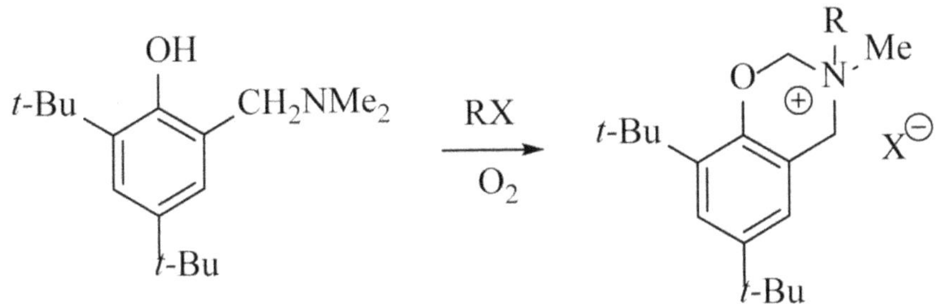

FIGURE 1.41 Spontaneous dehydroheterocyclization during quaternization of 6-dialkylaminomethyl-2,4-di-*tert*-butylphenol.

The largest amount of data on the unusual reactivity of HP was obtained using a pair of 3,6-Cat – 3,6-Q in synthesis, structural studies, and studies of reaction mechanisms. The powerful chelating effect of two adjacent hydroxyl (carbonyl) groups screened by *tert*-butyl substituents in this pair is manifested in the ease with which various metal complexes with many metals of the periodic system are formed [45]. The electronic state of the ligand in these complexes depends on the nature of the central element. Zero-valent metals, their oxides, and salts can participate in the formation of chelate complexes. The processes of complex formation are carried out both in the solid phase (see above) and in solutions. In the formation of complex compounds, there is a tendency toward the coordinative saturation of the central element with the replacement of the maximum number of coordination vacancies. Heterocyclic esters of 3,6-Cat with a central phosphorus atom form in the interaction with PCl_3 [46], other phosphorus compounds, and also in the interaction of 3,6-Q with white phosphorus [47]. A characteristic feature of these reactions is the formation of phosphoric structures (Figure 1.42).

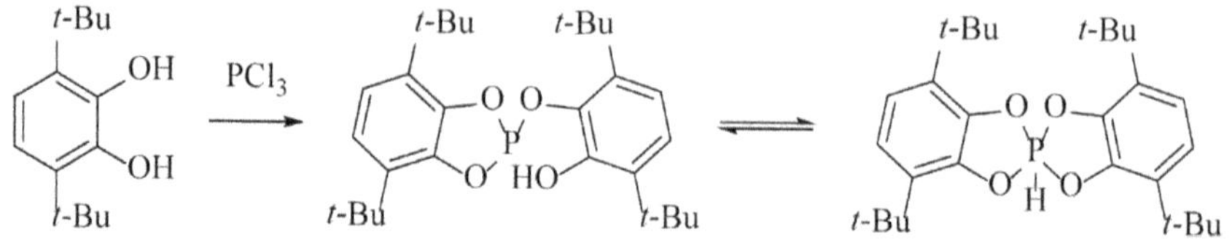

FIGURE 1.42 Formation of phosphoric structure in the interaction of the 3,6-di-*tert*-butyl catechol with PCl₃.

The same trend of coordinative saturation may be noted in the reactions of 3,6-Cat and 3,6-Q with silicon and its derivatives (Figure 1.43).

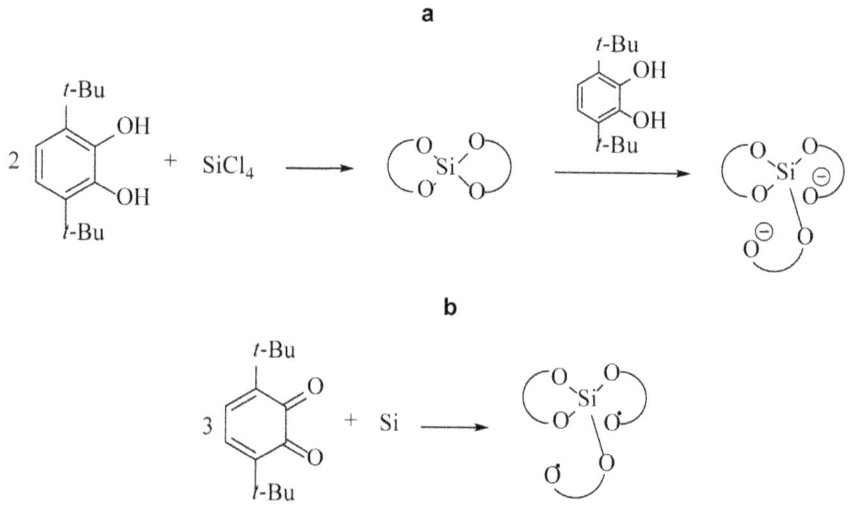

FIGURE 1.43 Formation of silicon complexes with catehate (a) and semiquinolate (b) ligands.

The complex of 3,6-Q with SiF₄ demonstrates the properties of the oxidant, although the components of the complex are not strong oxidants (see above) [37]. The method of direct synthesis of 3,6-Q complexes was developed on the basis of the studied transformations. Solid-phase 3,6-Q reactions with tungsten, molybdenum, and silicon with the formation of (SQ)₃W, (SQ)₃ Mo, (SQ)₂CatSi were carried out. Special studies were devoted to their structure [48–50]. The complex with silicon was identified as a biradical, the ESR parameters of which correspond to the structure obtained independently from silicon bis-3,6-di-*tert*-butylcatehate and 3,6-Q (Figure 1.44).

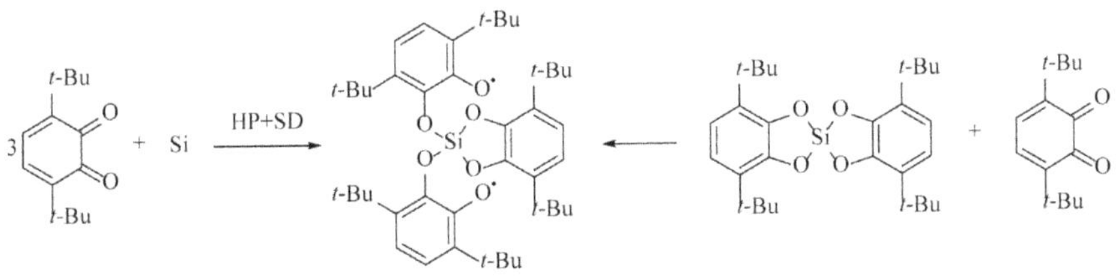

FIGURE 1.44   Two ways of synthesis of the biradical tris-ligand complex of silicon.

In the study of such complexes by the ESR method, the phenomenon of "wandering valence, " degenerate isomerization with the displacement of paramagnetic centers through ligands, was observed (Figure 1.45).

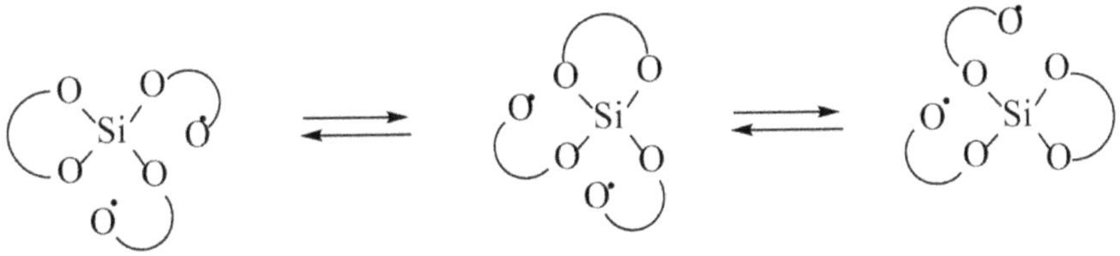

FIGURE 1.45   Tris-ligand complex of silicon with "wandering valence."

Metal complexes have high stability and do not undergo changes during storage. Paramagnetic compounds such as $(SQ)_3Al$, can be used as probes in the study of solid-phase processes. Their generating can occur in the system under study *in situ*. Thus, the joint grinding of low-density polyethylene and aluminum foil in the presence of 3,6-Q with the method of modified extrusion according to [51] gives rise to composite powder in which $(SQ)_3Al$ is registered. The ESR signal of the metal complex remained constant for several months.

The formation of metal complexes can be used for analytical purposes. Thus, the use of 3,6-Cat as a chelating agent allowed to detect natural micro-admixture of titanium compounds in SiO_2 (Figure 1.46).

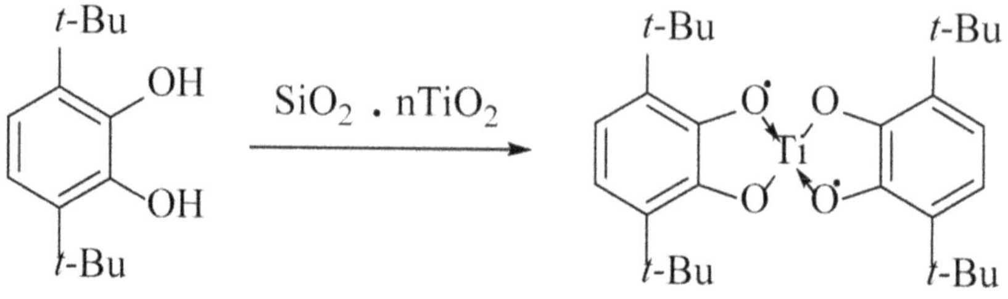

FIGURE 1.46 Formation of the titanium complex in the interaction of 3,6-di-*tert*-butyl catechol with an admixture of TiO$_2$ in SiO$_2$.

In general, the triad 3,6-Cat – 3,6–SQ – 3,6-Q is the most representative example of the use of HP in structural studies and special synthesis.

ACKNOWLEDGMENT

A significant part of the work was carried out by the support of Program of the Fundamental Research of the Presidium of the Russian Academy of Sciences "Development of methods for obtaining chemicals and creating new materials. Development of the methodology of organic synthesis and the creation of compounds with valuable applied properties" (grants 2011 7P, 2012–2013–8P).

KEYWORDS

- **2,6-di-*tert*-butylphenol**
- **3,6-di-*tert*-butyl catechol**
- **dextrans**
- **hydroxyethyl starch**
- **oligomeric polyethylene glycols**
- **phenosan-acid**

REFERENCES

1. Volodkin, A. A., & Zaikov, G. E., (2004). Patent No. 2231522. *A Method of Synthesis of Methyl Ester of β-(4-hydroxy-3,5-di-tert-buthylphenyl)-Propionic Acid.* Date of publication 20.01.2004, Bulletin 2. Int. C07C69/612 (2000.01) (in Russian).
2. Zaikov, G. E., Volodkin, A. A., & Lomakin, S. D., (2012). Fundamental research and method of synthesis of methyl ester of 3-(3'5'-di-*tert*-buthyl-4'-hydroxyphenyl)-propionic acid. *Chemical Physics and Mezoscopy, 14*(2), 271–284 (in Russian).
3. Volodkin, A. A., Efteeva, N. M., & Zaikov, G. E., (2012). Peculiarities of transesterification of methyl ester of 3-(3,5-di-*tert*-buthyl-4-hydroxyphenyl)-propionic acid with tetra (hydroxymetyl) methane and properties of reaction products. *Russian Chemical Bulletin, 9,* 1673–1677 (in Russian).
4. Volodkin, A. A., & Zaikov, G. E., (2007). Ion chain mechanism of catalytic alkylation of 2,6-di-*tert*-buthylphenol by methyl acrylate. *Reports of Russian Academy of Sciences, 414*(3), 349–351 (in Russian).
5. Tchasovskaya, T. E., Maltseva, E. L., & Palmina, N. P., (2013). Effect of potassium phenosan on structure of plasma membranes of mice liver cells in vitro. *Biophysics, 58*(1), 78–85.
6. Tchasovskaya, T. E., Platshina, I. G., & Palmina, N. P., (2013). Physico-chemical changes in liposomes, inducted with low concentrations of synthetic antioxidant–potassium phenosan. *Reports of Russian Academy of Sciences, 449*(6), 673–677 (in Russian).
7. Burlakova, E. B., Molochkina, E. M., & Nikiforov, G. A., (2008). Hybrid antioxidants. *Oxidation Communications Journal, 31*(4), 739–757.
8. Arefiev, D. V., Belostotskaya, I. S., Volieva, V. B., Domnina, N. S., Komissarova, N. L., Sergeeva, O. U., & Hrustaleva, R. S., (2007). Hybrid macromolecular antioxidants on the basis of hydrophilic polymers and hindered phenols. *Russian Chemical Bulletin, 4,* 751–761 (in Russian).
9. Arzamanova, I. G., Logvinenko, R. M., Grinberg, A. E., Gurvich, Y., Kumok, S. T., & Ribak, A. I., (1973). Influence of p-substitution on reaction capacity of biphenols. *Russian Journal of Physical Chemistry, 47,* 707–708.
10. Arefjev, D. V., Domnina, N. S., Komarova, E. A., & Bilibin, A. Y., (1999). Sterically hindered phenol-dextran conjugates: Synthesis and radical scavenging activity. *European Polymer Journal, 35*(2), 279–284.
11. Arefiev, D. V., Domnina, N. S., Komarova, E. A., & Bilibin, A. Y., (2000). Sterically hindered phenol-dextran conjugates: Radical scavenging activity in water and water-organic media. *European Polymer Journal, 36*(4), 857–860.
12. Rozantsev, E. G., & Scholle, E. G., (1979). *Organic Chemistry of Free Radicals* (p. 344). Moscow: Chemistry (in Russian).
13. Dneprovskii, A. S., & Temnikova, T. I., (1979). *The Theoretical Basis of Organic Chemistry* (p. 520). Leningrad: Chemistry (in Russian).
14. Yasushi, M., Noriaki, T., & Hiromi, K., (1993). Raman spectroscopic study on water in polymer solutions. *J. Phys. Chem., 97*(14), 13903–13906.
15. Zenkov, N. K., Lankin, V. Z., & Menshikova, E. B., (2001). *Oxidative Stress* (p. 343). Moscow: MAIK, (in Russian).
16. Voleva, V. B., Domnina, N. S., Sergeeva, O. Y., Komarova, E. A., Belostotskaya, I. S., & Komissarova, N. L., (2011). Structural factors responsible for the activity of

macromolecular phenolic antioxidants. *Russian Journal of Organic Chemistry, 47*(4), 480–485.

17. Domnina, N. S., Khrustaleva, R. S., Arefiev, D. V., Komarova, E. A., Sergeeva, O. J., & Tczyrlin, V. A., (2006). Patent No. 2273483. A Method of Synthesis of Methyl Ester of β-(4-hydroxy-3,5-di-tert-buthylphenyl)-Propionic Acid. Date of publication 20.01.2004 Bull. 2. Int. C07C69/612 (2000.01) (in Russian).

18. Filippov, S. K., Sergeeva, O. Y., Vlasov, P. S., Zavyalova, M. S., Belostotskaya, G. B., Garamus, V. M., Khrustaleva, R. S., Stepanek, P., & Domnina, N. S., (2015). Modified hydroxyethyl starch protects cells from oxidative damage. *Carbohydrate Polymers, 134*, 314–323.

19. Domnina, N. S., Sergeeva, O. Y., Komarova, E. A., Mikhailova, M. E., Voleva, V. B., Belostotskaya, I. S., & Komissarova, N. L., (2014). Indicator properties of oligoethylene glycol hybrids with sterically hindered phenols. *Russian Journal of Organic Chemistry, 50*(3), 371–375.

20. Sergeeva, O., Vlasov, P. S., Domnina, N. S., Bogomolova, A., Konarev, P. V., Svergun, D. I., Walterova, Z., Horsky, J., Stepanek, P., & Filippov, S. K., (2014). Novel thermosensitive telechelic PEGs with antioxidant activity: Synthesis, molecular properties and conformational behavior. *RSC Advances, 4*(79), 41763–41771.

21. Dobrun, L. A., Kuzyakina, E. L., Rakitina, O. V., Sergeeva, O. Y., Mikhailova, M. E., Domnina, N. S., & Lezov, A. V., (2011). Molecular characteristics and antioxidant activity of polyethylene glycols modified by sterically hindered phenols. *Journal of Structural Chemistry, 52*(6), 1161–1166.

22. Konstantinova, N. D., Didenko, L. V., Shustrova, N. M., Volieva, V. B., Belostotskaya, I. S., & Komissarova, N. L., (2004). Influence of antioxidant ICHPAN-16 on ultrastructure organization of Staphylococcus. *Proceeding of XX Russian Electron Spectroscopy Conference* (p. 236). Tshernogolovka, (in Russian).

23. Ovsyannikova, M. N., Volieva, V. B., Belostotskaya, I. S., Komissarova, N. L., Malkova, A. V., & Kurkovskaya, L. N., (2013). Antibacterial activity of substituted 1,3-dioxolanes. Dependence on substitution character. *Pharmaceutical Chemistry Journal, 47*(3), 18–21 (in Russian).

24. Volieva, V. B., Belostotskaya, I. S., Komissarova, N. L., Kurkovskaya, L. N., Pleshakova, A. P., & Prokofieva, T. I., (2007). Urotropin synthesis of 3,5-di-*tert*-buthylsalicylic acid derivatives. *Journal of Organic Chemistry, 43*(10), 1495–1498 (in Russian).

25. Volieva, V. B., Ovsyannikova, M. N., Belostotskaya, I. S., Komissarova, N. L., & Malkova, A. V., (2016). Dependence of antibacterial activity on phenol antioxidants structure and possibility transformation *in situ. Pharmaceutical Chemistry Journal, 50*(4), 96–100 (in Russian).

26. Belostotskaya, I. S., Komissarova, N. L., Djuaryan, E. V., & Erschov, V. V., (1972). Ortho-alkylation of catehol. *Russian Chemical Bulletin, 7*, 1594–1596 (in Russian).

27. Ecke, G. G., & Kolka, A. J., (1958). Patent USA No. 3075832. *Chem. Abstr., 59*, 12707d (1963).

28. Belostotskaya, I. S., Komissarova, N. L., Djuaryan, E. V., & Erschov, V. V., (1971). Identification of alkylcatehols with thin layer chromatography on silica gel. *Russian Chemical Bulletin,* 2816–2817 (in Russian).

29. Mazaletskaya, L. I., Karpuhina, G. V., Komissarova, N. L., & Belostotskaya, I. S., (1982). Inhibiting action of 3,6-di-*tert*-buthylcatecol at oxidation reaction of nonen-1. *Russian Chemical Bulletin, 3*, 505–509 (in Russian).

30. Volieva, V. B., Belostotskaya, I. S., Komissarova, N. L., Starikova, Z. A., & Kurkovskaya, L. N., (2006). Photooxygenolysis of 3,6-di-*tert*-buthyl-*o*-benzoquinone. *Journal of Organic Chemistry, 42*(2), 243–246 (in Russian).

31. Malisheva, N. A., Prokofiev, A. I., Bubnov, N. N., Solodovnikov, S. P., Prokofieva, T. I., Volieva, V. B., Ershov, V. V., & Kabatchnik, M. I., (1988). Cation-radicals of esters of 3,6-di-*tert*-buthylcatecol. *Russian Chemical Bulletin,* 1040–1047 (in Russian).

32. Varbanskaya, R. A., Pudov, V. S., Komissarova, N. L., Belostotskaya, I. S., Volieva, V. B., & Ershov, V. V. (1984). *Polymer Composition.* A.C.No.1143752 registered in the Federal service for intellectual property, patents and trademarks 08.11.1984.

33. Volieva, V. B., Zhorin, V. A., Ershov, V. V., & Enicolopyan, N. S., (1984). Direct synthesis of tris-(3,6-di-*tert*-buthylbenzoquone) chromium in high pressure and shear deformation conditions. *Russian Chemical Bulletin,* 1437 (in Russian).

34. Solodovnikov, S. P., Sarbasov, K., Tumanskii, B. L., Prokofiev, A. I., Volieva, V. B., Bubnov, N. N., & Kabatchnic, M. I., (1984). Investigation of electron structure of tris-(3,6-di-*tert*-buthylbenzoqunone) chromium. *Russian Chemical Bulletin, 8,* 1789–1794 (in Russian).

35. Volieva, V. B., Prokofieva, T. I., Ivanova, E. V., Zhorin, V. A., Ershov, V. V., & Enikolopov, N. S., (1986). Stimulation of intramolecular reaction with action of high pressure and shear deformation. *Russian Chemical Bulletin,* 2159 (in Russian).

36. Gonicberg, M. T., (1969). *Chemical Equilibrium and Rate of Reaction Under High Pressure* (p. 248). Moscow: Chemistry (in Russian).

37. Chekalov, A. K., Gvazava, N. G., Volieva, V. B., Prokofieva, T. I., Prokofiev, A. I., & Ershov, V. V., (1990). Redox transformations of 3,6-di-*tert*-buthyl-o-benzoqunone, catalyzed with tetra-fluorosilane. *Russian Chemical Bulletin,* 1105–1108 (in Russian).

38. Belostotskaya, I. S., Volieva, V. B., Komissarova, N. L., & Ershov, V. V., (1976). Reduction of 3,6-di-*tert*-buthyl-o-benzoqunone with hydrogen chloride. *Russian Chemical Bulletin,* 709 (in Russian).

39. Belostotskaya, I. S., Komissarova, N. L., Volieva, V. B., Ershov, V. V., Borisova, L. N., Glozman, O. N., & Smirnov, L. D., (1986). Oxidative geterocyclization of 6-bromo-2,4-di-*tert*-buthylphenol. *Russian Chemical Bulletin,* 709 (in Russian).

40. Volieva, V. B., Belostotskaya, I. S., & Ershov, V. V., (1996). Dehydratation of 2,2'-dihydroxy-3,3, '5,5'-tetra-*tert*-buthylphenyl under molecular bromine–a new way to dibenzofurane. *Russian Chemical Bulletin, 3,* 784–785 (in Russian).

41. Volieva, V. B., Belostotskaya, I. S., Komissarova, N. L., & Ershov, V. V., (1996). Solid phase bromination of hindered phenols 3,6-di-*tert*-buthyl. *Russian Chemical Bulletin, 3,* 1310–1312 (in Russian).

42. Belostotskaya, I. S., Komissarova, N. L., Prokof'eva, T. I., Kurkovskaya, L. N., & Voleva, V. B., (2005). New opportunities for Duff reaction. *Russian Journal of Organic Chemistry, 41*(5), 703–706.

43. Belostotskaya, I. S., Volieva, V. B., Komissarova, N. L., Decaprilevich, M. O., Hrustalev, V. N., Karmilov, A. U., & Ershov, V. V., (1997). Oxidation of 2-dialkylaminomethyl-4,6-di-*tert*-buthylphenols. *Russian Chemical Bulletin,* 1328–1335 (in Russian).

44. Voleva, V. B., Kurkovskaya, L. N., Belostotskaya, I. S., & Komissarova, N. L., (2003). ^{1}H NMR Study of quaternization of 2,4-di-*tert*-buthyl-6-dimethylaminomethylphenol. *Russian Journal of Organic Chemistry, 39*(1), 92–95.

45. Volieva, V. B., Prokofiev, A. I., Zhorin, V. A., Prokofieva, T. I., Belostotskaya, I. S., Komissarova, N. L., Karmilov, A. U., & Ershov, V. V., (1996). Reports of Russian

Academy of Sciences at investigation of organochemical reaction mechanisms, stimulated with action of high pressure and shear deformation. *Journal of Physical Chemistry B, 15*(1), 16–22.

46. Belostotskaya, I. S., Komissarova, N. L., Prokofieva, T. I., Volieva, V. B., & Ershov, V. V., (1978). Phosphorus acid derivations on the basis. of 3,6-di-*tert*.buthyl catechol. *Russian Chemical Bulletin,* 2385–2388 (in Russian).

47. Prokofiev, A. I., Hodak, A. A., Malysheva, N. M., Bubnov, N. N., Solodovnikov, S. P., Belostotskaya, I. S., Ershov, V. V., & Kabachnik, M. I., (1978). Interaction of white phosphorus with o-quinones. *Reports of Russian Academy of Sciences, 240,* 358–361 (in Russian).

48. Prokofiev, A. I., Volieva, V. B., Prokofieva, T. I., Bubnov, N. N., Solodovnicov, S. P., Ershov, V. V., & Kabachnik, M. I., (1988). Anion-radical tris-3,6-di-*tert*-buthyl-o-semiquinone complexes of molybdenum and tungsten. *Reports of Russian Academy of Sciences, 300*(5), 1139–1142 (in Russian).

49. Prokofieva, T. I., Malysheva, N. A., Volieva, V. B., Solodovnikov, S. P., Prokofiev, A. I., Bubnov, N. N., Ershov, V. V., & Kabachnik, M. I., (1987). ESR-investigation of vanadium tris-3,6-di-*tert*-buthyl-o-qunone complexes. *Reports of Russian Academy of Sciences, 295*(5), 1143–1147 (in Russian).

50. Prokofiev, A. I., Prokofieva, T. I., Belostotskaya, I. S., Bubnov, N. N., Solodovnikov, S. P., Ershov, V. V., & Kabachnik, M. I., (1979). Tautomerism in silicon-containing free radicals. *Tetrahedron, 35,* 2471–2482.

51. Akopyan, E. L., Karmilov, A. U., Nikolskii, V. G., Hachatryan, A. N., & Enikolopyan, N. S., (1986). Elastic deformation shredding of thermoplastics. *Reports of Russian Academy of Sciences, 291*(1), 133–136 (in Russian).

CHAPTER 2

Inhibitory Efficiency of the Plant Cell Components in the Model Oxidative Processes

LYUDMILA N. SHISHKINA, MIKHAIL V. KOZLOV,
LIDIYA I. MAZALETSKAYA, NATALYA V. KHRUSTOVA, and
NATALIYA I. SHELUDCHENKO

*Emanuel Institute of Biochemical Physics of Russian Academy of Sciences,
4, Kosygin St., Moscow 119334, Russia, E-mail: shishkina@sky.chph.ras.ru*

The inhibitory efficiency and antioxidant (AO) properties of the *Laminaria japonica* lipids, the alcohol extract of the *Sophora japonica* fruits, quercetin, dihydroquercetin (QH_2), "Serpisten" (a mixture of the plant sterols), β-carotene, "Vetoron," α-tocopherol and their mixtures were studied by the methyl oleate oxidation model, depending on the rate of the chain initiation, using the KINS computer program to analyze the participation of the plant cell components in the oxidation reactions at the different stages of the process. It is shown that all the investigated components of the plant objects have the ability to decompose the peroxides. It has been established that the inhibitory effect of the plant cell components is due to the physicochemical properties of the basic compounds in their composition and depends significantly on the oxidation conditions and their ability to participate in different stages of the oxidation process.

The data obtained allow us to suggest that the ability of the plant cell components to influence the intensity of oxidative processes underlies their biological activity.

2.1 INTRODUCTION

As known, antioxidants (AO, InH), which are predominantly the phenolic compounds [1–3], are the effective regulators of the oxidation processes in the food, cosmetic, and technical fats and oils, in polymeric materials and biological systems of varying complexity. It is plants that are the sources of the most natural biologically active substances (BAS) representing the different classes of the phenolic compounds and manifesting a variety of useful pharmacological properties along with AO properties [4–9]. The search for new effective tools to protect fats, oils, and lipid-containing medicines from oxidation, the development of a strategy and tactics of the prevention and therapy of various diseases are due to a necessity a detailed study of the mechanism of the inhibitory efficiency in the systems of varying degrees of complexity for selecting the regulators of the free radical oxidation processes that are the most effective and non-toxic for a organism.

At present, the natural compounds or extracts of the plants are actively enough used because of their low toxicity for the organism. However, the mechanism of the oxidative processes inhibition in the model systems is mostly studied for such natural phenolic AO as tocopherol and its homologs [10, 11], a number of flavonoids and related compounds [12–15], β-carotene [16]. It has been found that the natural phenolic AO is characterized by the high antiradical activity, i.e., the value of the rate constant of their interaction with peroxy radicals of the model oxidation substrate (k_7). However, their antioxidant activity (AOA), which is an integral parameter and represents the influence of the other factors on the inhibitory activity of compounds, depends essentially on the concentration of AO, the unsaturation degree of the oxidation substrate, the chemical structures of radicals which are leading in chain oxidation, the rates initiation of radicals in the system, the participation of AO in side reactions. These kinetic features of natural AO should be considered when selecting the most promising for use in a various sphere of the human action, depending on the goal and task of the practice.

The aim of this research was to study the effect of the oxidation conditions of the model substrate on the mechanism of its inhibition by the plant cell components and their mixtures.

2.2   MATERIALS AND METHODOLOGY

The objects of this study were lipids (LH) isolated from the brown algae *Laminaria japonica* by the Foch's method in the modification of Kates [17]; a 20% alcoholic extract of the *Sophora japonica* fruits; quercetin (Q) and dihydroquercetin (QH$_2$) from Sigma (USA); the "Serpisten" preparation (a mixture of the plant sterols, among which is 75% 20-hydroxyecdysone); β-carotene (Serva, Germany) and Vetoron, which is a mixture of 200 mg β-carotene+80 mg ascorbic acid+80 mg α-tocopherol (TP); mixture of TP (Sigma, USA) and "Serpisten" which was isolated from *Serratula coronata* by scientific workers in Institute of Biology of Komi Scientific Center of the Ural Branch of the Russian Academy of Sciences (head is Dr. Sc. V.V. Volodin) [18] and kindly placed us for investigations. All compounds were used without the additional purification.

AOA of the investigated objects was determined on the methyl oleate oxidation model [19], which is the thermal autooxidation of methyl oleate (RH). An autooxidation is carried out in a kinetic (constant bubbling of air through the oxidation cell, which ensures the independence of the oxidation rate on the oxygen concentration) and diffusion (oxidation in a thin layer, when the oxygen concentration is due to the rate of its diffusion into the methyl oleate layer) ranges. The initiation rate of radicals (W_0) under the autooxidation of RH was modified by carrying out the process at different temperatures or using RH of varying degrees of the oxidation. The course of oxidation was followed by the accumulation of hydroperoxides (ROOH), the concentration of which was determined by iodometric titration. The error of the measurement ROOH concentration determination does not exceed 1.5%. As the duration of the induction period (τ) during the autooxidation of RH in the kinetic range, we took the interval when the concentration of ROOH is equal to 0.02 mmol/g. When the oxidation of RH was performed in a thin layer, the value of τ is the interval from zero to the perpendicular dropped on the X-axis from the point of intersection of the linear plots of the kinetic curve of peroxide accumulation corresponding to the initial oxidation rate in the induction period and the maximum rate of the peroxide accumulation. The AOA value was calculated as the difference in the periods of the induction of the methyl oleate oxidation in the presence of additions and the methyl oleate itself, attributed to 1 g of the addition.

The inhibitory efficiency was evaluated by the difference in values of the induction period under the RH oxidation in the presence and the absence of test additions, attributed to the induction period of the RH oxidation (τ_{RH}), The kinetic curves were analyzed by the KINS computer software package [20].

Antiperoxide activity (APA), i.e., the ability preparations to decompose the peroxides, was evaluated by the difference in the peroxide concentrations in oxidized methyl oleate and after the addition of an additive attributed to 1 g of an addition [21].

The experimental data were processed by the commonly used variation statistics method using the Exel software package. The results are presented as the arithmetic mean values indicating root-mean-square errors of the arithmetic mean (M ± m).

2.3   RESULTS AND DISCUSSION

Various versions of chemiluminescent and photo-chemiluminescent methods, the model systems using initiators, EPR method, as well as autooxidation reactions of the fatty acids or their esters (for example, see [19, 22–25]) are used to evaluate the antiradical and AO properties of compounds. It was established that the oxidation of both individual organic compounds, and their oxidation in the presence of AO is a complex multi-stage process, the mechanism of which is usually represented as the following kinetic consequences of reactions (Scheme 2.1).

(0) $I \rightarrow 2I^\bullet \ (+RH) \rightarrow 2R^\bullet$ **initiation**

(1) $R^\bullet + O_2 \rightarrow RO_2^\bullet$

(2) $RH + RO_2^\bullet \rightarrow ROOH + R^\bullet$ **propagation of chain**

(3) $ROOH \rightarrow RO^\bullet + OH^\bullet$ **branching**

(6) $RO_2^\bullet + RO_2^\bullet \rightarrow ROOR + O_2$ **termination of chains**

(7) $InH + RO_2^\bullet \rightarrow ROOH + In^\bullet$ **inhibition**

(8) $In^\bullet + RO_2^\bullet \rightarrow$ **molecular products**

(9) $In^\bullet + In^\bullet \rightarrow$ **molecular products**   **(APA)**

(10) $In^\bullet + RH + O_2 \rightarrow RO_2^\bullet + InH$

(10') $In^\bullet + ROOH \rightarrow RO_2^\bullet + InH$

(11) $InH + ROOH \rightarrow$ **molecular products**

(12) $InH + O_2 \rightarrow In^\bullet + HO_2^\bullet$

SCHEME 2.1   Mechanism of the liquid-phase oxidation of the organic compounds in the presence of antioxidants.

However, it should be noted that a depending on the nature of the model substrate and the initiation rate of radicals in the system, the rate constants of the individual intermediate stages in the oxidation process have the different contributions to the evaluation of the inhibitory efficiency of the test compounds. This causes both the determination of the different parameters of their AOA and significant differences in the inhibitory efficiency of the compounds under using different model systems. The processes of lipid peroxidation (LPO) in the tissues of the organism and the processes of the oxidative degradation of the food predominantly proceed in the autooxidation range, which is an autoinitiated process with the positive feedback, which is realized through the hydroperoxide [22]. It is due to the necessity to use the model systems of the autooxidation of the fatty acids or their esters, in spite of their labor intensity and duration.

A detailed computer analysis of the kinetics of the hydroperoxide accumulation during the low-temperature autooxidation of RH, including in the presence of lipids from the biological objects, carried out in works [26, 27] by means of the software package KINS allows us to significantly expand the possibilities of using this model system for studying the mechanism of the inhibition of oxidative processes by various AO, the cell components and their mixtures. Indeed, as in case of the RH autooxidation at 37°C [27], the hydroperoxide accumulation during its autooxidation at 50°C (the diffusion oxidation region) and 60°C (the kinetic oxidation region) is described by the exponential dependence $[ROOH] = a \cdot \exp(kt)$ with a correlation coefficient of 0.94–1.0 (Figure 2.1a). Besides, the magnitude of the preexponential factor a of an exponential curve increases with increasing W_0 (Figure 2.1b). Earlier, an inverse correlation dependence was revealed between the values of the factor a of the kinetic curves of the peroxide accumulation during the RH autooxidation in the presence of lipids isolated from mammalian tissues and their AOA [27]. At the initial stages of the oxidation process, when the consumption of the substrate can be neglected, the value of the exponential index k is proportional to the total oxidation rate determined by the parameters $(k_2/\sqrt{k_6})[RH]_0)$ and $\sqrt{k_3}$ (see Scheme 2.1), and decreases with increasing W_0 [27].

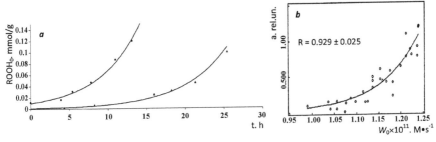

FIGURE 2.1 Kinetic regularities of the autooxidation of methyl oleate at 60°C: a – kinetic curves of the hydroperoxide accumulation during autooxidation; b – dependence of the preexponential factor a on the initiation rate of radicals.

It is believed that the main biologically active compounds of *L. Japonica* are 3,5-diiodotyrosine, fucoxanthin, and alginic acid, whose structural formulas are represented by Figure 2.2, and the color of brown algae is attributed to the presence of carotenoids (β-carotene and fucoxanthin). Rutin is considered as the main biologically active compound of *S. japonica* (Figure 2.2). In addition, the plant sterols [28] and TP belong to these compounds, which play an important role among the components of the plant cells. Hence, the investigated objects differ in the chemical nature of the constituent compounds; therefore the components of the plant cells can have the different inhibitory efficiency, including due one to the reciprocal influence of the compounds on the total AOA of the mixture. These assumptions resulted in the selection of the model compounds for the investigation, the structural formulas of which are given in Figure 2.3, and their mixtures for evaluation of the inhibitory efficiency mechanism of the plant cell components.

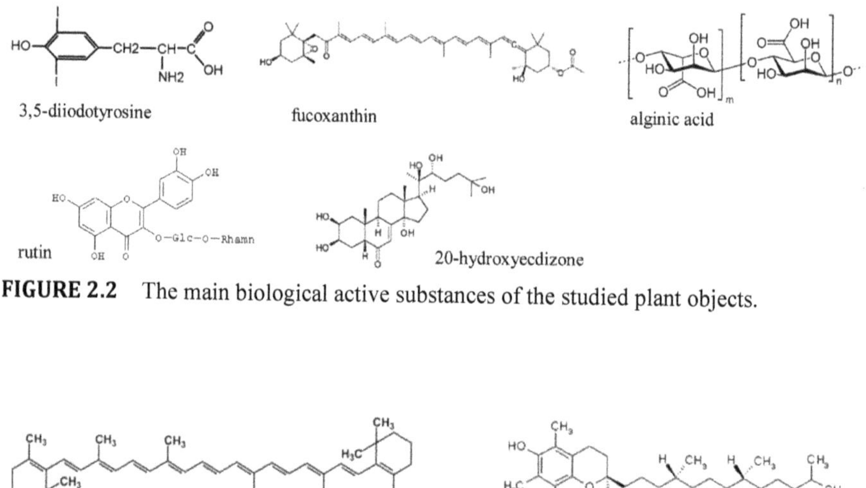

FIGURE 2.2   The main biological active substances of the studied plant objects.

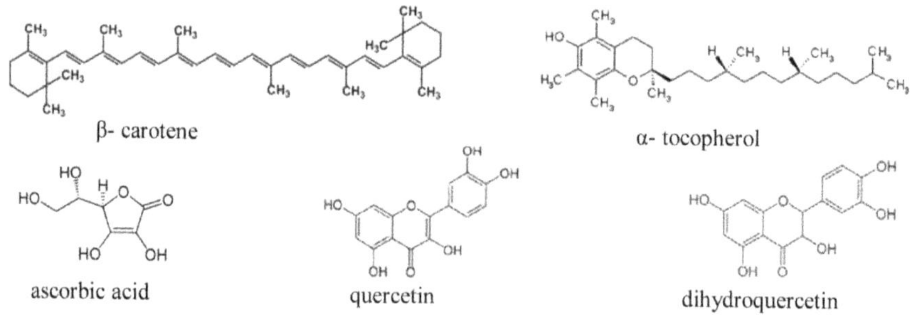

FIGURE 2.3   Structural formulas of the used model compounds.

Experiments have shown that 1 g of the powder of the dry brown algae contains 32.5 ± 3.4 mg (n = 10) of the total lipids. Based on the solubility of BAS contained in *L. Japonica* in organic solvents, it can be expected that they are extracted together with lipids. As shown earlier, in the processes of the low-temperature autooxidation of methyl oleate lipids extracted from animal tissues take part in the reactions of the radical initiation and the chain propagation already during the initial stages of the reaction, even if their concentration in the reaction medium does not exceed 3% [27, 29]. This causes a significant complication of the mechanism of the oxidation process and the appearance of additional intermediate steps, which is reflected in Scheme 2.2.

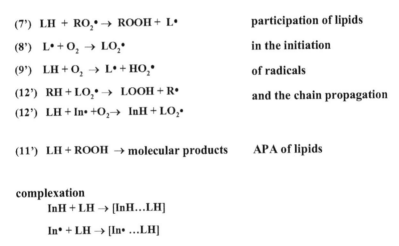

(7') $LH + RO_2^{\bullet} \rightarrow ROOH + L^{\bullet}$ **participation of lipids**

(8') $L^{\bullet} + O_2 \rightarrow LO_2^{\bullet}$ **in the initiation**

(9') $LH + O_2 \rightarrow L^{\bullet} + HO_2^{\bullet}$ **of radicals**

(12') $RH + LO_2^{\bullet} \rightarrow LOOH + R^{\bullet}$ **and the chain propagation**

(12') $LH + In^{\bullet} + O_2 \rightarrow InH + LO_2^{\bullet}$

(11') $LH + ROOH \rightarrow$ molecular products   **APA of lipids**

complexation
$InH + LH \rightarrow [InH...LH]$

$In^{\bullet} + LH \rightarrow [In^{\bullet}...LH]$

SCHEME 2.2 Mechanism of the participation of lipids in the autooxidation process.

Analysis of the physicochemical properties of lipids from *L. Japonica* revealed the complex character of the dependence of their ability to inhibit the RH autooxidation on the lipid concentration at $W_0 = (3.26 \pm 0.02) \cdot 10^{-10}$ M s^{-1}. In the concentration range from 1 to 20 mg/ml of RH, the brown algae lipids show both AO and prooxidant properties, accelerating the oxidation of methyl oleate (Table 2.1).

TABLE 2.1 The Ability of *L. Japonica* Lipids to Inhibit the Autooxidation of Methyl Oleate (Temperature 60°C)

[LH], mg/ml	1.0	1.75	2.7.	6.9	10.0	15.0	20.0
AOA, h×ml/g of LH	−3950	+975	+130	−200	+265 ± 40	55.7 ± 0	−35 ± 50

As shown, the prooxidant activity of lipids, despite the presence of AO in them, is due to their high degree of the unsaturation. Indeed, the lipids from *L. Japonica* contain TP, but the fatty acid composition of the lipids is predominantly represented by the unsaturated acids with a high content of the polyunsaturated fatty acids [30]. The participation of the brown algae lipids in the autooxidation processes at stages of the radical initiation and the chain propagation is indicated by the correlation between a decrease in the inhibitory efficiency of the *L. Japonica* lipids with an increase in the preexponential factor *a* of the kinetic curves of the peroxide accumulation during the oxidation of methyl oleate in their presence (Figure 2.4). However, the *L. Japonica* lipids do not exert a significant influence on the value of the exponential index: $k_{rel} = 1.125 \pm 0.058$, i.e., the overall oxidation rate of RH is increased by 12.5% ($n = 10$) in the entire range of the lipid concentrations studied.

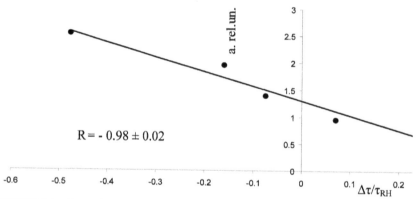

FIGURE 2.4 Dependence of the relative value of the preexponential factor *a* on the inhibitory efficiency of *Laminaria japonica* lipids during the autooxidation of methyl oleate (60°C).

It was mentioned above that one of the components of *L. japonica* is β-carotene. A study of its inhibitory efficiency in the autooxidation processes showed that under similar conditions β-carotene in the concentration range $(1.9 \div 3.7) \times 10^{-7}$ mol/l exhibits only prooxidant properties, more pronounced with decreasing W_0 (Table 2.2). However, its presence does not affect the value of the exponential index: $k_{rel} = 0.98 \pm 0.085$. Besides, the inhibitory efficiency of β-carotene (Table 2.2), as in the case of the lipids from *L. japonica*, linearly decreases with increasing value of factor *a*, the moat essentially at a lower initiation rate of radicals during autooxidation. This conclusion follows from the data presented in Figure 2.5.

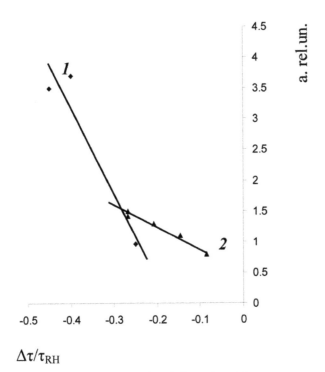

FIGURE 2.5   The interrelation between the relative value of the parameter a of the kinetic curves of the peroxide accumulation during autooxidation of methyl oleate at 60°C in the presence of β-carotene and its inhibitory efficiency depending on the initiation rate of radicals: $1 - W_0 = 2.92 \times 10^{-10}$ M s^{-1}; $2 - W_0 = 3.40 \times 10^{-10}$ M s^{-1}.

TABLE 2.2   Inhibitory Efficiency of β-Carotene and Vetoron in the Autooxidation of Methyl Oleate Depending on the Initiation Rate of Radicals in the Model System (Temperature 60°C)

$W_0 \times 10^{10}$, M s^{-1}	$\Delta\tau/\tau_{RH}$
β-Carotene	
2.92	-0.37 ± 0.06
3.40	-0.195 ± 0.035
Vetoron	
3.15	-0.223 ± 0.015
3.30	0.167 ± 0.075
3.40	0.037 ± 0.035

Similarly, to lipids of the brown algae, in the composition of Vetoron, in addition to β-carotene, ascorbic acid and TP are presented. This causes the appearance of its weak AO properties at a higher W_0, reaching a maximum value at $W_0 = 3.3 \times 10^{-10}$ M s^{-1}, and the preservation of the prooxidant activity of the preparation with a decrease in the initiation rate of radicals to $W_0 = 3.15 \times 10^{-10}$ M s^{-1} (see Table 2.2.). It was found that Vetoron does not take part in the reactions of the initiation of the radicals and the chain propagation at the early stages of the oxidation process: in the presence of Vetoron, the values of the preexponential factor a of the kinetic curves of the accumulation of ROOH corresponds to that in the autooxidation of methyl oleate in all variants of experiments and is equal $a_{rel.} = 0.97 \pm 0.07$ ($n = 14$). At the same time, the relative value of the exponential index k decreases from 1.435 ± 0.095 to 1.10 ± 0.08 as W_0 increases from 3.15×10^{-10} to 3.4×10^{-10} M s^{-1}. It indicates that with the rise of the initiation rate of radicals, the contribution of Vetoron's side reactions to the overall rate of the oxidation decreases. This conclusion agrees with the results of Ref. [31], in which it is shown that the inhibitory efficiency of Vetoron significantly exceeds the analogous ability of β-carotene in processes of the initiated oxidation.

A 20% alcoholic extract of *Sophora* japonica fruit in the reaction of the methyl oleate autooxidation at 60°C in concentrations at the range from 5 to 50 μl/ml has high AOA with the maximum level at a concentration of 5 μl/ml of RH: AOA = 24870 ± 870 h.ml/g. At concentrations of extract from 15 to 50 μl/ml the AOA value retained a high level (AOA = 14150 ± 640 h.ml/g, $n = 4$), although it reduced 1.7 times. Nevertheless, a linear correlation was found between the inhibitory efficiency of extract and its concentration in the reaction medium over the entire range of values studied (Figure 2.6a). Since the main BAS of the alcoholic extract of *S. japonica* fruit is believed to be rutin (glucorhamnoglycoside of quercetin), then Q and QH$_2$ were chosen as model compounds. When the autooxidation of methyl oleate was performed in a thin layer at a temperature 50°C a linear dependence was also revealed between the induction periods of the oxidation on the concentration of both flavonoids (Figure 2.6b). This allows us to conclude that both for the extract of *S. japonica* fruits, and for Q and QH$_2$ the contribution of side reactions to the RH autooxidation process is negligible.

However, a detailed analysis of the kinetic curves of the ROOH accumulation during the autooxidation of methyl oleate in the presence of the extract from *S. japonica* fruits and flavonoids revealed certain differences. So, with increasing the extract concentration from 5 to 50 μl/ml a significant decrease in the value a from 0.034 ±0.014 to 0.003 ±0.001 is revealed by compared with the corresponding value in the RH autooxidation. Earlier, a significant decrease in the relative value a has been shown for quercetin and QH$_2$ in the process of the RH autooxidation in a thin layer at 50°C [15].

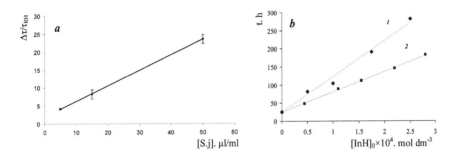

FIGURE 2.6 (a) The dependence of the inhibitory efficiency of the alcohol extract of *S. japonica* fruits on its concentration during the autooxidation of methyl oleate (60°C). (b) The dependence of the autooxidation induction period of methyl oleate (50°C) on the concentration of quercetin (1) and dihydroquercetin (2).

This allows us to conclude that all the studied preparations participate in reactions of the radical initiation and the chain propagation during the induction period of the oxidation. The presence of extract from the *S japonica* fruits in the reaction system is accompanied by a decrease in the relative value of the index k from 0.85 ± 0.15 to 0.30 ± 0.01 with an increase in its concentration from 5 to 50 µl/ml. Earlier, for quercetin and QH_2 under the RH autooxidation in a thin layer, their significant effect on the exponential index k for curves of the peroxide accumulation was not found in comparison with that for the kinetic curves of the peroxide accumulation under the RH autooxidation [15]. The present results suggest a preferential interaction of flavonoids with peroxy radicals in the autooxidation reactions, and a reduction of the RH autooxidation rate after the end of the induction period in the presence of an alcohol extract from the *S. japonica* fruits is due to its participation in the side reactions involved in the oxidation process.

One of the important side reactions that affect the oxidative processes is an interaction of AO and lipids with molecular oxidation products – peroxides (See Scheme 1, the reaction of 11 and Scheme 2, the reaction of 11'). Indeed, it has been established that all investigated components of the plant cells possess APA, i.e., an ability to decompose peroxides to molecular products (Figure 2.7). Among the studied model compounds APA only Vetoron possesses a high value of APA the magnitude of which under similar experimental conditions is 47.0 ± 2.3 mmol/(g×g$_{preparation}$).

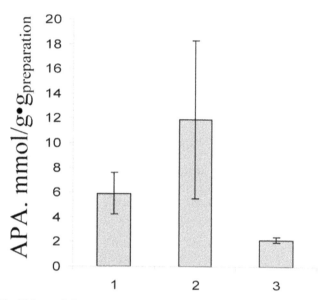

FIGURE 2.7 Values of the antiperoxide activity of lipids (1) and dry powder (2) of *L. japonica*, the 20% alcohol extract from *S. japonica* fruits (3).

The ability to decompose the peroxides is also preserved in a number isobornylphenol synthesized on the basis of the vegetable raw materials and characterized by relatively high antiradical activity [32, 33]. It is also shown that the magnitude of their APA essentially depends not only on the chemical structure of the compound, but also on the ability to form complexes with the natural phospholipids [33]. The mutual influence of components of the complex systems allows us to suggests that the high degree of unsaturation of β-carotene (the main component in Vetoron's composition) which has the prooxidant properties in the autooxidation reactions, eliminates the ability of Vetoron to decompose the peroxides and reduce the overall oxidation rate after an induction period. Besides, the high APA of Vetoron provides a presence, though weak, its AO properties in the autooxidation reactions. The absence of a linear dependence of the AO properties of *L. japonica* lipids on their concentration in the system, in our opinion, is also caused by the competition between the reactions of the radical initiation due to the high degree of the unsaturation of lipids themselves and carotenoids which are present in lipids and APA of the brown algae lipids.

Of course, it is necessary to bear in mind that among components of plant cells tocopherols, which play an important role in the regulation of oxidative processes are almost always present. The most common is α-tocopherol (TP), the kinetic characteristics of which have been well studied [10, 11]. It was established that TP is actively involved in side reactions, regardless of the oxidation conditions in the system, and this leads to a decrease in the induction period of the oxidation process with an increase in its concentrations [11, 15]. In the autooxidation of methyl oleate (temperature 60°C) TP at a concentration of 2.55×10^{-5} mol/l is characterized by an inhibitory activity of 0.36 and decreases the relative value of the factor a by 14%, and the exponent index is by less than 10% at the initiation rate of radicals W_0 = 3.44×10^{-10} M.s^{-1}. At the same time, the decrease of the initiation rate of radicals to $W_0 = 2.98 \times 10^{-10}$ M.s^{-1} and the TP concentration to 1.28×10^{-5} mol/l is accompanied by an increase of its inhibitory efficacy to 0.643 ± 0.043 with a significant reduction in the relative value and the factor a (0.415), and the exponential index k (0.69) as compared with the same values for RH. A more high inhibitory efficiency and a significant reduction in the values of a and k at low W_0 and at the lower of the studied TP concentrations correspond to the literature data. TP also has APA, the value of which is equal to 0.39 ± 0.13 mmol/(g×g$_{TF}$).

An important component of plant cells are sterols, which play a substantial role in their metabolism and have lately been used as a dietary supplement [28, 34]. It is shown that the biological activity of the plant sterols is based on their ability to participate in the regulation of LPO processes under the administration into an animal body [35]. In this regard, the inhibitory activity of "Serpisten" in a concentration range from 9.5×10^{-6} to 1.14×10^{-3} M was determined by the RH autooxidation both at physiological temperature of 37°C and at 60°C. It was found that at 37°C "Serpisten" has only prooxidant properties, and there is no linear dependence of the inhibition efficiency on its concentration. Moreover, at a low temperature of the oxidation "Serpisten" practically does not affect both the value of a in the whole range of the studied concentrations ($a_{rel} = 1.09 \pm 0.03$) and the value of k, if its concentration does not exceed 5.7×10^{-4} M ($k_{rel} = 1.04 \pm 0.03$). When the concentration is raised to 1.14×10^{-3} M "Serpisten" increases the exponential index by 1.55 times, which indicates its participation in the autooxidation process as a substrate.

The ability of "Serpisten" to participate in the autooxidation processes at the stages of the radical initiation and the chain propagation is even more clearly pronounced when the process of oxidation is carried out at 60°C.

Under these oxidation conditions, there is not a linear dependence of the inhibitory efficiency of "Serpisten" on its concentration, the preparation exhibits both AO and the prooxidant properties and practically does not change the value of the exponential index k in all variants of experiments. The value of the factor a decreases linearly with increasing the inhibitory efficiency of "Serpisten," however, the coefficient of the linear regression of this correlation dependence differ by 3.7 times, depending on the rate of the radical initiation in the system, which follows from the data presented in Figure 2.8.

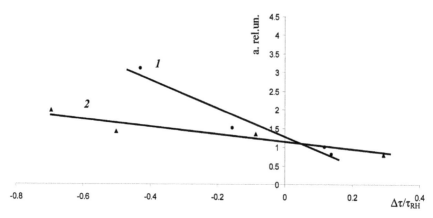

FIGURE 2.8 The relationship between the relative value of the preexponential factor a and the inhibitory efficiency of "Serpisten" in the autooxidation reaction of methyl oleate at a temperature 60°C and $W_0 = (2.96 \times 10^{-10} - 2.975 \times 10^{-10})$. M.s^{-1} (straight line 1) and $W_0 = (3.355 \times 10^{-10} - 3.38 \times 10^{-10})$ M.s^{-1} (straight line 2).

It should also be noted that "Serpisten," as well as other components of the plant objects, has the ability to decompose the peroxides (APA = 0.41 ± 0.09 mmol/(g×g$_{preparation}$) and may have a modifying effect on the inhibitory efficiency of the AO. The verification of this assumption has been verified by studying the inhibitory efficiency of mixture "Serpisten" with TP. A significant change in the parameters of the kinetic curves of the hydroperoxide accumulation of the mixture was revealed in comparison with the similar values for the individual components, including a decrease in the inhibitory activity of TP by 5.6 times in the presence of "Serpisten" (Figure 2.9).

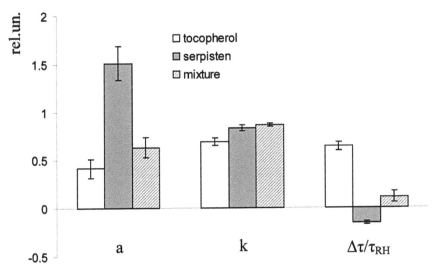

FIGURE 2.9 The relative value of the parameters of the exponential curves of the peroxide accumulation during the autooxidation of methyl oleate ($W_0 = 3.0 \times 10^{-10}$ M's $^{-1}$, temperature 60°C) in the presence of "Serpisten" at a concentration of 2.1×10^{-4} M, α-tocopherol at a concentration of 1.3×10^{-5} M and their mixtures and the inhibitory efficiency of the studied systems.

2.4 CONCLUSION

Thus, the results of the studying the inhibitory efficiency of components of the plant cells in the oxidative processes indicate a complex and ambiguous mechanism of the oxidation of the model substrate in their presence. All investigated compounds and preparations differ substantially both in their effect on the duration of the induction period of oxidation and the ability to participate in the reactions of the radical initiation and the chain propagation at the different stages of the oxidation process, which can be seen from the change in the parameters of the kinetic curves of the oxidation of the model substrate in their presence.

It should be noted that all the studied plant components have antiperoxide properties, i.e., the ability to decompose the peroxides, which persist in the compounds synthesized on the basis of the plant raw materials. It is also important that the inhibitory efficiency of such natural AO as α-tocopherol and flavonoids is essentially dependent on their participation in the formation of complexes with one of the basic structural components of biological membranes as phospholipids [33–37], and the total AO effect of the mixtures of the phospholipids with the natural AO is due to the rate of the radical initiation in the oxidation process [14, 38].

An analysis of the experimental data confirms the notion that the contribution of a separate intermediate stage of the oxidation process in the result depends essentially on the ratio of the reaction rates of all its components. It is therefore logical that the physicochemical properties of the compounds belonging to the plant object, in combination with the other components of cells determine the inhibitory efficiency of the plant extracts and tinctures, and their ability to influence the regulation of the oxidative processes when they are administrated into the organisms is the basis of their biological activity.

ACKNOWLEDGMENT

The authors are sincerely grateful to V. V. Volodin, PhD, for providing "Serpisten" preparation for our research.

KEYWORDS

- antioxidant activity
- antioxidants
- antiperoxide activity
- biologically active substances
- dihydroquercetin
- lipid peroxidation

REFERENCES

1. Burlakova, E. B., Shilov, A. E., Varfolomeev, S. D., & Zaikov, G. E., (2005). Chemical and biological kinetics: New aspects. *Chemical Kinetics* (Vol. 1, p. 528). Leiden – Boston: VSP.
2. Burlakova, E. B., & Varfolomeev, S. D., (2005). Chemical and biological kinetics: New aspects. *Biological Kinetics* (Vol. 2, p. 562). Leiden – Boston: VSP.
3. Zagoskina, N., & Burlakova, E., (2010). *Phenolic Compounds: Fundamental and Applied Aspects* (p. 400). Moscow: Scientific World, (in Russian).
4. Gupta, V. K., & Sharma, S. K., (2006). Plants as natural antioxidants. *Natural Product Radiance*, 5(4), 326–334.

5. Merkl, R., Hradkova, I., Filip, V., & Smidrkal, J., (2010). Antimicrobial and antioxidant properties of phenolic acids alkyl esters. *Czech. J. Food Sci., 28*(4), 275–279.

6. Sirisha, N., Sreenivasulu, M., Sangeeta, K., & Madhusudhana, C. C., (2010). Antioxidant properties of Ficus speicues – A Review. *Int. J. Pharm. Tech Research, 2*(4), 2174–2182.

7. Starlin, T., & Gopalakrishnan, V. K., (2013). Enzymatic and non-enzymatic antioxidant properties of *Tylophora pauciflora* Wight and Arn – an in vitro study. *Asian J. Pharm. Clin. Research, 6*(4), 68–71.

8. Okoh, S. O., Iweriegbor, B. S., Okoh, O. O., Nwodo, U. U., & Okoh, A. I., (2016). Bacterial and antioxidant properties of essential oils from the fruits *Dennettia tripetala* G. Baker. *BMC Complementary and Alternative Medicine, 16*, 486–498.

9. Misharina, T. A., Yerokhin, V. N., & Fatkullina, L. D., (2017). Effect of low doses of savory essential oil dietary supplementation of lifetime and the fatty acid composition of the aging mice tissues. In: Alexander, V., Kutchin, L. N., Shishkina, L., & Weisfeld, I., (eds.), *Chemistry and Technology of Plant Substances: Chemical and Biochemical Aspects* (pp. 211–226). Toronto – New Jersey: Apple Academic Press.

10. Burton, G. W., & Ingold, K. U., (1989). Vitamin E as an in vitro and in vivo antioxidant: Review. *Am. N.Y. Acad. Sci., 570*, 7–22.

11. Burlakova, E. B., Krashakov, S. A., & Khrapova, N. G., (1995). Kinetic Peculiarities of tocopherols as antioxidants, *Chem. Phys. Rep., 14*(10), 1657–1690.

12. Gody, V., Middleton, E., & Harborne, J. B., (1986). *Plant Flavonoids in Biology and Medicine: Biochemical, Pharmacological and Structure-Active Relationships* (p. 330). New York: Liss.

13. Belyakov, V. A., Roginsky, V. A., & Bors, W., (1995). Rate constants for the reaction of free radical with flavonoids and related compounds as determined by the kinetic chemiluminescence method. *J. Chem. Soc. Perkin Trans., 2*, 2319–2326.

14. Mazaletskaya, L. I., Sheludchenko, N. I., & Shishkina, L. N., (2012). Inhibitory efficiency of antioxidant and phospholipid mixtures under the different oxidation extent of methyl oleate. *Chemistry & Chem. Technology, 6*(1), 35–41.

15. Mazaletskaya, L. I., Sheludchenko, N. I., & Shishkina, L. N., (2012). Influence of the initiation rate of Radicals on the kinetic characteristics of quercetin and dihydroquercetin. In: Gennady, E. Z., & Kozilowsky, R. M., (eds.), *Chemical Reactions in Gas, Liquid and Solid Phases Synthesis, Properties and Application* (pp. 11–20). New York: Nova Science Publisher.

16. Kasaikina. O. T., (2005). Kinetic model of β-carotene/ influence of oxygen partial pressure. Burlakova, E. B., Shilov, A. E., Varfolomeev, S. D., & Zaikov, G. E., (eds.), *Chemical and Biological Kinetics: New Aspects: Chemical Kinetics* (Vol. 1, pp. 322–336). Leiden – Boston: VSP.

17. Kates, M., (1975). *The Technologue of Lipidology* (p. 322). Moscow: Mir, (Russian version).

18. Volodin, V. V., & Volodina, S. O. (2000). Method of the ecdysteroids production. Patent 2153346, Russia. MKIS A 61 K 53/7, Institute of Biology of Komi SC, Ural Branch of Russian Academy of Sciences, N 99106351/14, Publ. 27.07.2000, Bin 21.

19. Burlakova, E. B., Alesenko, A. V., Molochkina, E. M., Palmina, N. P., & Khrapova, N. G., (1975). *Bioantioxidants in the Radiation Damage and Tumor Growth* (p. 214). Moscow: Nauka (Science, in Rus.) Publishing House.

20. Brin, E. F., & Travin, S. O., (1991). Modeling mechanisms of the chemical reactions. *Chem. Phys. Reports, 10*(6), 830–837.

21. Menshov, V. A., Shishkina, L. N., & Kishkovskii, Z. N., (1994). Effect of biosorbents on the composition, content, and antioxidative properties of lipids in the medium. *Applied Biochemistry and Microbiology, 30*(3), 359–369.

22. Denisov, E. T., & Azatjaian, V. A., (1996). *The Inhibition of Chain Reactions* (p. 268). Chernogolovka, (in Russian).

23. Emanuel, N. M., Zaikov, G. E., & Maizus, Z. K., (1984). *Oxidation of Organic Compounds: Effect of Medium* (p. 650). Oxford: Pergamon Press.

24. Emanuel, N. M., & Gal, D., (1984). In: Berezin, I. V., (ed.), *Oxidation of Ethylbenzene (Model Reaction)* (p. 376). Moscow: Nauka (Science, in Rus.) Publishing House.

25. Burlakova, E. B., Kruglyakova, K. E., & Shishkina, L. N., (1992). *Investigation of Synthetic and Natural Antioxidants In Vitro and In Vivo* (p. 110). Moscow: Nauka (Science, in Rus.) Publishing House.

26. Shishkina, L. N., Menshov, V. A., & Brin, E. F., (1996). Prospects of utilization of the model system of methyl oleate oxidation for studies of kinetic properties of lipids. *Biology Bull., 3*, 240–244.

27. Shishkina. L. N., & Khrustova, N. V., (2006). Kinetic characteristics of lipids of mammalian tissues in Autooxidation reactions. *Biophysics, 51*(3), 292–298.

28. Volodin, V. V, (2003). *Phytoecdysteroids* (p. 293). St. Peterburg: Nauka (Science, in Rus.) Publishing House. (in Russian).

29. Khrustova, N. V., & Shishkina, L. N., (2004). The role of peroxides in the mechanism of low-temperature autooxidation of methyl oleate and its solution with lipids. *Kinetic Catal., 45*(6), 799–809.

30. Honya, M., Kinoshita, T., Ishikawa, M., Mori, H., & Nishizawa, K., (1994). Seasonal variation in the lipid content of cultured *Laminaria japonica*: Fatty acids, sterols, β-carotene and tocopherol. *J. Appl. Phycol., 6*, 25–29.

31. Kasaikina, O. T., Kartashova, Z. S., Lobanova, T. V., & Sirota, T. V., (1998). Effect of surround on the reactivity of β-carotene with respect to oxygen and free radicals. *Biochemistry (Moscow)/ Suppl. Ser. A. Membrane and Cell Biology, 15*(2), 168–176 (in Russian).

32. Mazaletskaya, L. I., Sheludchenko, N. I., Shishkina, L. N., Kutchin, A. V., & Chukicheva, I. Y., (2011). Kinetic characteristics of isobornylphenols with peroxyl radicals. *Petroleum Chemistry, 51*(5), 348–355.

33. Marakulina, K. M., Kramor, R. V., Lukanina, Y. K., Shishkina, L. N., Fedorova I. V., Kutchin, A. V., & Chukicheva, I. Y., (2013). Complexation of lecithin with phenolic antioxidants. In: Rafiq, I., (ed.), *The Science and Engineering of Sustainable Petroleum* (pp. 215–226). New York: Nova Science Publisher.

34. Volodin, V. V., Pchelenko, L. D., Volodina, S. O., Kudyasheva, A. G., Shevchenko, O. G., & Zagorskaya, N. G., (2006). Pharmacological estimate of new containing ecdysteroid substance "Serpisten." *Plant Resources, 42*(3), 113–129. (in Russian).

35. Shishkina, L. N., Shevchenko, O. G., & Zagorskaya, N. G., (2011). Influence of ecdysteroid-containing compounds on the oxidation processes regulation. *Oxidation Commun., 34*(3), 711–725.

36. Burlakova, E. B., Mazaletskaya, L. I., Sheludchenko, N. I., & Shishkina, L. N., (1995). Inhibitory effect of the mixtures of phenol antioxidants and phosphatidylcholine. *Russian Chemical Bulletin, 44*(6), 1014–1020.

37. Xu, K., Liu, B., Ma, Y., Du, J., Li, G., Gao, H., Zhang, Y., & Ning, G., (2009). Physicochemical properties and antioxidant activity of luteolin-phospholipid complex. *Molecules, 14*, 3486–3493.
38. Mazaletskaya, L. I., Sheludchenko, N. I., & Shishkina, L. N., (2010). Lecithin influence on the effectiveness of the antioxidant effect of flavonoids and α-tocopherol. *Applied Biochemistry and Microbiology, 46*(2), 135–139.

CHAPTER 3

Potassium Salt of Phenosan Influences on the State of the Soluble and Membrane-Bounded Proteins

OLGA M. ALEKSEEVA,[1] YURY A. KIM,[2] and ELENA A. YAGOLNIK[3]

Department of Plant Science, Bharathidasan University, Tiruchirappalli–620024, India

[1]*Emanuel Institute of Biochemical Physics of Russian Academy of Sciences, 4, Kosygin St., Moscow 119334, Russia, E-mail: olgavek@yandex.ru*

[2]*Institute of Cell Biophysics of Russian Academy of Sciences, Pushchino, Moscow Region 142292, Russia*

[3]*Tulsky State University, 92, Prospect Lenin, Tula 300012, Russia*

ABSTRACT

By two biophysical methods – the spectral measuring of intensity of intrinsic protein's fluorescence and registration of changes of protein microdomain organization by adiabatic differential scanning calorimetry (DSC) – we examined biological effects of biological activity substance – antioxidant (AO) potassium salt of phenosan, on free and membrane proteins structure: bovine serum albumin (BSA) and proteins microdomains of erythrocyte ghosts.

We show by spectral method (on fluorescence intensity of two trypto-phane residues at BSA molecule) that potassium salt of phenosan (potassium phenosan) at 10^{-4} M reduces the fluorescence intensity and shifts the wavelengths of emission. These indicate that the environment around the tryptophane is hydrophilizated when large concentrations of phenoksan, due to globule loosening of molecule BSA and quenched by oxygen, dissolved in water. Potassium phenosan solutions at concentrations of $10^{-17} - 10^{-6}$ M increase the intensity of intrinsic BSA fluorescence, as compared to control without AO. This phenomenon may be explained by the preservation of tryptophane residues emission from quenching by the adsorbed potassium phenosan molecules.

By the applauding of adiabatic process, DSC has been shown the shifts of temperature denaturation of proteins microdomains at erythrocyte ghosts in the presence of AO. The proteins of erythrocytes membranous cytoskeleton intimately mutually operate with each other and form the membranous domains, the denaturation of which forms structural heat transitions: five identified structural transitions of proteins family: spektrin, ancyrin, bands proteins 4.1 and 4.2 and band protein 3, are mutual of membranous cytoskeleton in practice of all cells at animal organism. The general DSC ghost melting parameter Tmax shifts were determined. Tmax were changed without shape changing of protein microdomains peaks. Potassium phenosan causes restructuring in protein microdomains organization; however, its domains were remaining.

The obtained *in vitro* data suggest that influencing of potassium phenosan to free soluble protein (BSA) and proteins, which are membrane-bounded (erythrocyte ghosts), is in strong dependence on potassium phenosan concentration at experimental mediums.

3.1 INTRODUCTION

The potassium salt of phenosan-acid (potassium phenosan) is hindered phenol, [β-4-oxy-(3,5-ditretbutyl-4-oxiphenyl) potassium propionate], was synthesized in Institute of Chemical Physics of USSR Academy of Sciences and Moscow petroleum processing plant [1]. It is suggested as an antioxidant (AO). The potassium phenosan is amphiphilic, which is primarily being distributed in an exterior layer of the membrane. It changes the bilayer thickness and regularity at multilayers.

The creation of phenosan-acid preparation is the successful experience by work's cooperation of N. N. Semenov Institute of Chemical Physics of Academy of Sciences of USSR.

Two main points of route "idea – thing": academic institution – petroleum processing plant, allowed the creating the highly effective stabilizer of polymers phenosan-acid by the success for 5 years instead of 10–15 years. Phenosan-acid was named by "plant fenols and academy of science" coworking. Phenosan-acid is white crystalline powder with the melting temperature 85°C. It is dissolved in acetone, benzene, toluene, diethyl ether, and heptane; but it is not dissolved in water. Esters of acid phenosan-acid and diethylene glycol and others are belonging among no staining heat stabilizers for polyolefins and other polymeric materials at dosages 0.1–1%.

At the base of phenosan-acid many substances were synthesized with variable tasks for certain targets from stabilizations of polymers to influencing to biological objects of different levels of the organization. For more bio-available, the hydrophilic potassium salt of phenosan-acid, named potassium phenosan, was created at Institute of Biochemical Physics of Russian Academy of Sciences (IBCP RAS) [1]. The potassium phenosan is dissolved in water very well. This substance is not toxic. When it was tested on biologic objects actions, its favorable biological influence was found (by works of IBCP RAS) [2]. Thus, the liposome's bilayers, formed by phospholipids, changed its organization when treated by large doses of phenosan-acid and potassium phenosan. Application of 5% phenosan-acid and potassium phenosan destroyed the phosphatidylcholine bilayer. The solutions of these substances (5–10%) exchange for 0.7–0.5 nm the recurrence interval of membranes in multilamellar liposomes, formed from phosphatidylcholine. These biologically active substances (BAS) increase the membrane thickness on 0.3 nm [3].

The architectonics of erythrocytes was change under phenosan-acid actions. Phenosan-acid as the amphiphilic substance was distributed at the outer leaf of membrane's lipid bilayer [2]. Phenosan-acid is the strong AO, which affects the structure and the composition of the membrane's lipid phase. But phenosan-acid hasn't the certain target for its action. Potassium phenosan at 10^{-5}M was used, because under this concentration this BSA was not destructed the dimyristoyl phosphatidylcholine (DMPC) lipid microdomains organization at bilayers.

The group of hybrid AOs – ICPANs was synthesized at IBCP RAS. These substances were created for the best binding with the first target of biological objects – biomembranes [4]. Such approaches allow receiving substances with a certain target of the action. For the penetrating to deep locuses of biomembranes, the group of more hydrophobic substances was synthesized. So, the "float" ICPANs, keep in its composition charged the onium group ("anchor") with lipophilic long-chain alkyl tail ("float"), that makes it possible effectively interact with charged lipid bilayer of cell membranes, needing AO defense for the maintenance of normal level of lipid peroxidation (LPO) [4]. AO properties of ICPANs were so much increased in compared to phenosan-acid properties [5]. Multitarget properties of ICPANs were realized at first by inhibition of acetylcholine esterase [6]. This enzyme plays a key role at neurodegenerative deceases. And second ICPANs can penetrate to hydrophobic inner locuses of biomembranes and thus stabilized the membrane structure. The next actions of ICPANs are the binding with the membrane surface. But its influences are in the great dependences of ICPANs concentrations. Large concentrations disrupt the membrane structures [7] and change cellular morphology [8].

But phenosan-acid and potassium phenosan are more bioavailable substance, and its influences are not so disrupting [9]. At experiments in vivo, the investigations of phenosan-acid acid metabolism at rabbit body were occurred. The next derivative of phenosan-acid and potassium phenosan under its metabolism cinnamic acid is more bioavailable for hepatic tissue treatments [10].

Some derivatives of phenosan-acid are accumulated at membrane and create the local subfraction at membrane. And for uniform distributions at biomembranes surfaces the hybrid macromolecular antioxidant (HMA) were created at IBCP RAS and Saint-Petersburg University. In the synthesis of HMA, as of macromolecular basis, containing hydrophilic polymers – the dextran and hydroxyl-ethylated starch with different fractional substitutionality of polymer link by screened phenols were used. HMA are the synthetic macromolecular conjugates with hydrophilic bio- and synthetic polymers. As the base dextran, hydroxyethylated starch and polyvinyl alcohol were used. As bounded materials Phenosan-acid and its dehydrogenated analog 4-hydroxy-3, 5-di-*tert*-butil-cinnamic acid were used [11]. HMA were created for uniform distributions of AOs at biomembranes surfaces. These properties are very important for preserve the bio-object from large local concentrations of AOs. Thus, the AO stress may be is prevented.

But in this chapter, we deal with only one derivative of phenosan-acid – potassium phenosan, which used as aqua solutions at the wide range of concentrations. Its influences to biological objects, which are state at different levels of organization, are so variable. And we tested the structural effects of potassium phenosan to simple bio-objects from animal's blood. In Figure 3.1, the structural formula of phenosan-acid acid and its potassium salt are presented.

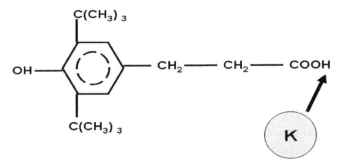

FIGURE 3.1 Structural formula of phenosan-acid and its potassium salt, which is hindered phenol [β-4-oxy-(3,5-ditretbutyl-4-oxiphenyl) potassium propionate], synthesized as antioxidant for polymer stabilization in IBCP RAS [1].

At our works with testing of BAS actions, we usually used the experiment protocols by the scheme, presented in Figure 3.2. But now only two points we showed: blood serum protein BSA (bovine protein) and erythrocytes ghost.

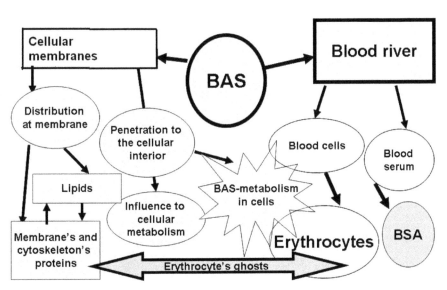

FIGURE 3.2 Scheme of investigations of the structural role of exogenous biological active substances (BAS) at bio-objects – first targets at our work: BSA and erythrocytes ghosts. These experimental points of scheme are marked by gray.

One of the first targets for BAS at animal' body is the blood river. At this tissue, there are two objects for our investigations. These are the albumin at blood serum and red blood cells – erythrocytes. We prepared the model objects for the investigations of our biological active substance influences to these two targets. At first, bovine serum albumin (BSA) and at second, erythrocytes ghosts were our experimental objects. BSA was used for testing of structural changing of free proteins when potassium phenosan presented at experimental medium. Ghosts were used for testing of structural changing of membrane-bounded proteins when potassium phenosan presented at experimental medium. Potassium phenosan was applauded as aqua solutions at a wide range of concentrations.

Some changes of serum albumins conformation were registered on change of extent of quenching its intrinsic fluorescence [12]. So many works are performed by this time using this approach for testing of actions of any biological activity substances on albumins [13].

The schematic picture of the serum albumin molecular structure is presented in Figure 3.3.

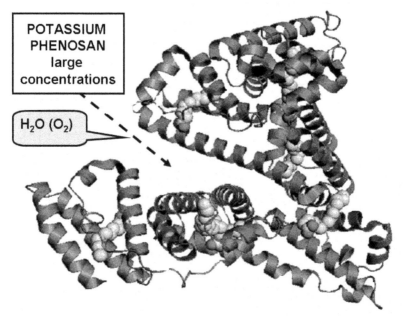

FIGURE 3.3   Scheme of serum albumin molecular structure. (Scheme was modified by us from our earlier paper [14]).

The albumin's molecules rapidly change its conformation at the expense of specific architecture of molecule: intermittent motives of hard spirals α- and β-folded structures with mobile loops Figure 3.2. Some changes of serum albumins conformation were registered on change of extent of quenching its intrinsic fluorescence [13]. We use this approach for the test of potassium phenosan influences on albumins, as experimental object "free soluble protein." We tested the albumin's binding with exogenous synthetics materials by using the registration of the intensity of intrinsic fluorescence of the BSA.

BSA contains two tryptophane residues in hydrophobic regions of its molecule. There is fluorescent emission of two tryptophane residues in hydrophobic regions of molecule BSA after excitation of tryptophane. The first residue is located with close to a surface, a second residue located at the deep inside of the protein globule. When BSA molecule loosening, or unfolding, the availability of tryptophane residues for quencher – oxygen, which was dissolute in water, increase greatly. The quenching of tryptophane fluorescence was observed in this case. The wavelengths of a maximum of fluorescence change too.

The next object for investigation of potassium phenosan actions to animals' cells were erythrocytes – the next first target at Blood River. At our work, we tested the potassium phenosan influence to membrane-bounded proteins of the cellular envelope and intracellular cytoskeleton. For this task, we treated erythrocytes with osmotic shock [15]. Hemoglobin went out from cells fast. These events are presented in Figure 3.4.

And by cooling differential centrifugation, we pelleted the clean closed cellular envelopes with cellular cytoskeleton – erythrocytes ghosts. The preparations of erythrocytes ghosts are free from hemoglobin completely. These experimental objects are a simple and adequate model of cellular envelopes with cellular cytoskeleton for many types of animal's cells at different types of tissues. The main proteins structural components of ghosts are similar to components of another animal's cell [16].

Structural organization of protein microdomains was investigated by adiabatic differential scanning calorimetry (DSC) method [17]. It is known that under the temperature increasing or decreasing the rearrangements of lipids and proteins microdomains at biomembranes are changed too so much [18]. Any pathological events at the body may be registered by measuring of structural transitions of erythrocyte ghost membranes by DSC method. Thus the lipid and protein microdomain organization of human erythrocyte ghost membranes are changed under oxidative stress factors actions that are mutual characteristics of many diseases [19]. The structural rearrangements of lipids microdomains are registered at low-temperature interval under DSC melting, and changes of organizations protein microdomains occurred when 10–90°C DSC melting [20].

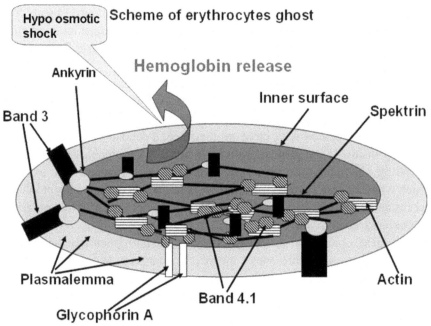

FIGURE 3.4   Scheme of proteins content of erythrocytes ghosts. (Scheme was modified by us from our earlier paper [14]).

For example, these approaches with DSC melting of erythrocyte ghost membranes were used and for investigations of the different effects of other BAS to animal's body. The synthetic plant growth regulator melafen, which kills good native seeds when was applied at large concentrations. And melafen greatly stimulates the plant development and growth from the bed sleep seeds at low and ultra-low concentrations. By these facts of strong concentration-depended regulation, the influences melafen to animal's cells were very interesting. At first for this study, the DSC melting of erythrocyte ghost membranes under melafen treatment were provided with native and aging ghost. This model of fairly simple to be used, and on thermograms are clearly mirrored all peaks of thermo induced denaturation transitions of cytoskeleton's proteins [21]. There are at DSC melting curves five peaks that mirrored the five temperature induced transitions of protein microdomains in ghost membranes of rat erythrocytes [16, 17]. These five temperatures inducted transitions of protein microdomains were submitted in control and after preliminary incubation with testable the aqueous solutions of BAS. Melafen didn't exert these peaks and the great destructive actions to the isolated erythrocytes membranes (with cytoskeleton) under the concentrations that activate the plant growth [22].

Hypo-osmotic hemolysis of isolated erythrocytes is used for testing of BAS influence. Thus this approach is not used only for prepared of erythrocytes ghost. By hypo- (for biological active substance – melafen) [23] and hyper-hemolysis (for BAS – saponins faction and triterpens glycosides) we may test the biological active substance effects to membrane permeability, that is in strong dependence from lipid and protein organizations at biomembranes [24].

3.2   MATERIALS AND METHODOLOGY

1. **Materials:** KCl, KH_2PO_4, NaCl, $MgCl_2$ (Merck); BSA, (Sigma).
2. **Technology:** Potassium phenosan actions to BSA were measured by standards spectroscopic methods [12] by using spectrofluorimeter Perkin-Elmer-44B. The conditions for measurements of fluorescence quenching: protein: wavelengths λ_{ex} = 286 nm (slit 3 nm), λ_{em} = 345 nm (slit 4 nm). BSA 66µg/ml at medium, contained: 0.1 M KCl, 0.02 M HEPES, pH 7.1, 24°C. The quartz cells (1 cm) were used for registration of fluorescence quenching by fluorescent spectrophotometer "Perkin-Elmer MPF-44B." For measurements of optical density was used the spectrophotometer "Specord M 40."

The preparations of erythrocytes ghosts from the blood of rats were performed by the osmotic shock method [25]. The erythrocyte membrane raised its permeability to hemoglobin when hypo-osmotic medium. The maintaining of the cold thermal regime (4°C), without freezing, when centrifugation permits to receive the colorless draft of ghost membranes. Ghost membranes were suspended at phosphate buffer easily.

The registration of the potassium phenosan actions to protein microdomain organization at erythrocytes ghosts were made by the method of DSC [26]. The DSC melting of samples of erythrocytes ghosts treated by this AO was occurred from 10°C to 90°C with the step of 1 grad/min with points 701, concentrations of ghost's protein 1 mg/ml, the volume of sample 0.73 mL heating velocity K/min.

3.3   RESULTS

At first, our investigation deals with the influence of potassium phenosan (at the wide concentration range of aqua solutions) to structural properties of soluble protein – BSA. Potassium phenosan actions were tested by the spectral methods. We found some changing of BSA structure.

As we see in Figure 3.5, the small increase of intrinsic fluorescence intensity occurred, when BSA treated by AO under low and ultra-low concentrations of its aqua solutions. That correlated with increasing of the potassium phenosan concentrations. Maybe this AO associated with surface of BSA as an envelope. Thus these molecules of potassium phenosan preserve the BSA globule from loosing. And the water doesn't penetrate to entertaining of BSA. And the strong quencher – O_2, which was soluted at water volume usually, don't penetrate to deep locuses of BSA molecule, where tryptophane residues exist, and thus O_2 don't quench the tryptophane fluorescence.

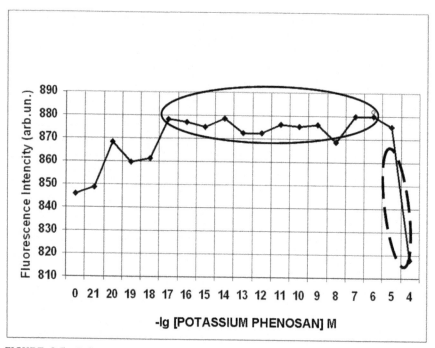

FIGURE 3.5   Influence of the potassium phenosan concentrations to the fluorescence intensity of BSA triptophanil emission (arbitrary units). Region of large concentrations of substance is marked by the oval with the continual line. Region of the low concentrations of potassium phenosan is marked by the oval with a dotted line (Figure was modified by us from our earlier paper [7]).

But large concentrations of potassium phenosan: 10^{-5}, 10^{-4} M, greatly loosed the BSA globule; thus triptophanils become available for O_2, and fluorescence of triptophanils decreased greatly. This phenomenon we may observe, because the great quenching of intrinsic fluorescence occurred under the BSA treatment by large concentrations of AO. Under the increasing of potassium phenosan concentrations, the wavelengths of a maximum of BSA triptophanil emission shifted and increased (Figure 3.6). But when large concentrations of this substance: 10^{-5}, 10^{-4} M, the wavelengths of a maximum of BSA triptophanil emission decreased. Those facts supported the changing of the environment around of BSA tryptophane residues in depending of changing of the potassium phenosan concentrations.

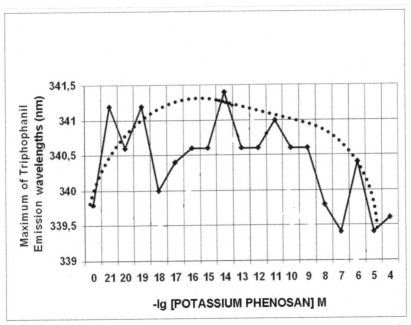

FIGURE 3.6   Influence of the potassium phenosan concentrations to the wavelengths of a maximum of BSA triptophanil emission. It is revealed a shift of triptophanil emission wavelengths (nm). The line, which results in the tendency of potassium phenosan at the low and large concentrations, is marked by the dotted line (Figure was modified by us from our earlier paper [7]).

The shapes and views of emission spectra were similar (not presented), only the maximal value of fluorescence intensity BSA was a change in dependence from the potassium phenosan concentrations at the experimental medium. When large concentrations of potassium phenosan were presented, there were the noticeable tendency of the fluorescence quenching by the potassium phenosan. And increasing of fluorescence intensity there were, when low and ultra-small concentrations of potassium phenosan presented. Evidently, conformational rearrangements occurred in BSA molecules. These rearrangements were small, and had the different directions.

Potassium phenosan molecules affected to the BSA so that under the low and ultra-low concentrations, there was the preserving of the protein tryptophane residues from quenching from oxygen, dissolved in water. At the large potassium phenosan concentrations, the change of protein conformation became essential. At this case, the tryptophane residues that lying at deep locus of molecule became more available for water (and oxygen, respectively) that was indicated by fluorescence quenching. The "loosening" of BSA molecule structure occurs. So, soluble proteins that unhardened of the membrane lipids were under the essential potassium phenosan actions. Taking into account that it is the hydrophilic substance; it can change the water environment.

At this case we may suppose that potassium phenosan influence to BSA by two ways: mediated through the water, or directly to the influence to hydrophilic sites of BSA molecules. The mechanism was unknown. These influences were mainly changed in dependence on the potassium phenosan concentration presented in the surrounding solution. There were not clear evidences of BSA-AO linkage existence. However, this biological active substance may be acting immediately through the change the properties of water medium surrounding of the protein's molecules.

At the second part of our study, we investigated the potassium phenosan influencing to structural rearrangements membrane-bounded proteins at experimental model – erythrocyte ghosts. By the using of DSC melting has been shown the shift of temperature of proteins microdomains denaturation at erythrocyte ghosts under the presence of potassium phenosan. All proteins of erythrocytes membrane and cytoskeleton intimately mutually interact with each other and form the membranous protein microdomains. The temperature depended denaturation of these protein microdomains under DSC melting forms the structural heat transitions, which mirrored as peaks at DSC melting curves. It is known five identified structural transitions of protein's microdomains that named *A, B1, B2, C, D*. **A**-peak corresponds of denaturation of proteins – spektrin and actin. That is followed by the total loss of erythrocytes deformability and ghost membranes [27, 28].

B1-peak corresponds of denaturation of proteins – ancyrin and demantin, and proteins of bands 4,1 and 4,2. B-peak corresponds of denaturation of proteins of band 3, which are the cytoplasm fragments of proteins of ion's channels. C-peak is linked with denaturation of membrane fragment 55 kDa of proteins band-3, which are ion-channels microdomains; *D*–peak corresponds of denaturation of unidentified another proteins, which assure the membrane's blabbing [16–19, 21].

One important parameter of the DSC, when erythrocytes ghost are heated, is Tmax that is the temperature, at which the maximum of heat capacity occurs. At the picture that corresponds of the thermodenaturation kinetic cures, there are five peaks that corresponded to protein microdomains thermodenaturation. Five peaks are seen for main endothermic phase transitions at certain temperatures. Respectively, when Tmax (°C) increasing, protein's microdomains stay more resistance to destruction actions due to temperature increases. But, when Tmax (°C) decreases, thermostability of these proteins decreases too.

In Figure 3.7, the data of erythrocytes ghosts DSC melting thermograms are presented: the control samples of ghost denaturation and under the potassium phenosan treatment. We may see five peaks of protein's microdomains thermo-inducted denaturation transitions. The treatment of erythrocytes ghosts by potassium phenosan influences to thermodenaturation characteristics of membrane-bounded proteins. The Tmax (°C) of *A*-peak, *B1, B2, D*-peaks are shifted under the potassium phenosan treatment. But the shapes of all these peaks are not changed.

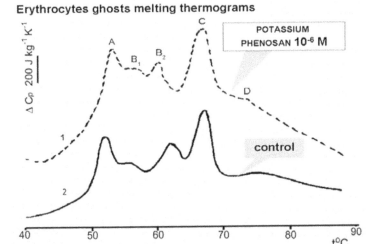

FIGURE 3.7 Potassium phenosan influences to thermodenaturation characteristics of membrane-bounded proteins. Erythrocytes ghosts DSC melting thermograms. Peaks of protein's microdomains thermo-inducted denaturation transitions. (Figure was modified by us from our earlier paper [7]).

The digital data of these DSC experiments at model biomembrane – erythrocytes ghosts, with potassium phenosan influences to the structural changing of membrane-bounded proteins, are presented in Table 3.1.

TABLE 3.1 Influence of AO to Maximum of Temperature of Protein's Microdomains Thermodenaturation Transition (°C) at Erythrocytes Ghost Membranes*

Peaks of protein's microdomains thermo-inducted denaturation transitions	Temperature of maximum of transition (°C) + potassium phenosan	Temperature of maximum of transition (°C), control
A (spektrin, actin)	53.2	52.0
B1 (ancyrin, demantin, band 4.1 and 4.2)	56.4	55.2
B2 (band three cytoplasm channel's fragment)	60.2	62.0
C (band 3 – ion channels)	67.0	67.4
D (membrane's blabbing)	73.6	75.0

*Table was modified by us from our earlier paper [9].

At the base of data presented in Figure 3.7 and Table 3.1, we may conclude that at erythrocyte's ghosts potassium phenosan raised the thermoresistance of membranous and cytoskeleton's proteins: spektrin, actin, ancyrin, and demantin. AO shifted the maximum of protein denaturation temperature Tmax. But the rearrangements of proteins microdomains *A, B1, B2, C, D* did not occur. The membrane organizations of erythrocytes ghosts became more stability.

3.4 DISCUSSION

Phenosan-acid is amphiphilic synthetic AO [1]. Its derivatives are interesting substances for investigation [4–6, 11]. Our work is devoted to the biology study by using model objects of the animals' origin. We deal with hydrophilic derivative of phenosan-acid – its potassium salt, which is called potassium phenosan. All derivatives, formed at the base of phenosan-acid, exhibit the strong AO activity [4, 5]. This activity was tested on the large number of biological objects. But potassium phenosan is the AO of nonspecific action. It acts as structural effector on enzymes and biomembranes. At animals' body, phenosan-acid, and its potassium salt exhibit the specific activity offered the properties as adaptogen [10]. At investigations which were performed of the scientific workers of Emanuel Institute of Biochemical Physics of RAS, it has been noted that the phenosan-acid and potassium phenosan have no certain targets for its actions at biomembranes. As the amphiphilic agent, it operates in all areas of surface layers of biomembrane, as in exterior and in internal list of bilayer appears to penetrating through biomembrane defects to internal surface of bilayer plasma membrane. This has been shown by means of spin-labeled probes on erythrocyte membranes [2].

Phenosan-acid and its derivative potassium phenosan change the: bilayer thickness and regularity at multilayers [3]. At the presented work we show by spectral method (on intrinsic fluorescence intensity of 2 tryptophane residues in BSA molecule [13]) that potassium phenosan at 10^{-4} M reduces the intrinsic fluorescence intensity of BSA and shifts the wavelengths of emission maximum. These facts indicate that the environment around of the tryptophane is hydrophilizated, when large concentrations of AO are presented in medium. Due to potassium phenosan actions, the globule of molecule BSA are loosening. And tryptophane residues became available for water entering. That results to intrinsic fluorescence quickening by O_2, which was dissolved in water. But low and ultra-low concentrations of potassium phenosan ($10^{-17} - 10^{-6}$ M) increase the fluorescence intensity of BSA compared to control. This phenomenon may be explained by the preserve of tryptophane residues from quenching by the adsorbed AO molecules.

Also, the water solutions of potassium phenosan may be the regulator of transporting function of albumins, as it will be introduced to the animal's body. And albumins may be taken part in extracting fatty acids from any molecules, or in binding of free fatty acids. As is known, the water solutions of potassium phenosan change the fatty acid's content of membranes and mitochondrial electron transport [29, 30].

Our obtained data suggests to the fact that BSA, which is one of the first targets in "Blood River, " in particular, in blood plasma, was exposed to potassium phenosan, that was used over at a wide range of concentrations. And the albumin structure was varied under the presence of potassium phenosan. Respectively, the presupposition was arizing that if some properties of albumin change, a carrier of biologically active substances and as osmotically active substance. Albumins maintain the colloid-osmotic pressure at blood plasma and at other fluids (in cerebrospinal fluid, for example). It can be assumed that the transport effectiveness will be decreased and can be unbalanced of osmotic pressure in compartments with biological fluids. This is why the application of potassium phenosan demands the great cares and the observances of concentration limitations, because the soluble protein of animal origin – BSA, changes its structural properties in their attendance so much. Biological activity substances influence to structural and functional properties of animal cells. These influences of biological active substance on many cellular properties is may be mediated by BAS effects on the structural properties of bilayers of cellular membranes. Biological activity substances change structural parameters of heating or cooling of cellular components. As result, biological activity substances cause some variances of many functions during of animal life: slipping, hibernation, and others. Besides, the action of BAS in the structural changes in membranes when biological objects become complicated, that is transition from the model object to whole cell.

At the second part of our investigation, we deal with the membrane-bounded proteins properties. Membranes as the major components of cells form numerous compartments with local parameters within cells. Membranes lend the base for concentration or creation of determinate orders of molecules, or for its posttranslational modifications. Structural basis of membranes are the lipids and proteins. In our earlier works, the influence of biologically active substances, as operative factors of environment, on lipid-lipid interactions in membranes structure of multilamellar liposomes has been examined [31]. As experimental object of this presentation with lipid-protein composition, the erythrocyte ghosts were used. To study the effect of potassium phenosan on protein-lipid interactions in membranes structure, it was investigated by DSC. The obtained DSC data suggest that low and middle concentrations of aqua solutions of this substance stabilized the cytoskeleton proteins structural organization.

3.5 CONCLUSION

Biological active substance – potassium phenosan, change the structures of both free soluble proteins and membrane proteins. At our case, free protein serum albumin BSA from bovine blood serum was tested. Potassium phenosan, when all large concentrations, changes at great station the structure of albumin. Thus in this case, potassium phenosan may change the transporting properties of serum albumin. And for investigations of the potassium phenosan influence to the membrane-bounded proteins, we used the cytoskeleton's proteins of erythrocytes ghosts. Erythrocytes ghosts contain the plasmalemma and intracellular proteins, which are firmly linked with each other. These dissoluble non-free proteins compose between them cellular cytoskeleton. Erythrocytes ghosts are the liberated from hemoglobin cellular envelopes with plasmalemma and cytoskeleton proteins (actin, spektrin, demantin, channels fragments, incorporated to membrane: glico-porin A, proteins of band 3, and interiors channel fragments–ancyrin). All these components are common for many cells of different animal's tissues. Thus, biologically active substance potassium phenosan may influence as to properties of blood serum, as to blood erythrocytes that may be followed by changing of cellular fate at the animal body.

KEYWORDS

- **differential scanning calorimetry**
- **erythrocyte ghost**
- **fluorescence**
- **triptophanil emission**

REFERENCES

1. *Ershov, V. V., Nikiforov, G. A.,* & Volodkin, *A. A., (1972). Space Hampered Fenols (p.* 352*).* Moscow: Chemistry (in Russian).
2. Gendel, L. J., Kim, L. V., Luneva, O. G., Fedin, V. A., & Kruglakova, K. E., (1996). Changes of cursory architectonics of erythrocytes under the impact of synthetic antioxidant Fenosan-1. *Reports of the Russian Academy of Science. Series. Biol., 4,* 508–512 (in Russian).
3. Archipova, G. V., Burlakova, E. B., Krivandin, A. V., & Pogoretskaya, I. L., (1996). Phenosan-acid influence to phospholipid membrane. *Neurochemistry, 13,* 128–132 (in Russian)*.*
4. *Nikiforov, G. A.,* Belostotskaya, I. S., Vol'eva, V. B., Komissarova, N. L., & Gorbunov, D. B., (2003). Bioantioxidants "float types" at base of derivatives 2,6 ditret butyl phenyl. *Scientific Bulletin of the Tyumen Academy of Medicine: Bioantioxidants. 1,* pp. 50, 51 (in Russian).
5. Perevozkina, M. G., Storozhok, N. M., & *Nikiforov, G. A., (2005).* Correlation between *chemical* structure and inhibiting properties of sterically hindered fenols of the ichfan group. *Biomed. Chemistry, 51, 413–423* (in Russian).
6. Ozerova, I. B., Molochkina, E. M., Nikiforov, G. A., & Shishkina, L. N., (1996). Influences of new fenolic antioxidants (AO) of group of space hampered fenols to soluble acetylcholine esterase (ACE). In: *First Scientific-Practical Conference Alzheimer Disease: Progress at Neurobiology, Diagnostic and Therapy* (pp. 48–49)*.* Moscow. Russia (in Russian).
7. Alekseeva, O. M., Kim, Y. A., Rykov, V. A., Golochshapov, A. N., & Mill, E. M., (2010). Influences of space hampered *fenols to lipid structures, and to free and membrane-bounded proteins. In: Fenolic Substances: Fundamental and Aspects for Application*

(pp. 125–136). Scientific space. Part 1: *Fenolic substances: Structure, properties, and biology activity* (in Russian).

8. Albantova, A., Binyukov, V. I., Alekseeva, O. M., & Mill, E. M., (2012). The investigation of influence of Phenosan, ICHPHAN-10 on the erythrocytes in vivo by AFM method. In: Varfolomeev, S. D., Burlakova, E. B., Popov, A. A., & Zaikov, G. E., (eds.), *Modern Problems in Biochemical Physics* (pp. 45–48). Horizonts, Nova Science Publishers, New York, Chapter 5.

9. Alekseeva, O. M., Rykov, V. A., Kim, Y. A., Golochshapov, A. N., & Burlakova, E. B., (2009). Influences of perspective neuroprotector *phenosan to structures, and functions of biological membranes. In:* Zinchenko, V. P., Kolesnikov, S. S., & Beregnov, A. V., (eds.), *International Scientific Conference "Reception and Intracellular Signaling"* (pp. 317–321). Pushchino. Moscow region (in Russian).

10. Prokopov, A. A., Berland, A. C., & Shukil, L., V., (2006). Investigation of Phenosan-acid acid metabolism at rabbit body. *Pharmaceutical Chemistry Journal, 2*, pp. 3, 4 (in Russian).

11. Arefev, D. V., Belostotskaya, I. S., Voleva, V. B., Domnina, N. S., Komissarova, N. L., Sergeeva, O. Y., & Khrustaleva, R. S., (2007). Hybrid macromolecular antioxidant on the basis of hydrophilic polymers and fenols. *Russian Chemical Bulletin, 56*(4), 781–790 (in Russian).

12. Lakowicz, J. R., (1983). *Principles of Fluorescence Spectroscopy* (p. 389). New York: Plenum Press.

13. Diaz, X., Abuin, E., & Lissi, E., (2003). Quenching of BSA intrinsic fluorescence by alkylpyrimidinium cations its relationship to surfactant-protein association. *J. Photochem. Photobiol., 155*, 157–162.

14. Alekseeva, O. M., Fatkullina, L. D., & Shatalova, O. V., (2014). Melafen influences to structure and function of some proteins of animal origin. In: Fattachov, S. G., Kuznetsov, V. V., & Zagoskina, N. V., (eds.), *Melafen: Mechanism of Actions and Regions of Using* (pp. 343–358). Kazan. Print-Service-XXI Century, Part 4.2 (in Russian).

15. Sato, Y., Yamakose, H., & Suzuki, Y., (1993). Mechanism of hypotonic hemolysis of human erythrocytes. *Biol. Pharm. Bull., 16*(5), 506–512.

16. Brandts, J. F., Eryckson, L., Lysko, K. A., Shchwartz, A. T., & Taverna, R. D., (1977). Calorimetric studies of the structural transitions of the human erythrocyte membrane. The involvement of spectrin in the A transition. *Biochemistry, 16*, 3450–3454.

17. Jackson, W. M., Kostyla, J., Nordin, J. H., & Brandts, J. F., (1973). Calorimetric study of protein transitions in human erythrocyte ghosts. *Biochemistry, 12*(19), 3662–3667.

18. Gulevsky, A. K., Riazantcsev, V. V., & Belous, A. M., (1990). Structural rearrangements of lipids and proteins of erythrocytes membrane under low temperatures. *Scientific Topics of High School: Biologist Sciences, 29–36.*

19. Akoev, V. R., Matveev, A. V., Belyaeva, T. V., & Kim, Y. A., (1998). The effect of oxidative stress on structural transitions of human erythrocytes ghost membranes. *Biochim Biophys Acta., 1371*(2). 284–294.

20. Akoev, V. R., Sherbinina, S. P., Matveev, A. V., Tarachovsky, Y. S., Deev, A. A., & Shnirov, V. L., (1997). Investigations of structural transitions at erithricytes membrane when hereditary hemochromatosis. *Bulletin of Experimental Biology and Medicine, 123*(3), 279–284.

21. Shnyrov, V. L., Zhadan, G. G., & Salia, C. H., (1986). Thermal transitions in rat erythrocyte ghosts. *Biomed. Biochim. Acta., 45.* 1119–1121.

22. Alekseeva, O. M., Fatkullina, L. D., Kim, Y. A., & Zaikov, G. E., (2014). The melafen influence to the erythrocyte's proteins and lipids. *Herald of Kazan Technology University, 17*(9), 176–181.

23. Alekseeva, O. M., Fatkullina, L. D., Goloschapov, A. N., & Zaikov, G. E., (2015). Influence of solution ion strange to dependence of blood bioprocesses from biological active substance. *XII Russian Conference with International Part: Solvatation and Complexion Problems at Solutions. From Effects at Solutions to New Materials* (pp. 40, 41). Ivanovo. Russia (in Russian).

24. Kim, Y. A., Elemesov, R. E., & Akoev, V. R., (2000). Hyperosmotic hemolysis of erythrocytes and antihemolytic activity of saponins faction and triterpens glycosides from Panax Ginseng C.A. Meyer. *Biological Membrane, 17*(2). 15–26 (in Russian).

25. Dodge, J. T., Mitchell, C., & Hanahan, D. J., (1963). The preparation and chemical characteristics of hemoglobin-free ghost of human erythrocytes. *Arch. Biochem. Biophys., 100*. 199–130.

26. Privalov, P. L., & Plotnikov, V. V., (1989). Three generations of scanning microcalorimeters for liquids. *Therm. Acta., 139*, 257–277.

27. Mohandas, N., Greenquist, A. C., & Shohet, S. B., (1978). Effect of heat and metabolic depletion on erythrocyte deformability, spectrin extractability, and phosphorylation. *The Red Cell* (pp. 453–472). New York, Alan R. Liss, Inc.

28. Heath, B. P., Mohandas, N., Wyatt, J. L., & Shohet, S. B., (1982). Deformability of isolated red cells. *Biochim. Biophys Acta., 691*(2), 211–219.

29. Zhigacheva, I. V., & Kaplan, E. Y., (1985). Fatty acid contains liver mitochondria under inserting of antioxidants of ionol type to white rats. *Biochemistry, 50*, 1582–1586 (in Russian).

30. Zhigacheva, I. V., Evseenko, L. S., & Burlakova, E. B., (2008). Potassium Phenosan and electron transport at the terminal (cytochromoxidase) site of respiratory chain of liver mitochondria. *International Scientific Conference Biological Motility: Achievements and Perspectives* (Vol. 1, pp. 188–190). Pushchino, Moscow region.

31. Alekseeva, O. M., Narimanova, R. A., Yagolnik, E. A., & Kim, Y. A., (2011). Phenosan-acid and its derivatives influence to membrane components. *Russian Conference Bio-Stimulators at Agriculture and Medicine* (pp. 15–19). Ufa. Russia (in Russian).

CHAPTER 4

Antioxidants as Adaptogens and Plant Growth Regulators

IRINA V. ZHIGACHEVA

Emanuel Institute of Biochemical Physics of Russian Academy of Sciences, 4, Kosygin St., Moscow, 119334, Russia, E-mail: zhigacheva@mail.ru

ABSTRACT

In this chapter, we studied the influence of spatially hindered phenols as adaptogens and plant growth and development regulators (PGRs) on the intensity of free radical processes, an indication of whose is the intensity of lipid peroxidation (LPO) in biological membranes, particularly the membranes of mitochondria stressful effects lead to a shift in antioxidant pro-oxidant equilibrium in the direction of increasing the generation of reactive oxygen species (ROS) by mitochondria. As a result, a 3–4-fold increase in the fluorescence intensity of LPO products is observed in the membranes of these organelles. Spatially hindered phenols reduce the intensity of LPO to control values, which contributes to the preservation of high functional activity of the mitochondria. Prevent dysfunction of mitochondria is associated with an increase in the sustainability of plant and animal organisms to the action of stress factors.

4.1 INTRODUCTION

Mitochondria, being "energy stations" of cells, play one of the main roles in the organism's response to the action of stress factors. As a result of incomplete reduction of oxygen, these organelles form a reactive oxygen species (ROS), which involved in cellular signaling [1, 2]. Normally, the steady-state level of ROS in organs and tissues is very low (of the order of $10^{-10} - 10^{-11}$ M) due to the presence in them of enzymatic and non-enzymatic systems of regulation of accumulation and elimination of ROS [3, 4]. The shift of antioxidant – pro-oxidant balance towards increased production of ROS leads to the development of oxidative stress. Stress factors modify the membranes of mitochondria [5, 6]. Activation of free radical processes, which is an indicator of the intensity of lipid peroxidation (LPO), affects the lipid composition of membranes. First of all, the fatty acid composition of mitochondrial membranes changes [7, 8], which, possibly, leads to a change in protein-lipid interactions. As a result, the activities of enzymes that form a single complex with the membrane changes and, first of all, the activity of enzymes of the mitochondrial electron transport chain [6]. In this case, the rates of electron transport and the efficiency of oxidative phosphorylation changes [9, 10]. The transformation of mitochondrial energy can lead to both cell death and adaptation to changing environmental conditions. The resistance of organisms to stress factors, probably, will depend on the compliance of the functional state of mitochondria to the energy needs of the cell and the whole organism. In this regard, studies of the mechanisms redox regulation of different physiological processes in normal conditions and under the action of stress factors are actual, and the mitochondria are the main organelles regulating the redox potential of the cell [10, 11].

It can be assumed that biological active substance (BAS) – adaptogens, possessing antioxidant properties, can influence both the production of ROS in the respiratory chain of mitochondria, and the functional state of these organelles. Introduction to the body before the stressful action of antioxidants may lead to a decrease in the generation of ROS by mitochondria under stress, as a result of which dysfunction of the mitochondria, is likely, to be prevented, which, apparently, will contribute to an increase in the nonspecific resistance of organisms to stress factors. Antioxidants primarily have a claim on the role of adaptogens, particularly, claim to this role the synthetic phenolic antioxidants, which have rather high coefficients of interaction with peroxyl radicals (k_7) [12, 13]. The choice of this class of antioxidants is primarily due to the simplicity of their production, high efficiency of chains breaking, the possibility of changing properties over a wide range due to variation of substituents and low toxicity [14].

Since plant growth and development regulators (PGRs) increase plant resistance to stress factors [15], it can be assumed that they can influence the generation of ROS by mitochondria. Based on this, we hypothesized that antioxidants, probably, can be used as PGRs. The object of the study was chosen antioxidants from the class of sterically hindered phenols: potassium phenosan (PHEN) (3, 5-di-*tert*-butyl-4-hydroxyphenyl propionate potassium) (Figure 4.1) and sodium anphen (ANPH) (1-carboxy-1-(N-methylamide)-2-(3,5-di-*tert*-butyl-4-hydroxy-phenyl-propionate   sodium) (Figure 4.2).

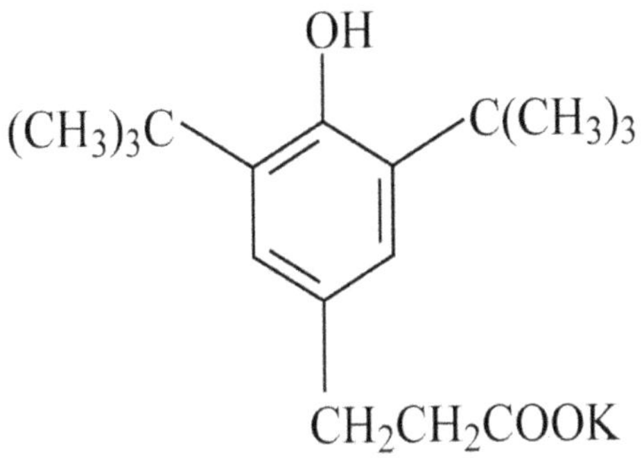

FIGURE 4.1   Potassium phenosan (PHEN).

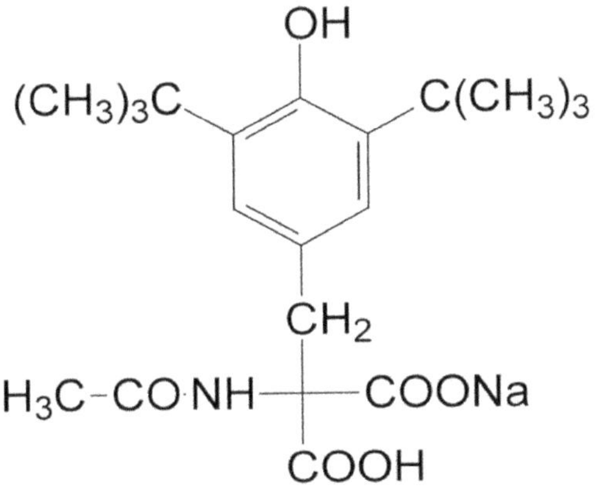

FIGURE 4.2   Sodium anphen (ANPH).

The effectiveness of antioxidants, as well as other BAS that show their activity as a function of the dose, apparently depends on the concentration [16]. The goal of the study, in connection with this, was also to study the functional state of rats liver mitochondria under stress conditions and the effect of various concentrations of the investigated antioxidants on the functional state of these organelles. In died the effect of different concentrations of the investigated antioxidants on the bioenergetic characteristics of mitochondria of 6-day etiolated seedlings of pea (*Pisum sativum* L., cv. Alpha).

4.2 MATERIALS AND METHODOLOGY

Tests were performed on male rats of Wistar line weighing 120–140 g and on pea seedlings (*P. sativum*), cv. Alpha, which was grown in standard conditions and in the conditions of water deficiency (WD).

4.2.1 REGULATORY STANDARDS

The study was performed according to the Rules of Laboratory Practice in the Russian Federation, in accordance with the rules adopted by the European Convention for the protection of vertebrate animals used for experimental and other scientific purposes [17]. The research was performed according to the approved protocol, in accordance with standard operating procedures of the researcher (SOPR), as well as with the Guidelines at laboratory animals and alternative models in biomedical researches on laboratory animals [18].

4.2.2 GERMINATION OF PEA SEEDS

Pea (*Pisum sativum* L., cv. Alfa) seeds were washed with soapy water and 0.01% $KMnO_4$. Control seeds were then soaked in water, experimental seeds – in a 10^{-6} M solution of ANPH or in a 10^{-13} M solution of PHEN for 30 min. Thereafter, seeds were transferred into covered trays on moistened filter paper in darkness for 2 days. After 2 days, half of the control (WD) and seeds treated with ANPH (WD + ANPH) or PHEN (WD + PHEN) were transferred to dry filter paper. After 2 days of WD, seedlings were transferred to wet filter paper WD or paper moistened with 10^{-6} M ANPH solution, or 10^{-13} M solution of PHEN, where they were in the next two days at 24°C. The second half of the seeds in the control group remained on moist filter paper for 6 days. On the sixth day, mitochondria were isolated from the epicotyls of the seedlings of all the study groups.

4.2.3 ISOLATION OF RAT LIVER MITOCHONDRIA

It was performed by differential centrifugation [18]. The first centrifugation at 600 g for 10 min, the second at 9000 g, 10 min. The pellet was re-suspended in the medium, containing: 0.25 M sucrose, 10 mM HEPES, pH 7,4.

4.2.4   *ISOLATION OF PEA SEEDLING MITOCHONDRIA*

Isolation of mitochondria from 6-days-old epicotyl of pea seedlings (*P. sativum*) was performed by differential centrifugation (at 25,000 g for 5 min and at 3,000 g for 3 min) [19]. Precipitation of the mitochondria was carried out for 10 min at 11000 g. The pellet was resuspended in 2–3 mL of media containing: 0.4 M sucrose, 20 mM KH_2PO_4 (pH 7.4), 0.1% BSA (free from fatty acids (FA)), and the mitochondria were again deposited at 11,000 g for 10 min.

4.2.5   *RESPIRATION OF MITOCHONDRIA*

We recorded polarographically (Polarograph LP-7 (Czech), using Clarke oxygen electrode. Pea sprout mitochondria were incubated in a medium containing: 0.4 M sucrose, 20 mM HEPES-Tris buffer (pH 7.2), 5 mM KH_2PO_4, 4 mM $MgCl_2$, and 0.1% BSA. The rat liver mitochondria were incubated in a medium containing 0.25 M sucrose, 10 mM Tris-HCl, 2 mM $MgSO_4$, 2 mM KH_2PO_4, and 10 mM KCl, pH 7.4 (28°C). The rate of respiration was expressed in ng-moll O_2/mg protein min (for rat liver mitochondria) or ng-atom O_2/mg protein min (for pea seedling mitochondria).

The protein was determined by the Biuret method.

The level of LPO was evaluated by the fluorescence method [20]. Lipids were extracted by a mixture of chloroform and methanol (2:1). Lipids of mitochondrial membranes (3–5 mg of protein) were extracted in the glass homogenizer for 1 min at 10°C. Thereafter, an equal volume of distilled water was added to the homogenate, and after rapid mixing, the homogenate was transferred into 12 mL centrifuge tubes. Samples were centrifuged at 600 g for 5 min. The aliquot (3 mL) of the chloroform (lower) layer was taken, 0.3 mL of methanol was added, and fluorescence was recorded in 10-mm quartz cuvettes with a spectrofluorometer (Fluoro Max Horiba Yvon, Germany). The excitation wavelength was 360 nm; the emission wavelength was 420–470 nm. The results were expressed in arbitrary units of fluorescence/mg of protein.

The double bond index (DBI), characterizing the degree of unsaturation of lipids, was calculated by the formula: $DBI = \Sigma P_j n_j / 100$, where P_j is the FA content (in%), n_j is the number of double bonds in each acid. Also, we used the unsaturation coefficient as the ratio of the sum of unsaturated FA to the sum of saturated FA.

The protective activities of the preparations were investigated using the models of acute hypobaric hypoxia (AHH) and acute alcohol intoxication (AAI).

(i) **Model of acute hypobaric hypoxia (AHH)** in rats was carried out in the hyperbaric chamber in a low-pressure atmosphere (230.40 mm Hg), which corresponds to the height of 9000 m above sea level. In the first minute in the chamber created the rarefaction corresponding to the 5 thousand meters (corresponding to the atmospheric pressure of 405 mm Hg) above sea level. In each subsequent minute "ascent" carried out on the further one thousand meters. Time staying of rats at a height" of 9.0 thousand meters above sea level – 5.0 min.

(ii) **Model of acute alcohol intoxication (AAI).** Mice of the BALB/c line weighing 20–25 grams are subcutaneously injected with ethanol 8 g/kg.

To all test animals, the preparation was introduced intraperitoneally at a chosen dose 45 min before the event.

The following reagents were used: methanol, chloroform (Merck, Germany), and sucrose, Tris, EDTA (Ethylenediaminetetraacetic acid), FCCP (carbonyl cyanide-*p*-trifluoromethylphenyl-hydrazone), malate, glutamate, succinate ADP, (Sigma, Aldrich, USA), HEPES (4-(2-hydroxyethyl)-1-piperazineethanesulfonic acid) (Biochemica Ultra, for molecular biology) (MB Biomedicals, Germany).

4.3   RESULTS AND DISCUSSION

It is known that under stress conditions, mitochondria are one of the main sources of ROS [21, 22]. To study the antistress properties of antioxidants, it was necessary to develop a model, that simulated stress, i.e., to find conditions under which the production of ROS by mitochondria will increase and, consequently, LPO will be activated. We solved this problem by developing a model of "aging" of the mitochondria (15 min incubation in a hypotonic medium containing 1 mM KH_2PO_4). The "aging" of the mitochondria caused the activation of free radical oxidation, which was expressed in a 3-fold increase in fluorescence intensity of the final products of LPO (Schiff bases) (Figure 4.3).

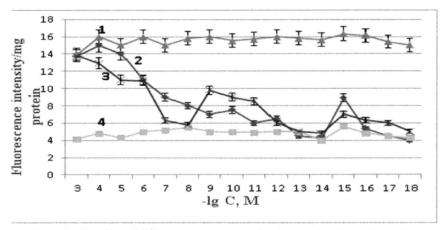

FIGURE 4.3 The effect of different concentrations of sodium anphen (ANPH) or potassium phenosan (PHEN) and "aging" on the intensity of fluorescence of LPO products.
(*Note:* 1 – "aging" of rat liver mitochondria; 2 – "aging" of rat liver mitochondria + ANPH; 3 –"aging" of rat liver mitochondria + PHEN; 4 – control. X-axis – lg C, M; Y-axis – fluorescence intensity in arbitrary units/mg protein [23, Significantly modified]).

The introduction of ANPH or PHEN into the mitochondrial incubation medium led to a decrease in the fluorescence of the final products of LPO in the membranes of mitochondria and had a dose dependence. PHEN at a concentration of $10^{-8} - 10^{-14}$ M and $10^{-16} - 10^{-18}$ M reduced the fluorescence intensity of the final LPO products in mitochondrial membranes of rat liver to almost the control values. ANPH effectively reduced the fluorescence intensity of the final LPO products in mitochondrial membranes of rat liver at concentrations $10^{-6} - 10^{-8}$ M and $10^{-13} - 10^{-14}$ M. Similar data were obtained for mitochondria isolated from etiolated pea seedlings. These data may indicate the presence at drugs antistress properties, the presence of which was tested on the model of AHH and AAI.

AHH led to the activation of LPO in membranes of liver mitochondria. At the same time, the fluorescence intensity of the final products of LPO increased almost 3-fold. Introduction to rats 10^{-6} M ANPH or 10^{-13} M PHEN 45 min prior to exposure prevented the activation of LPO (Figure 4.4).

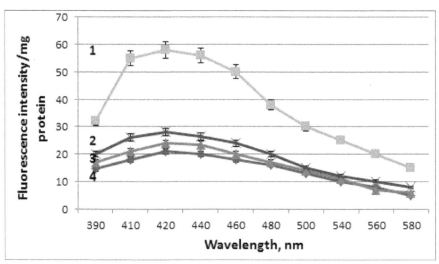

FIGURE 4.4   Effect of acute hypobaric hypoxia (AHH), potassium phenosan (PHEN) or sodium anphen (ANPH) on the spectra of fluorescence of LPO products. (*Note:* 1 – AHH; 2 – AHH + ANPH; 3 – AHH+. PHEN; 4 – control. X-axis: the wavelength in nm; Y-axis: fluorescence intensity in arbitrary units/mg protein).

AAI was accompanied by an almost 2-fold increase in fluorescence intensity of the final products of LPO (Figure 4.5). This is consistent with the literature data, since one of the main hepatotoxic effects of acetaldehyde, formed in the liver cells from ethanol as a result of the activation of alcohol dehydrogenase and the system of microsomal oxidation, is the intensification of LPO processes [24]. Injection to mice 10^{-6} M, sodium anphen or 10^{-13} M potassium phenasan 45 min before exposure reduced the fluorescence intensity of the final products of LPO. In this case, the fluorescence intensity when administered to mice ANPH was above the control values by 30%, and when the PHEN was administered, it was above the control values by 20%.

Changes in physicochemical properties of mitochondrial membranes, probably had an impact on the activity of enzymes associated with the membrane, namely on the activity of enzymes of the respiratory chain of mitochondria. Indeed, AHH resulted in a reduction in the maximum rates of oxidation of NAD-dependent substrates (Table 4.1).

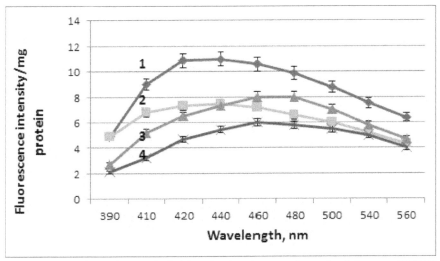

FIGURE 4.5 Effect of acute alcohol intoxication, the potassium salt of phenosan (PHEN) or sodium anphen (ANPH) on the spectra of fluorescence of LPO products. (*Note:* 1 – AAI; 2 – AAI + ANPH; 3 – AAI + PHEN; 4 – control. X-axis – the wavelength in nm; Y-axis – fluorescence intensity in arbitrary units/mg protein).

TABLE 4.1 Effect of Acute Hypobaric Hypoxia, Potassium Phenosan and Sodium Anphen on the Rate of Oxidation of NAD-Dependent Substrates by Rat Liver Mitochondria (Oxidation Rates are Presented in ng moles of O_2/mg Protein min) (The Number of Experiments is 10)

Method of processing	Vo	V_3	V_4	V_3/V_4	FCCP
Control	7.2 ±1.0	30.2±0.9	8.6±0.5	3.52±0.03	29.2±1.3
AHH	7.8±1.4	22.8±1.3	10.7±0,2	2.13±0.05	21.3±1.2
AHH + potassium phenosan	8.1±1.9	30.8±1.5	8.6±0.9	3.58±0,05	31.8±1.9
AHH + sodium anphen	8.6±1.8	31.2±1.4	8.8±1.0	3.55±0,04	32.6±2.4

(*Note:* The incubation medium contained 0.25 M sucrose, 10 mM Tris-HCl, 2 mM KH_2PO_4, 5 mM $MgCl_2$ and 10 mM KCl, pH 7.5 Other additives: 1 mM malate, 4 mM glutamate, 125 µM ADP, 10^{-6} M FCCP – carbonylcyanide-p-trifluoromethoxy-phenylhydrazone. AHH – acute hypobaric hypoxia; Vo – the rates of oxidation of substrates; V_3 – rates of oxidation of substrates in the presence of ADP; V_4 – rates of oxidation in a state of rest [rates of oxidation of the substrate with the exhaustion of ADP]).

At the same time, the efficiency of oxidative phosphorylation RCR (respiratory control rates) decreased from 3.52 ± 0.03 to 2.13 ± 0.05. However, the rate of oxidation of succinate even slightly was increased in the presence of ADP, which agrees with the data obtained in the laboratory of M.N. Kondrashova [25, 26] (Table 4.2).

TABLE 4.2   Influence of Short-Term AHH, Potassium Phenosan and Sodium Anphen on the Rates of Succinate Oxidation by Mitochondria of Rat Liver (Oxidation Rates are Given in ng moles of O_2/mg Protein min) (The number of experiments is 10)

Group	Vo	V_3	V_4	V_3/V_4	FCCP
Control	7.5 ±1.0	30.4±1.1	9.4±1/2	3.23±0.1	28.7±1.6
AHH	9.0±1.8	34.2±1.0	11.0±1.5	3.11±0.2	31.2±1.4
AHH + potassium phenosan	9.8±2.1	33.1±1.3	10.2±1.0	3.25±0.2	30.0±1.9
AHH + sodium anphen	10.8±1.8	34.3±2.2	10.7±3.3	3.21±0.03	32.5±2.8

(*Note:* The incubation medium contained 0.25 M sucrose, 10 mM Tris-HCl, 2 mM KH_2PO_4, 5 mM $MgCl_2$ and 10 mM KCl, pH 7.5 Other additives: 5 mM succinate, 125 µM ADP, 10^{-6} M FCCP (carbonylcyanide-p-trifluoromethoxy phenylhydrazone). AHH – acute hypobaric hypoxia; Vo – the rate of oxidation of substrates; V_3 – rate of oxidation of substrates in the presence of ADP; V_4 – rates of oxidation in a state of rest (rates of oxidation of the substrate with the exhaustion of ADP). FCCP – maximum oxidation rates in the presence of uncoupler FCCP).

Changes in the bioenergetic characteristics of rats liver mitochondria under conditions of AHH are probably due to inactivation of the NAD-dependent oxidation pathway [25], which leads to inhibition of electron transfer in the NADH-CoQ region, accumulation of reduced pyridine nucleotides, and inhibition of ATP synthesis in the first stage of oxidative phosphorylation. In this case, occur a switching of substrate region from complex I to complex II, i.e., to the oxidation of succinate [26]. Introduction to rats 10^{-13} M potassium phenasan or 10^{-6} M sodium anaphen 45 min before the exposure prevented changes in the functional characteristics of liver mitochondria: the oxidation rates of NAD-dependent substrates both in the presence of ADP and in the presence of FCCP had not differ from the control values (see Table 4.1). At the same time, the rate of respiratory control was comparable with the corresponding values in the control group of animals. The oxidation rates and the efficiency of oxidative phosphorylation during oxidation of succinate were not different from the control values (see Table 4.2), which indicated the correction of the functional characteristics of the mitochondria by the drug. Since it is known that the restoration of oxidation rates of NAD-dependent substrates by mitochondria is important in the formation of resistance to hypoxia [27, 28], it is possible to conclude about the presence of adaptogenic properties of studied antioxidants. It should be noted that the activation of LPO under stress conditions, which caused changes in the functional characteristics of mitochondria, has also affected physiological parameters and, above all, the survival of animals in conditions of hypoxia and AAI. The life span of animals (mice) under conditions of various types of hypoxia increased 1.8–4.5 times and 3.9 times in acute alcohol poisoning, and the survival rate of animals increased by 12–40% (Tables 4.3 and 4.4).

TABLE 4.3 The Protective Activity of Potassium Phenosan (Presents the Results of 10 Experiments)

Action	Measured parameter	Control	Potassium phenosan, 10^{-13} M
Raising to an altitude of 11.5 thousand meters (hypobaric hypoxia)	Lifetime in minute	4.0±1.1	15.0± 2.8
	Survivors	20%	50%
Injection of sodium azide, 20 mg/kg (cytotoxic hypoxia)	Lifetime in minutes	3.0±0.6	11.0±1.1
	Survivors	0%	50%
Injection of sodium nitrite, 250 mg/kg (hemic hypoxia)	Lifetime in minutes	15.1±2.5	3.5±3.0
	Survivors	20%	60%
Swimming with a load at a temperature of 2°C (low-temperature stress in combination with muscular load)	Lifetime in minutes	3.2±0.6	14.3±2.4
	Survivors	0%	0%

TABLE 4.4 Protective Activity Sodium Anphen (Presents the Results of 10 Experiments)

Action	Measured parameter	Control	Sodium anphen, 10^{-6} M
Injection of sodium azide, 20 mg/kg (cytotoxic hypoxia)	Lifetime in minutes	2.3±0.5	5.2±1.2
	Survivors	0%	40%
Injection of sodium nitrite, 250 mg/kg (hemic hypoxia)	Lifetime in minutes	20.5±3,1	35.7±6.4
	Survivors	0%	15%
Injection of ethanol, 8 g/kg	Lifetime in minutes	35.4±6.1	137.4±31.1
	Survivors	0%	12%

Based on these data, we can conclude about the presence of antistress properties ANPH and PHEN.

The assumption that the main feature of the preparations adaptogens, including regulators of growth and development of plants, is to reduce the excessive production of ROS, and consequently, the decrease of intensity of processes of LPO in biological membranes and mainly in the mitochondrial membranes was checked by us on the mitochondria of six-days etiolated pea seedlings using the antioxidants studied. The stress model was used for the water shortage. As you know, WD reduces functional activity, as chloroplasts and mitochondria [29]. In this connection, it was interesting to find out whether the bioenergetic characteristics of mitochondria will change under conditions of WD and in the treatment of seeds and seedlings with antioxidants in these conditions. The WD resulted in the activation of free radical oxidation in mitochondria membranes of etiolated pea seedlings, as evidenced by a 3-fold increase in fluorescence intensity of LPO products (Figure 4.6).

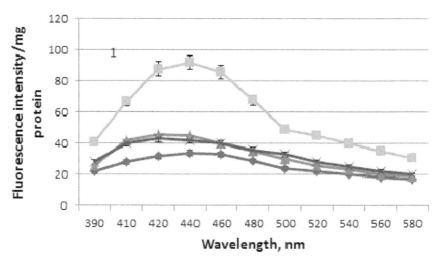

FIGURE 4.6   Effect of water deficiency (WD), potassium phenosan (PHEN) or sodium anphen (ANPH) on the fluorescence of LPO products. (*Note:* 1 – WD; 2 – Wnd+ ANPH; 3 – WD + PHEN; 4 – control. X-axis – wavelength, nm; Y-axis – fluorescence intensity in arbitrary units/mg protein).

The data obtained are consistent with the literature data on the effect of WD on the activation of free radical oxidation in membranes of wheat seedlings [30, 31]. It should be noted that the treatment of seeds and seedlings of peas by the tested antioxidants reduced the fluorescence intensity of LPO products to almost control values.

Activation of LPO, possibly, could lead to a change in the FA composition of the membranes. In this connection, in the next series of experiments, we studied the effect of WD and antioxidants on the FA composition of the total lipid fraction of the mitochondria membranes. The WD caused an increase in the relative content saturated FA and a decrease in the content of unsaturated FA in the membranes of pea seedlings mitochondria. Significant changes were observed in the content of FA with 18 carbon atoms. The index of double bonds of FA containing 18 carbon atoms decreased from 1.45 ± 0.02 to 1.28 ± 0.01 (Figure 4.7), and the unsaturation factor of FA containing 18 carbon atoms decreased from 23.54 ± 0.07 to 15.15 ± 0.22 [32]

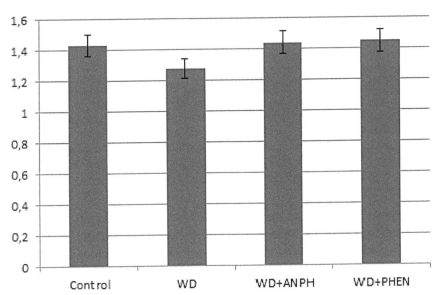

FIGURE 4.7 Effect of water deficiency, potassium phenosan or sodium anphen on the index of double bonds of FA, which contains 18 carbon atoms, in membranes mitochondria of pea seedlings. (*Note:* X-axis – different groups of plants; Y-axis – the index of the double bonds of FA containing 18 carbon atoms).

Changing the content of FA with 18 carbon atoms due to dehydration are also observed in membrane lipids of potato, suspension of cells, and membrane lipids of the leaves of *Arabidopsis thaliana* [33–35]. In all these cases, WD was accompanied by a decrease in the contents of linoleic and linolenic acids and an increase in the content of stearic acid in mitochondrial membranes. The modification of the physicochemical properties of mitochondrial membranes, probably, could lead to changes in lipid-protein interactions, and, consequently, the activity of the enzymes of the mito-chondrial respiratory chain. Indeed, the WD caused a 40% decrease in the maximum oxidation rates of NAD-dependent substrates and a 30% decrease in the respiratory control rates in the oxidation of these substrates by pea seedlings mitochondria (Table 4.5). At the same time, the oxidation rates of succinate decreased by only 10–15%.

TABLE 4.5 Effects of Water Deficiency and a Treatment of Pea Seeds with Potassium Phenosan (PHEN) or Sodium Anphen (ANPH) on the Rate of NAD-Dependent Substrate Oxidation by Mitochondria, Isolated from Pea Seedlings, ng-atom/mg Protein min

Group	Vo	V_3	V_4	V_3/V_4	FCCP
Control	20.0±2.5	70.0±5.4	30.0±2.0	2,33±0.01	72.0±5.0
WD	11.8±1.9	47.6±3.2	38.2±1.0	1.25±0.02	49.9±4.8
WD+PHEN	21.8±1.9	72.0±3.5	29.1±1.4	2.48± 0.02	73.4±3.5
WD+ANPH	19.8±1.9	69.0±3.2	29.4±2.1	2.35±0.02	70.0±5.4

Note: The incubation medium contained: 0.4 M sucrose, 20 mM HEPES-Tris buffer, 5 mM KH_2PO_4, 4 mM MgCl, and 0.1% BSA, 10 mM malate, 10 mM glutamate. Other additives: 125 µM ADP, 10^{-6} M FCCP (carbonyl cyanide-p-trifluoromethoxy phenylhydrazone). The results of the 10 experiments are presented. WD – water deficiency, Vo – the rate of oxidation of substrates; V_3 – rate of oxidation of substrates in the presence of ADP; V_4 – maximum oxidation rates in the presence of uncoupler FCCP.

Treatment of seeds and seedlings of pea 10^{-6} M, ANPH or 10^{-13} M potassium phenasan prevented changes in the efficiency of oxidative phosphorylation caused by WD and promoted the preservation of high oxidation rates of NAD-dependent substrates in the presence of ADP or FCCP. Changes in the bioenergetic characteristics of mitochondria are apparently related to the physicochemical state of the membranes of these organelles. PHEN or ANPH, preventing the oxidation of unsaturated C_{18} FA, main linoleic acid (see Figure 4.5), which is one of the basic FA included in the composition of cardiolipin, providing the effective functioning of the respiratory chain of mitochondria, probably due to the formation of supercomplexes respiratory carriers [36]. It is known that pea seedling is particularly sensitive to drought stress. As has been shown previously, that earlier growth stages were more sensitive to WD than subsequent ones [37]. In our experiments, we used the most sensitive to WD the growth stage of pea seedlings (two-day seedlings). WD inhibits the growth process (Figure 4.8), which is consistent with literature data.

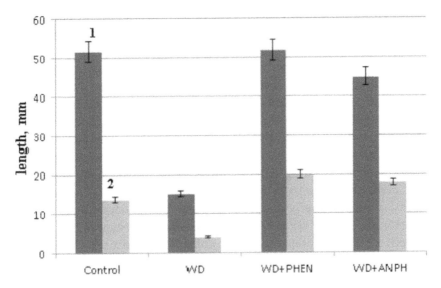

FIGURE 4.8 Effect of water deficiency, potassium phenosan (PHEN) or sodium anphen (ANPH) on the length of shoots and roots of 6-days-old pea seedling; (1) sprout (2) root. (*Note:* X-axis – different groups of plants; Y-axis – length, mm).

Treatment of pea seeds with studied antioxidants prevented inhibition of growth of roots and shoots under these conditions. In this case, the roots of the seedlings treated with potassium phenasan were longer by 45% than in the control samples, and the treated with the anphen, by 31%. It should be noted that the studied antioxidants prevented inhibition of growth of shoots in conditions of WD. However, the length of shoots of pea seedlings treated with ANPH was below the control values by 13%. This difference in the effect of the studied antioxidants on growth processes is apparently due to the difference in the FA composition of mitochondria membranes of pea seedlings treated with PHEN or ANPH.

4.4 CONCLUSION

Based on these data, it can be concluded that the investigated antioxidants can be used as adaptogens, and as regulators of growth and development of plants. Their adaptogenic properties may be due to antiradical activity. According to M. N. Perevozkina [13], the effective coefficient of PHEN interacting with peroxyl radicals during the oxidation of methyloleate (60°C) k_7 is 2.2×10^4 $(Ms)^{-1}$. Similar values observed for k_7 of

ANPH. However, the protective effect of the preparations, conditional to antiradical properties probably could only be manifested in a concentration of 10^{-6} M, i.e., in the concentration at which ANPH was used [38]. As shown in the work of Palmina N. P. [39], PHEN in small and ultra-small concentrations (10^{-9}–10^{-10} and 10^{-14}–10^{-16} M) had changed the microviscosity of the annular lipid bilayer (~20 A) membranes of the endoplasmic reticulum (ER). These changes correlated with the inhibition of LPO ($r = 0.551$, $p = 0.033$). Note that the administration of the preparation to animals (mice) also influenced the activity of the enzymes of the antioxidant system: the drug reduced the activity of Mn-SOD, and in 1.5 times increased the activity of Cu, Zn-SOD, by 1.5 times increased the activity of glutathione peroxidase (GP) and glutathione reductase (GR). However, the ratio of GP/GR activities remained constant [40]. Thus, the protective effect of preparations is probably related to the antioxidant activity of drugs and the effect of PHEN on the activity of antioxidant enzymes.

The protective activity of PHEN and ANPH as regulators of growth and development of plants, apparently, also related to their antioxidant activity. Preparations, preventing the activation of LPO, probably, protect unsaturated FA, which is part of the phospholipids of mitochondrial membranes, from peroxidation. Preventing changes in the fatty acid composition of mitochondrial membranes is reflected in the bioenergetic characteristics: preserved a high activity of NADH-dehydrogenase complex of the mitochondrial respiratory chain. Note that the mitochondria of germinating seeds are characterized by relatively low rates of oxidation of NAD-dependent substrates. The increase in the activity of NAD-dependent dehydrogenases activates energy processes in the cell, which increases the resistance of the plant organism to changing environmental conditions.

ACKNOWLEDGMENTS

I am very grateful to Tamara Misharina, Mariya Terenina, and Natalya Krikunova for help in the study of the effect of WD on the fatty acid composition in the lipid fraction of mitochondria membranes of pea seedlings.

KEYWORDS

- **lipid peroxidation**
- **mitochondria**
- **mitochondrial dysfunction**
- **peroxyl radicals**
- **spatially hindered phenols**
- **stress-factors**

REFERENCES

1. Skulachev, V. P., (2009). New data on the biochemical mechanism of programmed aging of the organism and antioxidant protection of mitochondria. *Biochemistry, 74*(12), 1718–1721 (in Russian).
2. Kunduzova, O. R., Escourrou, G., & Seguelas, M. H., (2003). Prevention of apoptotic and necrotic cell death, caspase-3 activation, and renal dysfunction by melatonin after ischemia/reperfusion. *FASEB J., 17*(8), 872–874.
3. Tailor, N. L., Day, D. A., & Millar, A. H., (2003). Targets of stress-induced oxidative damage in plant mitochondria and their impact on cell carbon/nitrogen metabolism. *J. Exp. Bot., 55*(394), 1–10.
4. Taylor, N. L., Heazlewood, J. L., Day, D. A., & Millar, A. H., (2005). Differential impact of environmental stresses on the pea mitochondrial proteome. *Molecular & Cell Proteomics, 4*, 1122–1133.
5. Palmina, N. P., (2009). The mechanism of action of ultra-low doses. *Chemistry and Life, 2*, 10–13 (in Russian).
6. Brand, M. D., Affourtit, C., Esteves, T. C., Green, K., Lambert, A. J., Miwa, S., Pakay, J. L., & Parker, N., (2004). Mitochondrial superoxide: Production, biological effects, activation of uncoupling proteins. *Free Radical Biology and Medicine, 37*, 755–767.
7. Falcone, D. L., Ogas, J. P., & Somerville, C. R., (2004). Regulation of membrane fatty acid composition by temperature in mutants of Arabidopsis with alterations in membrane lipid composition. *BMC Plant Biology, 4*(17), 4–17.
8. Stupnikova, I., Benamar, A., Tolleter, D., Grelet, J., Borovskii, G., Dorne, A. J., & Macherel, D., (2006). Pea seed mitochondria are endowed with a remarkable tolerance to extreme physiological temperatures, *Plant Physiology, 140*, 326–335.
9. Le Bras, M., Clement, M. V., Pervaiz, S., & Brenner, C., (2005). Reactive oxygen species and the mitochondrial signaling pathway of cell death. *Histol. Histopathol., 20*, 205–219.
10. Møller, I. M., (2001). Plant mitochondria and oxidative stress: Electron transport, NADPH turnover, and metabolism of reactive oxygen species. *Annu. Rev. Plant Physiol. Plant Mol. Biol., 52*, 561–591.
11. Grabelnych, O. I., Sumina, O. N., Funderat, S. P., Pobezhimova, T. P., Voinikov, V. K., & Kolesnichenko, A. V., (2004). The distribution of electron transport between the

main cytochrome and alternative pathways in plant mitochondria during short-term cold stress and cold hardening. *J. Thermal Biol.*, *29*, 165–175.

12. Kovtun, G. A., (2000). Reactivity of the interaction of phenolic antioxidants with peroxy radicals. *Catalysis and Petrochemistry*, *4*, 1–9 (in Russian).

13. Perevozkina, M. G., (2003). Kinetics and mechanism of inhibitory action of derivatives of phenozan, salicylic acid and their sinteticheskih mixtures with α-tocopherol and phospholipids. *PhD Thesis in Chemistry* (p. 24). Tyumen, (in Russian).

14. Roginsky, V. A., (1988). *Phenolic Antioxidants: Reactivity and Efficacy* (p. 247). Moscow, Science (Nauka in Rus.) (in Russian).

15. Chalova, L. I., & Ozeretskovskaya, O. L., (1984). Biological inducers of plant protective reactions and possible ways of their practical use. In: Berezin, I. V., (ed.), *Biochemistry of Immunity, Rest, Aging of Plants* (pp. 41–57). Moscow, Science (Nauka in Rus.) (in Russian).

16. Burlakova, E. B., Konradov, A. A., & Maltseva, E. L., (2007). Effect of ultralow doses of biological active substance and low-intensity physical factors. In: Rubin, A. B., (ed.), *Problems of Regulation in Biological Systems: Biophysical Aspects* (pp. 390–423). Moscow, Izhevsk, Research Center Regular and chaotic dynamics, Institute of computer studies Publ. (in Russian).

17. Karkishchenko, N. N., & Grachev, S. V., (2010). *Guidance on Laboratory Animals and Alternative Models in Biomedical Research* (2nd edn., p. 358). Moscow: Profile (in Russian).

18. Mokhova, E. N., Skulachev, V. P., & Zhigacheva, I. V., (1977). Activation of the external pathway of NADH oxidation in liver mitochondria of cold-adapted rats. *BBA*, *501*, 415–423.

19. Popov, V. N., Ruge, E. K., & Starkov, A. A., (2003). Effect of inhibitors of electronic transport on the formation of reactive oxygen species during the oxidation of succinate by pea mitochondria. *Biochemistry*, *68*(7), 910–916 (in Russian).

20. Fletcher, B. I., Dillard, C. D., & Tappel, A. L., (1973). Measurement of fluorescent lipid peroxidation products in biological systems and tissues. *Analytical Biochemistry*, *52*, 1–9.

21. Plotnikov, E., Chupyrkina, A., Vasileva, A., Kazachenko, A., & Zorov, D., (2008). The role of reactive oxygen and nitrogen species in the pathogenesis of acute renal failure. *BBA*, *1777*, S58–S59.

22. Zorov, D. B., Isaev, N. K., Plotnikov, E. Y., Zorova, L. D., Stelmashuk, E. V., Vasilyeva, A. K., Arkhangelskaya, A. A., & Khryapenkova, T. G., (2007). Mitochondria as the many-faceted Janus. *Biochemistry*, *72*, 1371–1384 (in Russian).

23. Zhigacheva, I., & Mil, E., (2016). Potassium phenosan as an adaptogen to stress. *World Journal of Pharmaceutical and Life Science*, *2*(6), 556–566.

24. Dorkina, E G., Sergeeva, E. O., Oganesyan, E. T., Parfentieva, E. P., Sadzhaya, L. A., Terekhov, A. Y., School, I. V., et al., (2007). Influence of bioflavonoids on lipid peroxidation and antioxidant systems of liver of rats in acute alcohol poisoning. *Bulletin VolGMU*, *3*(23), 50–52 (in Russian).

25. Kashuro, V. A., Dolgo-Saburov, V. B., Basharin, V. A., Bonitenko, I. Y., & Lapina, N., (2010). Some mechanisms of disorders of bioenergy and optimizing their approaches to pharmacotherapy. *Pharmacology*, *11*, 611–632 (in Russian).

26. Maevsky, E. I., Grishina, E. V., Rosenfeld, A. S., Zyakun, A. M., Vereshchagin, I. M., & Kondrashova, M. N., (2000). Anaerobic formation of succinate and facilitation of

its oxidation. Possible mechanisms of cell adaptation to oxygen starvation. *Biomedical Journal, 1*, 32—36 (in Russian).

27. Maevsky, E. I., Rosenfeld, A. S., Grishina, E. V., & Kondrashova, M. N., (2001). *Correction of Metabolic Acidosis by Maintaining the Function of Mitochondria* (p. 155). Pushchino: Institute of Theoretical and Experimental Biophysics RAS, (in Russian).

28. Nikonov, V. V., & Pavlenko, A. Y., (2009). Metabolic therapy of hypoxic conditions. *Medicine of Emergency States, 3*(4), 22–23 (in Russian).

29. Shugaeva, N. A., Vyskrebentseva, E. I., Orekhova, S. O., & Shugaev, A. G., (2007). Influence of water deficiency on respiration of conducting bundles of a leaf petiole of sugar beet. *Plant Physiology, 54*(3), 373–380 (in Russian).

30. Selote, D. S., Bharti, S., & Khanna-Chopra, R., (2004). Drought acclimation reduces O2•- accumulation and lipid peroxidation in wheat seedlings. *Biochem. Biophys. Res. Commun., 314*(3), 724–729.

31. Miller, G., Suzuki, N., Ciftci-Yilmaz, S., & Mittler, R., (2010). Reactive oxygen species homeostasis and signaling during drought and salinity stresses. *Plant Cell & Environment, 33*(4), 453–467.

32. Zhigacheva, I., Generozova, I., Shugaev, A., Misharina, T., Terenina, M., & Krikunova, N., (2015). Organophosphorus plant growth regulators provides high activity complex I mitochondrial respiratory chain *Pisum sativum* L seedlings in conditions insufficient moisture. *Annual Research & Review in Biology, 5*(1), 85–96.

33. Gigon, A., Matos, A. R., Laffray, D., Fodil, Y. Z., & Pham-Thi, A. T., (2004). Effect of drought stress on lipid metabolism in the leaves of *Arabidopsis thaliana* (Ecotype Columbia), *Ann. Bot., 94*(3), 345–351.

34. Guo, Y. P., & Li Jia-rui, (2002). Changes of fatty acids composition of membrane lipids, ethylene release and lipoxygenase activity in leaves of apricot under drought. *J. Zhejiang Univ., (Agric. & Life Sci.), 28*, 513–517.

35. Makarenko, S. P., Konstantinov, Y. M., Kotimchenko, S. V., Konenkina, T. A., & Arziev, A. S., (2003). Fatty acid composition of mitochondrial membrane lipids in Cultivated (*Zea mays*) and Wild (*Elymus sibiricus*) Grasses. *Plant Physiology, 50*(4), 487–492 (in Russian).

36. Paradies, G., Petrosillo, G., Pistolese, M., Venosa, N., Federici, A., & Ruggiero, F. M., (2004). Decrease in mitochondrial complex I activity in ischemic/perfused rat heart. Involvement of reactive oxygen species and cardiolipin. *Circulation Research, 94*, 53–59.

37. Generosova, I. P., & Shugaev, A. G., (2012). Respiratory metabolism of mitochondria of sprouts of peas of different ages under conditions of lack of moisture and rewatering. *Plant Physiology, 59*(2), 262–273 (in Russian).

38. Burlakova, E. B., (2006). Bioantioxidants: Yesterday, today, tomorrow. Oxidation, oxidative stress, antioxidants. *Proceedings of All-Russian Conference of Young Scientists and II School Academician N.M. Emanuel.* Moscow. RUDN, 5–37 (in Russian).

39. Palmina, N. P., Chasovskaya, I. E., Belov, V. V., & Maltseva, E. L., (2012). Dose dependences of the changes in microviscosity of lipid of biological membranes induced by the synthetic antioxidant potassium phenosan. *Reports of Russian Academy of Sciences, 443*(4), 511–515 (in Russian).

40. Gurevich, S. M., Kozachenko, S. M., & Nagler, L. G., (2006). Change of activities of antioxidant enzymes under the effect of potassium phenosan in small doses. *Proceedings of VII International Conference Bioantioxidant.* Moscow. RUDN, 100–102 (in Russian).

CHAPTER 5

How the Antioxidant Melafen Acts to the Structural and Functional Properties of Erythrocytes

OLGA M. ALEKSEEVA,[1] ANNA V. KREMENTSOVA,[1] ALEXANDER N. GOLOSHCHAPOV,[1] and YURY A. KIM[2]

[1] *Emanuel Institute of Biochemical Physics of Russian Academy of Sciences, 4, Kosygin St., Moscow 119334, Russia,*
E-mail: olgavek@yandex.ru

[2] *Institute of Cell Biophysics of Russian Academy of Sciences, Pushchino, Moscow Region 142290, Russia*

ABSTRACT

This investigation deals with the influence of melafen (melamine salt of bis(oximethyl) phosphinic acid) on the structural properties of membrane-bounded proteins. Melafen is applied in agriculture as a strong plant growth regulator. Melafen was used as aqua solutions, at a wide concentration's region at all our investigations. The erythrocyte's ghost that was isolated from blood erythrocytes were used as the experimental objects. The melafen interactions with membrane-bounded proteins at the erythrocyte's ghost were tested by the differential scanning microcalorimetry (DSC) method. The application of DSC allowed us to receive the structural data of protein microdomains organization at erythrocyte's ghost membranes. We did not reveal any noticeable structural changes of the protein's microdomain organization at erythrocyte ghost membranes under the large concentrations of melafen. The melafen concentrations, which are presented at our experiments, were so much bigger, than concentrations of melafen, which are used at crop production. The melafen actions to the protein-lipid interrelationships were tested by registrations of the microviscosity of isolated native erythrocytes. The using of electron spin resonance method (ESR) with the introduction of spin-labeled probes to erythrocytes allowed us to receive some structural data of protein-lipid organization at the different depths of erythrocyte's membranes (2–4 Å; 4–8 Å). The application of the spectral method for registration of hemolysis degree, allowed us to receive some data about melafen influences to membrane resistance under the bed environment, when the hypo- and hyper-osmolarity existed. We concluded that melafen has not any destruction's influences to the membranes bounded proteins under the concentrations that activate the plant growth. But melafen had essential influencing to the structural and functional properties of membrane-bounded proteins when we used the bigger concentrations.

5.1 INTRODUCTION

Melafen is a plant growth regulator – heterocyclic organic phosphor compound, synthesized at the A. E. Arbuzov (Institute of organic and physical chemistry of the Russian Academy of Sciences). Melafen is the salt of bis (oximethyl) phosphinic acid with a heterocyclic base of melamine (Figure 5.1).

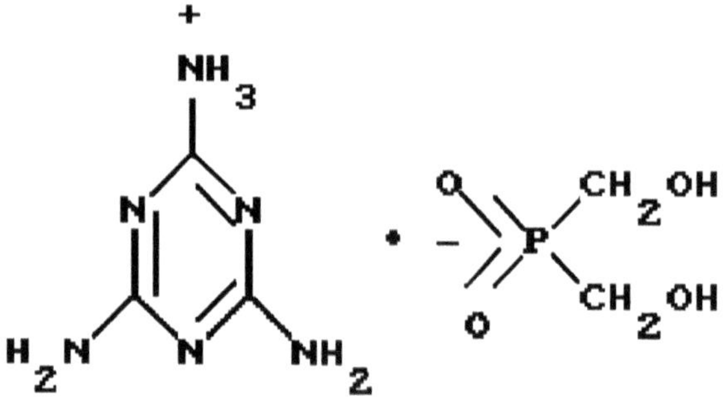

FIGURE 5.1   The structural formula of melafen.

It was received with high yield by one stage from industrially available products. Melafen is the synthetic hydrophilic preparation, which is effectual when it is used at aqua solutions with ultra-low concentrations of melafen ($10^{-10} – 10^{-7}$ M) [1]. Melafen is a polyfunctional substance. As we may see at our text later, melafen is a stimulator of agricultural plants and alga growth

[1, 2]. And it is a strong activator of revitalizing of slept agricultural seeds. But it is the strong and dangerous substance when it is used at large concentrations – the viable agricultural seeds die irrevocably even. But in the study with using of organelles of vegetation and animal origin, as experimental objects, have been found the evidences of variations of the lipid composition of organelles, and their structure and functions were changed too. When low concentrations were used, melafen act as an adaptogen. When in large concentration it was used – toxic substance. The last property has been used in our examining on the insulated animal cells and for the whole animal organism study. We have found an oppression of calcium homeostasis of transformed and normal cells [3]. At the level of the animal's organism, melafen influenced as an inhibitor of the growth of solid malignancy [4].

As we may see at the structural formula of melafen, we found two components: heterocyclic base melamine with a positive charge at Nitrogen atom and bis (oximethyl) phosphinic acid with a negative charge. Melamine has not toxically properties. But the melamine in practice is not metabolized, and it is not cleared from the body with urine. At the same time, when high concentrations of melamine exist in food, melamine can begin the crystallization in the urine [5]. As a result, it becomes apparent in crystal-urine and when very high concentrations leading to rocks education in kidneys. Thus, for these reasons, so many animals and 6 China babies died. At its food, the gluten melamine was found.

The second compound is phosphinic acid. It is known that their different derivatives (the salts with metals Na, Mg, Zn, Ca, Co, Cu) are the plant growth regulators [6]. However, these compounds often are very toxic, because the concentrations or doses of preparations are sufficiently high, when compounds are employed for the seed treatment. The methods of its preparing are very difficult and complex. These compounds are not stable, and are insoluble at the water. By these reasons, the stable hydrophilic substance was synthesized from available compounds by one stage. It was the substance called melafen [1].

It is more likely, that the key role at melafen molecule belongs to the active phosphinic group. This role, is likely, shall be concluded on contact with outer membranes of vegetable cells. As a result, the functional state of membranes is activated, and the trigger of physiological and genetical programs can follow. It exerts the wide action spectrum on the plant body by regulations of energetic processes during all ontogenesis of plants.

Melafen at certain concentrations (3×10^{-10} M) influences (increase slightly) on the generation of superoxide anion radical at chlorella cells [2].

As was supported by many works [7–9] the education of various forms of active oxygen plays a great role not only in damage of plants. But various forms of active oxygen have the most important functions for activation of physiological processes in plant species. The superoxide anion radical may be a signal molecule, which influences the regulatory mechanisms and to the physiology answers of vegetable cells.

Melafen increases the intensity of oxidizing and renovates processes at the expense of activating of such ferments as the peroxidase, the amylase, and the katalase. Melafen facilitates the earlier start of metabolism in germinating seeds and raises the power of germination and increases the field emergence on 4–5%. Melafen stimulates the respiration of vegetable cells; it increases the effectiveness of cyclical phosphorylation [10].

Melafen increases the plants stress-resistance for the bed environment: overcooling and drought. The reasons for these effects are that the melafen increases the effectiveness of the plant's energy metabolism. At these cases, melafen changes of the fatty acid composition and the microviscosity of membranes of microsomes and mitochondria membranes at plant's cells [11, 12]. Melafen, as was shoved at several investigations, exhibits the antioxidant properties. It indirectly influences antioxidant systems. Melafen, together with phytohormone IAA (indole-3-acetic acid) shows synergism in the action that appeared in a unidirectional increase of concentration of carotenoids, and activation of catalase and peroxidase [13]. Another group of authors [14] held the experiments for investigations of melafen effects to the plant and animal mitochondria. They found that the aqua solutions of antioxidant melafen at low and super low concentrations prevented any changes in the fatty acid composition of mitochondria. These solutions also affected to the microviscosity essentially, and influenced to the peroxide oxidizing of lipids. Zhigacheva I. V. et al., showed that electron transport was activated by melafen, when oxidizing the NAD-dependent substrates was occurred, and also changed the structural characterizations of mitochondria membranes of pea [12]. The direct and indirect relationships between the functional state of mitochondria from bodies of vegetation and animal origin and multiple resistances to stressors factors of bed environment were shown [11–14]. Thus, melafen is adaptogen because it can take part in decreasing of the free radical processes, which activate peroxide oxidizing of lipids.

But there are the great concentration limitations, when melafen is used. Large concentrations of melafen cause the immediate death of seeds and plants. And low and ultra-small melafen concentrations activate seed germination and increase the plant resistance under the stress conditions:

the overcooling and drought. However, we suggested that for organisms of animal origin, such melafen effects might have another vector. This vector direction may be in strong dependence from biological active substance (BAS) concentrations. This is why the investigation of aqua-melafen solutions at wide concentrations region has a great significance. The testing of influences of solutions with different concentrations of melafen to some first targets at animal organisms was our main task. The first targets for BAS are the cellular membranes and its components. This is why we used some model objects, which were imitated the real objects with different organization degree. As earlier, we showed the influences of melafen, at the wide concentration range, to the structural properties of lipid membranes with different composition [15]. The lipid-melafen interactions were tested with the aid of DSC-method and small-angle X-ray diffraction. We did not reveal (by the method of X-ray diffraction) of any noticeable structural changes at the egg lecithin membranes under the melafen concentrations that used at crop production. But on the basis of data, obtained by using differential scanning microcalorimetry (DSC), we concluded that the domain structure of dimyristoyl phosphatidylcholine membrane was changed greatly by melafen treatment. The next model objects for our investigations were the cellular membranes, which were isolated from the animal's cells, and the native cells, which were isolated from the animal body. These membranes consist of lipids, proteins, glycoproteins, and other components. Now we tested the melafen influences to the membrane-bounded proteins and lipids-proteins interrelationships.

Melafen is the strong and dangerous regulator of plants growth. Aqueous solutions of melafen at concentration $10^{-11} - 10^{-10}$ M intensified the plant development and the stress resistance under the bed environment. But under the concentration of melafen to 10^{-8} M, the 10^{-7} M plant's seeds quickly died. Therefore, our studies were carried out in a wide range of melafen concentrations ($10^{-21} - 10^{-2}$ M).

Considering the close interdependence of plant and animal bodies in nature, it was necessary to investigate the action of plant growth regulator to any objects of animal origin. The primary targets for BAS in animal's cells are the membranes and their components. When BAS appeared into the blood-vascular system, the first targets are the blood cells. This is why for carrying out the tests of melafen actions as of simple models of primary targets, erythrocytes, which were isolated as the whole cells, and the insulated ghosts that were the erythrocyte's membrane, emancipated from hemoglobin, were selected and were used. The main purpose was to determine how the aqueous solutions of melafen in a wide range of concentrations influence the structure of proteins with animals originated. Due to the present work, the melafen action as on membrane proteins, as well as on protein-lipid interactions has been examined.

Structural studies when influences aqueous solutions of melafen over a wide range of concentrations (10^{-21} – 10^{-2} M) were held on two models: erythrocyte's ghosts and insolated native erythrocytes. Conformational rearrangements of protein's domains at membranes of erythrocyte ghosts were examined by the method of adiabatic DSC. The structural organization of the whole membrane considering as the lipid and protein's phase transitions was characterized by the registrations of the membrane microviscosity. The functional changes of native cells were investigated by testing of melafen actions to the degree of erythrocyte's hemolysis. The degrees of spontaneous and induced hemolysis by variations of experimental medium ion strange were registered by the spectral method.

In order to verify our assumption, how melafen influencing to the cellular membranes, may be mediated by the actions to membrane-bounded protein components, for a description of this problem the microcalorimetry scanning of erythrocyte membranes was held. The testing of melafen actions on the thermostability of protein microdomains in erythrocyte's membrane was performed by using of erythrocyte ghosts. The erythrocyte ghosts are the plasmatic erythrocyte membranes with all elements of the cytoskeleton. These erythrocyte ghosts are fully emancipated from hemoglobin by means of hypo-osmotic hemolysis [16, 17]. In Figure 5.2, the scheme of erythrocytes ghost is submitted, where revealing the erythrocyte conversion to the ghost, when hemoglobin was exited.

The main protein elements of erythrocyte's cytoskeleton and the external plasmalemma are shown.

The erythrocyte membrane increased its permeability to hemoglobin when hypo-osmotic medium. The maintaining of the cold thermal regime (4°C), without freezing, when centrifugation permits to receive the colorless draft of ghost membranes. Ghost membranes were suspended at phosphate buffer easily. It should be noted that erythrocyte ghost is the fine model for studying of protein-lipid interactions at cellular membrane and the cellular cytoskeleton. The main components are typical for most of the cells of the animal body (see Figure 5.2).

For explorations on such model comfortable practice a method of adiabatic differential scanning microcalorimeter (DSC) by means of high-sensitive microcalorimeter DASM-4 [19].

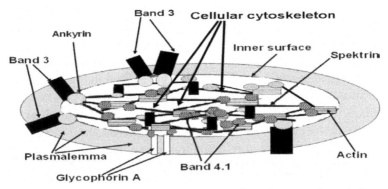

FIGURE 5.2 Scheme of erythrocyte ghost structure: plasmalemma and cytoskeleton, after hemoglobin releasing due to hypo-osmotic hemolysis. The basic proteins microdomains in erythrocyte ghost are denoted by riflemen's — there are plasmalemma (erythrocyte envelope) and cellular cytoskeleton components: ancyrin, spektrin, actin, proteins of band 4.1. Glycophorin A, proteins of band 3 are the integral proteins, which are incorporated to plasmalemma. (Figure was modified from our works [18]).

The microcalorimetry research at isotonic conditions discovers the five structural thermo denaturation transitions into ghost membranes (Figure 5.3), which are called *A*-transition, B$_1$-transition, B$_2$-transition, *C*- transition and *D*-transitions [20, 21].

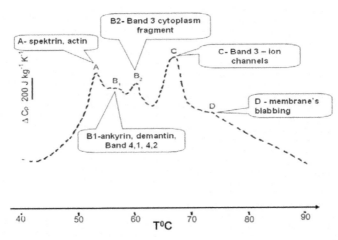

FIGURE 5.3 Thermograms of erythrocyte ghosts were registered by means of DASM-4. Identified endothermic phase transitions of protein's microdomains (A, B1, B2, C, D) in erythrocyte ghost's membranes, when registered by the DSC method, are noted. The designations: the axis of ordinates – ΔCp change of relative heat capacity (J/K) at the peak of a maximum of heat absorption (transition intensity); the X-axis – temperature °C. (Figure was modified from our work [18]).

The A-transition is determined by the microdomains denaturation of the cytoskeleton, set up of complex α- and β-spectrin and actin. Denaturation of spectrin-actin complex microdomain results in the disappearance of A-transition. That is followed by the total loss of erythrocytes deformability [22] and ghost membranes too [23]. B_1-transition is linked to denaturation of membranous microdomain, set up of ancyrin and proteins of bands 4.1, 4.2 and demantin. B_2-transition is linked with denaturation of cytoplasm fragment of protein band-3 microdomains. C-transition is linked with denaturation of membrane fragment 55 kDa of proteins band-3, which are the ion-channels microdomains. D-transition has linked with unidentified proteins denaturation and membrane bubbling microdomains [20, 21].

It should be noted that pathology damaged erythrocytes (e.g., oxidized, etc.) in the bloodstream are lacking, since used by macrophages. However, we used the native erythrocytes isolated from a healthy organism. All treatments were provided without macrophages in experimental test glass. The model of erythrocyte ghosts for DSC studies of cell membranes is quite adequately because the same changes occur in membranes and cytoskeletons of other cells. The proteins composition of cytoskeletons: spectrin, ancyrin, bands proteins 4.1 and 4.2 are developing and band protein 3 is submitted consisting of membranous skeletal in practice all cells of an organism. The DSC method, together with polyacrylamide protein electrophoresis allows investigators to define arizing when pathologies changes occurred in the cell membrane and erythrocyte cytoskeleton [24–26].

5.2   MATERIALS AND METHODOLOGY

5.2.1   MATERIALS

Melafen (melamine salt bis (oximethyl) phosphinic acid) was synthesized at Arbuzov A. E. Institute of Organic and Physical Chemistry of RAS Kazan. Melafen was used as the aqua solutions at the wide concentration range ($10^{-21} - 10^{-3}$ M).

The erythrocytes and its ghost were prepared by the method [24]. The melafen interactions with membrane-bounded proteins at the erythrocyte's ghost were tested by the DSC method [19].

The alcoholic solutions of probes were inputted to 5%-suspensions of erythrocytes as 30 min before up to experimental objects measurements. Registrations were occurred by the ESR spectrometer ER-200D SRC of company Bruker (Germany). On obtaining the electron paramagnetic resonance spectrums, employing formula for high turning probes, calculated the time of rotational correlation of probe $\tau_c \times 10^{-10}$s, which is the reorientation time of probe at the angle $\sim\pi/2$ of and describes the lipids microviscosity in the membrane [27]. The results were expressed in relative units. Control served the membranes models without the addition of melafen solutions.

The hemolysis was being conducted during 45 min at room temperature. The erythrocytes stability to spontaneous hemolysis studied by the method of Jager [28], based on photometry ($\lambda = 540$ nm) of ectoglobular hemoglobin, which incoming on medium owing to snap lyzes of erythrocyte membranes, induced by the lipid oxidation by the air oxygen [28]. To operate, we used the 5%-suspension of erythrocytes on medium, containing 0.1 M Tris-HCl buffer (pH 7.4) and the saline solution in correlation 1:1. All the examining held with fresh, just insulated erythrocytes. The hemolysis level identified on optical absorption of the supernatant at the wavelength 540 nm by spectrophotometer "Specord M40." For 100% the hemolysis level supernatant absorption after hemolysis, caused by 0.1% of Triton $\times100$ is elected.

5.3   RESULTS AND DISCUSSION

For investigations of effects of melafen aqueous solutions to the protein's microdomains organizations at cellular membrane, we used the ghosts of rat erythrocytes. This model is fairly simple for using. The DSK thermograms are clearly mirrored all peaks of thermo-inducted denaturation transitions of cytoskeleton's proteins [29]. In Figure 5.4, five temperatures inducted transitions of protein microdomains in ghost membranes of rat erythrocytes were submitted as control and after preliminary incubation with the testable aqueous solutions of melafen.

At Figures 5.4 and 5.5, the ghost's thermograms in the presence of melafen aqueous solutions at first and second days after receiving of a ghost from rat erythrocytes are presented.

In Figure 5.5, five identified endothermic phase transitions of proteins (A, B1, B2, C, D) in erythrocyte ghost's membranes when registered by the DSC method are noted.

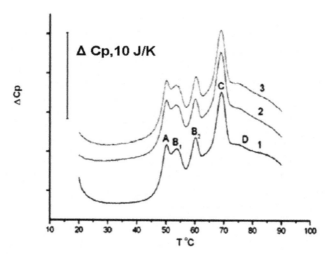

FIGURE 5.4 The melafen influences to the endothermic thermodenaturation of protein's microdomains at erythrocyte's ghost membranes on the first day after the receiving of ghost. (Note: The designations: the axis of ordinates – ΔCp change of relative heat capacity (J/K) at the peak of a maximum of heat absorption (transition intensity); the X-axis – T °C temperature °C. Five identified endothermic phase transitions of protein's microdomains (A, B1, B2, C, D) in erythrocyte ghost's membranes when registered by the DSC method are noted. Thermograms are named as 1 – control; 2 – with melafen 10^{-5} M; 3 – with melafen 10^{-3} M. (Figure was modified from our work [18, 30])).

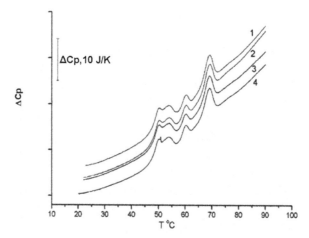

FIGURE 5.5 Effect of melafen to the thermodenaturation of proteins at erythrocyte's ghost membranes on the second day after the receiving of ghost. (*Note:* the axis of ordinates ΔCp – change of relative heat capacity (J/K) at the peak of a maximum of heat absorption (transition intensity); the X-axis – T°C. On thermograms are named: 1 – control; 2 – with melafen 10^{-8} M; 3 – with melafen 10^{-6} M; 4 – with melafen 10^{-4} M. (Figure was modified from our works [18, 30])).

As it is seen from Figures 5.4 and 5.5, the structure of ghost after aging was changed essentially, but the addition of melafen solution before scanning was not affected by views of curves. The dislocations of thermo-denaturation peaks and any essential changes in peak's amplitudes did not happen. It indicates that the aqueous solutions of melafen were not affected directly on the structural organization of protein components of the membrane. But the melafen aqueous solutions caused some small restructuring of protein's microdomains as on freshly isolated preparations (3–9%), and after the process of aging of the erythrocyte's ghost. In Table 5.1, the data presented in the calculation of relative enthalpy.

TABLE 5.1 Melafen Influence on Temperature Dependence of Relative Enthalpy (Thermograms – Peak A) of Membranes Suspensions of Erythrocyte Ghosts at the First Day of After Receiving of Erythrocyte Ghosts*

Concentration of melafen	ΔC_p	Δ (%)
Without melafen	$17.1 \pm 0{,}01$	–
Melafen 10^{-5} M	$17.6 \pm 0{,}01$	+3%
Melafen 10^{-3} M	18.6 ± 0.01	+9%

Note: ΔC_p – change of relative heat capacity (J/K) at the peak of a maximum of heat absorption (transition intensity); Δ (%) – the different quantity, where 100% – the quantity of control answer.
*Table was modified from our work [18, 30].

As it can be seen from data of Table 5.1, the melafen, aqueous solutions when concentrations 10^{-5} M and 10^{-3} M, did not cause of important changes in relative heat capacity (J/K) in peak maximum of heat absorption and (the A transition intensity) (3–9%).

A-transition is the characteristic of therm-induced endothermic dena-turation of two most important protein's cytoskeleton components, which are contained in practice of all cells of animal origin – spectrin and related actin (refer to Figures 5.2 and 5.3). Other endothermic transitions, which are the characteristics of thermo-denaturation properties of cytoskeleton components and channel fragments also, in practice, didn't change its ampli-tude. The same data have been received and for thermograms, registered in the second day after the ghost preparation, or on "aged" ghosts. These results suggest that the microdomains organization of cytoskeleton proteins and plasma membrane did not show the major changes in the presence of melafen aqueous solutions of high concentrations.

It indicates that melafen haven't any considerable affects to thermal denaturation properties of protein components at cellular membrane and cytoskeleton directly. Melafen does not cause any changes, which may lead to proteins loss, or to the restructuring of protein's domains.

The structural research of influencing of melafen aqueous solutions of high concentration to animal's cellular membranes had been continued on insulated entries erythrocytes by the measurements of microviscosity by means of electron spin resonance method (ESR).

The lipids, which are the surrounding of proteins, play a major role in the structural organization and in the maintaining of activity of the membrane protein components. The lipid's microviscosity in areas of lipid-lipid and lipid-protein, which were mirrored at the interactions at membranes, has a great significance. The microviscosity of erythrocyte membranes was estimated on rotational correlation time two materials, which are included in the lipid phase of membranes. These two materials are the two spin probes: 2,2,6,6-tertmethyl-4-capriloil-oxipiperidin-1-oxil (probe 1) and 5,6-benzo-2,2,6,6-tertmethyl-1,2,3,4-tetra hydro-γ-carbolin-3-oxil (probe 2) [31]. Two spin probes vary in its hydrophobic properties. Their structural formulas are introduced in Figure 5.6.

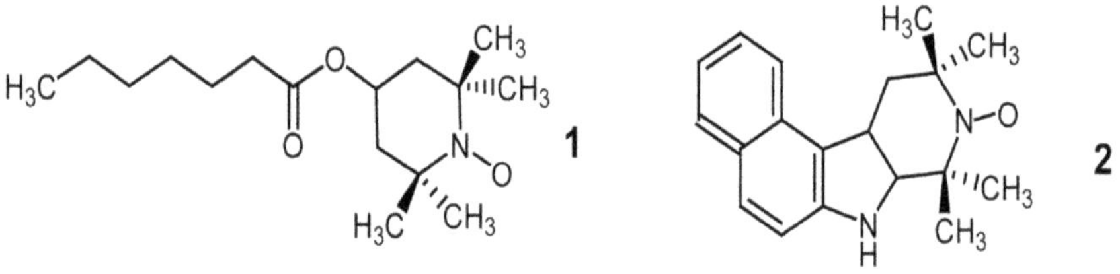

FIGURE 5.6   Structural formulas of nitroxyl radicals – spin-labeled probes 1 and 2.

It is known that the probe 1 primarily localizes in surface regions of lipid membrane in the area of lipids to a distance 2–4 Å, and probe 2, in the zone, which are near to protein and annular lipids at a depth of 6–8 Å from the surface [32]. The influences of melafen aqueous solutions to membranes microviscosity of insulated erythrocytes were measured by using two probes (Table 5.2).

TABLE 5.2 Influence of Melafen on Relative Changes of Erythrocyte Membrane Microviscosity*

Concentration of melafen	Probe 1, arb. un.	Probe 2, arb. un.
Without melafen	1.0	1.0
Melafen 4×10^{-5} M	0.98	0.93
Melafen 4×10^{-3} M	0.74	0.90

Note: Probe 1–2,2,6,6-tertmethyl-4-capriloil-oxipiperidin-1-oxil; Probe 2–5,6-benzo-2,2,6,6-tetramethy-l-1,2,3,4-terthydro-γ-carbolin-3–oxil; arb. un. – arbitrary unity.
*Table was modified from our work [18].

From Table 5.2 data, it is possible to see the suggestions that melafen, when the biggest doses, reduced the microviscosity of both regions of bilayer: at cursory lipid region at a depth of 2–4 Å and at the area of lipid in near protein's regions 6–8 Å. The microviscosity of both regions of the erythrocyte membrane was tapered off, when large concentration 10^{-3} M of melafen aqueous solutions. In lipid region at a depth of 2–4 Å was tapered of maximally – up to 25%. In earlier works [33], by using of aqueous solutions of melafen, when smaller concentrations ($10^{-10} – 10^{-7}$ M), was also shown the microviscosity change of erythrocyte membranes at near protein's regions on 7%, in lipid surface regions, low concentrations of melafen in practice unaltered the bilayer microviscosity. Melafen aqueous solutions with higher concentrations are influenced on lipid-lipid interactions in surface regions of membranes essentially (up to 25%). And the aqueous solutions of melafen that acted to membranes, have the insignificant influence (up to 10%) on lipid-protein interactions in deeper areas of the membrane of erythrocytes, when very large, and low concentrations of melafen.

When we tested the action of melafen aqueous solutions on cellular membranes, the functioning was some important changes in membrane permeability were found. This part of our work was accomplished to insulated entries erythrocytes. For these tests, we used the model of hemolysis of erythrocytes in hypo and hyperosmotic medium in contrast with spontaneous hemolysis. Such a model reflects the probability of melafen influence on the degree of membrane stability, and cell resistance to damaging factors of bed environment. The hemolysis, when immediate by addition of small amounts of melafen aqueous solutions with different concentrations to erythrocytes suspension, will describe the damaging action of Melafen to plasmalemma.

As it was disclosed that melafen in all range of concentrations ($10^{-13}–10^{-3}$ M) do not cause additional hemolysis and do not protect against spontaneous hemolysis. So, if the value of 100%, it is the hemolysis, in which the hemoglobin from erythrocytes is released fully, when the addition of Triton ×100, then the spontaneous hemolysis compiled 3%, and the hemolysis in the presence of melafen 3% ± 0,3%.

The measurements of erythrocytes hemolysis in hypo and hyperosmotic conditions (0–4 mM NaCl) were held with using of approaches and foundation, which were obtained by Kim, Yu. A. and his co-workers at work [34]. We held these experiments because we wanted to answer to the main question: what the explored preparation melafen can render some protective effect on erythrocyte's membrane? Or also melafen can speed up the erythrocyte's membrane destruction, when local unfavorable factors presented at medium. Obtained data are given in Figure 5.7.

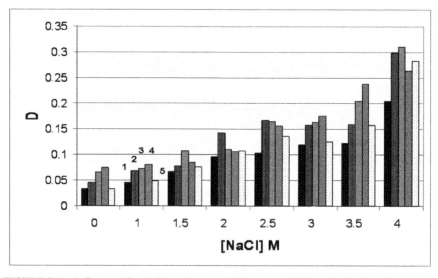

FIGURE 5.7   Influence of melafen on hypo- and hyperosmotic hemolysis of erythrocytes. (Note: The hemolysis level was identified on optical absorption (D_{540}) of supernatant at the wavelength 540 nm. For 100% the hemolysis level supernatant absorption after hemolysis, caused by 0.1% of Triton ×100 is elected. The designations of columns: 1 – control; 2 – 10^{-9} M melafen; 3 – 10^{-7} M melafen; 4 – 10^{-5} M melafen; 5 – 10^{-3} M melafen; the axis of ordinates – D – optical density; the X-axis – [NaCl] M – concentration of NaCl. (Figure was modified from our work [18, 30]).

As it can be seen from data at Figure 5.7, the general trend of increasing of hemolysis degree of erythrocytes varied significantly when melafen presence over a wide range of concentrations (10^{-9}–10^{-3} M). And these trends became so bigger under the increasing of environment osmolarity. But the melafen addition in concentration 10^{-12} M and 10^{-11} M don't change of general picture of hemolysis of erythrocytes (data were not shown).

5.4 CONCLUSION

At this part of our investigations, we tested the influence of plant growth regulator melafen to organisms of animal origin, that we used as the test object the erythrocytes. Erythrocytes were one of the first targets when some exogenous substances appeared in blood riverbed. This model is very simple for preparation, and fairly, it is stable for measurements. At the same time, the erythrocytes have the cytoskeleton protein composition, which is similar to most animals' cells to a considerable extent. By this, the investigations that were provided with this experimental object had the essential significance.

The composition and the structure of structural elements of the erythrocyte membrane are representative for many cells of an organism. The obtained data about measurements of proteins thermostability, their relationships with lipids, the changing of membrane permeability indicated that melafen didn't exert the great destructive actions to the insolated erythrocytes ant its isolated membranes (with cytoskeleton) under the concentrations that activate the plant growth. But melafen influencing was more essential to the structure and functioning of membrane-bounded proteins, when the bigger melafen concentrations presented at an experimental medium.

KEYWORDS

- **differential scanning microcalorimetry**
- **erythrocyte ghost**
- **hemolysis**
- **microviscosity**

REFERENCES

1. Fattakhov, S. G., Reznik, V. S., & Konovalov, A. I., (2002). Melamine salt of bis (hydroxymethyl) phosphinic acid (Melaphene) as a new generation regulator of plant growth regulator. *Proceedings of the 13th International Conference on Chemistry of Phosphorus Compounds.* St. Petersburg. 80. (in Russian).

2. Kashina, O. A., & Loseva, N. L., (2011). Influence to the growth and energetic processes of chlorella cells. *Phosphor Organic Substance Melafen* (p. 116). Lambert Academic Publishing.

3. Alekseeva, O. M., Fatkullina, L. D., Kim, Y. A., Burlakova, E. B., Fattakhov, S. G., Goloshchapov, A. N., & Konovalov, A. I., (2009). Melafen influence on structural and functional state of liposome membrane and cells of ascetic Ehrlich carcinoma. *Bulletin of Experimental Biology and Medicine. 147*(6), 684–688 (in Russian).

4. Krementsova, A. V., Erokhin, V. N., Semenov, V. A., & Alekseeva, O. M., (2015). The aqua melafen solutions influences on the parameters of the mice's solid carcinoma Luis development. In: Berlin, A. A., & Joswik, R., (eds.), *The Chemistry and Physics of Engineering Materials, Modern Analytical Methodologies* (Vol. 1. pp. 45–63). Apple Academic Press.

5. Li, G., Jiao, S., Yin, X., Deng, Y., Pang, X., & Wang, Y., (2009). The risk of melamine-induced nephrolithiasis in young children starts at a lower intake level than recommended by the WHO. *Pediatric Nephrology, 25*(1), 135–141.

6. Melnikov, N. N., Novozhilov, K. V., & Belan, S. R., (1995). The pesticides and the plant growth regulators, handbook, Moscow. *Chemistry*, p. 576 (in Russian).

7. Skulachev, V. P., (1989). *Energetic of Biological Membranes* (p. 564). Moscow. Nauka (Science) (in Russian).

8. Merzlyak, M. N., (1999). Activated oxygen and the life activity of plants. *Soros Educational Journal, 9*, 20–26 (in Russian).

9. Bolwell, G. P., (1999). Role of active oxygen species and NO in plant defense responses. *Opin. Plant. Biol.,* 287–294.

10. Loseva, N. L., Kashina, O. A., & Tribunskih, V. I., (2003). Potential mechanism of melafen action to physiological processes of growth of plant cell. *Proceedings of the International Conference on Plant Physiology is Base of Fitobiotechnology* (pp. 409–410). Penza (in Russian).

11. Zhigacheva, I. V., Fatkulina, L. D., Burlakova, E. B., Shugaev, A. I., Generosova, I. P., Fattahov, S. G., & Konovalov, A. I., (2008). Influence of phosphorogenic plant growth regulator to the structural characteristics of membranes plant's an animal's origin. *Biological Membrane, 25*(2), 150–156 (in Russian).

12. Zhigacheva, I. V., Misharina, T. A., Trenina, M. B., Krikunova, N. N., Burlakova, E. B., Generosova, I. P., Shugaev, A. I., & Fattahov, S. G., (2010). Fatty acids content of mitochondrial membranes of pea seedlings in conditions of insufficient moistening and treatment by the phosphor organic plant growth regulator. *Biological Membranes, 27*, 256–261 (in Russian).

13. Kirillova, I. G., & Sorokina, G. N., (2012). The influence of the regulator of the growth of melafen and copper microelement on the components of the antioxidant system and productivity of *Solanum tuberosum* L. Scientific notes of Turgenev I.A. State University. *Series of Natural, Technological and Medical Science, 3*, 103–106 (in Russian).

14. Zhigacheva, I. V., Fatkulina, L. D., Rusina, I. F., Shugaev, A. I., Generosova, I. P., Fattahov, S. G., & Konovalov, A. I., (2007). Antistress properties Influence of substance melafen. *Reports of Russian Academy of Sciences, 414*(2), 263–265 (in Russian).

15. Alekseeva, O. M., Krivandin, A. V., Shatalova, O. V., Rykov, V. A., Fattakhov, S. G., Burlakova, E. B., & Konovalov, A. I., (2009). The melafen-lipid – interaction determination in phospholipid membranes. *Reports of Russian Academy of Sciences, 427*(6), 218–220 (in Russian).

16. Dodge, J. T., Mitchell, C., & Hanahan, D. J., (1963). The preparation and chemical characteristics of hemoglobin-free ghost of human erythrocytes. *Arch. Biochem. Biophys., 100*, 199–130.

17. Sato, Y., Yamakose, H., & Suzuki, Y., (1993). Mechanism of hypotonic hemolysis of human erythrocytes. *Biol. Pharm. Bull., 16*(5), 506–512.

18. Alekseeva, O. M., Fatkullina, L. D., & Shatalova, O. V., (2014). Melafen influence to structure and function of several proteins of animal's origin. In: Fattachov, S. G., Kuznetsov, V. V., & Zagoskina, N. V., (eds.), *Melafen: Mechanism of Action and Regions of Using Part 4.2* (pp. 343–358). Print-Servis-XXI century, Kazan (in Russian).

19. Privalov, P. L., & Plotnikov, V. V., (1989). Three generations of scanning microcalorimeters for liquids. *Therm. Acta, 139*, 257–277.

20. Jackson, W. M., Kostyla, J., Nordin, J. H., & Brandts, J. F., (1973). Calorimetric study of protein transitions in human erythrocyte ghosts. *Biochemistry, 12*, 3662–3667.

21. Brandts, J. F., Lysko, K. A., Schwartz, A. T., Eryckson, L., Carlson, R. B., Vincentelli, J., & Taverna, R. D., (1976). Structural studies of the erythrocyte membrane. *International Colloquium C. R. S., (Colloques. Internationaux. du. C. R. S.), 246*, 169–175.

22. Mohandas, N., Greenquis, A. C., & Shohet, S. B., (1978). Effect of heat and metabolic depletion on erythrocyte deformability, spectrin extractability and phosphorylation. *The Red Cell* (pp. 453–472). New York. Alan R. Liss. Inc.

23. Heath, B. P., Mohandas, N., Wyatt, J. L., & Shohet, S. B., (1982). Deformability of isolated red cells. *Biochim. et Biophis. Acta, 691*(2), 211–219.

24. Akoev, V. R., Sherbinina, S. P., Matveev, A. V., Tarachovsky, Y. S., Deev, A. A., & Shnirov, V. L., (1997). Investigations of structural transitions at erythrocytes membrane when hereditary hemochromatosis. *Bulletin Experimental Biology and Medicine, 123*(3), 279–284 (in Russian).

25. Gulevski, A. K., Riazancev, V. V., & Belous, A. M., (1990). Structural transitions of lipids and proteins of erythrocytes membrane when low-temperature action. *Scientific Reports of High School. Biol. Sciences*, 29–36 (in Russian).

26. Akoev, V. R., Matveev, A. V., Belyaeva, T. V., & Kim, Y. A., (1998). The effect of oxidative stress on structural transitions of human erythrocyte ghost membranes. *Biochim Biophys Acta, 1371*(2), 284–294.

27. Burlakova, E. B., & Goloshchapov, A. N., (1975). Study of thermoinduced transitions in membranes of animal organs during injection of antioxidants and malignant growth. *Biophysics, 25*(1), 97–191 (in Russian).

28. Jager, F. C., (1968). Determination of vitamin E requirement in rats by means of spontaneous hemolysis in vitro. *Nutrition and Diets. (Nutr. et Diets), 10*(3), 215–223 (in Dutch).

29. Shnyrov, V. L., Zhadan, G. G., & Salia, C. H., (1986). Thermal transitions in rat erythrocyte ghosts. *Biomed. Biochim. Acta, 45*, 1119–1121.

30. Alekseeva, O. M., & Kim, Y. A., (2018). Bed environment actions to erythrocyte's membranes. New information technology in medicine, pharmacology, biology and ecology. *Proceedings of the XXVI International Conference and Discussion Scientific Club: New Information Technologies in Medicine, Biology, Pharmacology and Ecology* (pp. 151–155). Crimea, Yalta–Gurzuf (in Russian).

31. Binukov, V. I., Borunova, S. F., & Goldfeld, M. G., (1972). Paramagnetic probes at EPR-spectroscopic. *Biochemistry, 36*(6), 1149–1153 (in Russian).

32. Vasserman, A. M., & Kovarsky, A. L., (1986). *Spin Probes and Zones at Physicochemistry and Polymers* (p. 255). Moscow, Nauka (Science in Russ) (in Russian).

33. Zhigacheva, I. V., Fatkulina, L. D., Burlakova, E. B., Shugaev, A. I., Fattahov, S. G., & Konovalov, A. I., (2006). Preparation Melafen and physical-chemical state of membrane. *Reports of Russian Academy of Sciences, 409*(4), 547–549 (in Russian).

34. Kim, Y. A., Elemesov, R. E., & Akoev, V. R., (2000). Hyperosmotic hemolysis of erythrocytes and antihemolytic activity of saponins faction and triterpenes glycosides from Panax Ginseng, C. A. Meyer. *Biological Membrane, 17*(2), 15–26 (in Russian).

CHAPTER 6

Role of Water-Soluble Compounds of Reduced Sulfur in the Formation of the Toxic Properties of the Aquatic Environment

ELENA V. SHTAMM,[1] VIACHESLAV O. SHVYDKIY,[1] YURII I. SKURLATOV,[2] and LIUDMILA V. SEMENYAK[3]

[1]*Emanuel Institute of Biochemical Physics, Russian Academy of Science, 4, Kosygin St., Moscow 119334, Russia, E-mail: slavuta58@gmail.com*

[2]*Semenov Institute of Chemical Physics of Russian Academy of Sciences, 4–1, Kosygin St., Moscow 119334, Russia*

[3]*Russian Research Institute of Fisheries and Oceanography, 17, VerhnyayaKrasnoselskaya St., Moscow 107140, Russia*

ABSTRACT

The review of own and literary data confirming essentially important role of oxidation-reduction processes with participation of hydrogen peroxide and reduced sulfur water-soluble compounds for formation of the biological full value of the natural water environment and normal functioning of aerobic aquatic ecosystems is submitted. Hydrogen peroxide is the integral participant chemical-biological circulation of molecular oxygen: it is formed as an intermediate product at the 4th an electronic photo-oxidation of molecules of water in the processes of photosynthesis and at the 4th electronic reduction of O_2 to water in oxidation processes (including, in processes of cellular respiration). Reduced sulfur compounds (RSH) are formed in the anaerobic environment, mainly, as a result of microbiological processes of a sulfate-reduction, or are dumped in natural waters as a part of sewage.

As established by us, among the reduced-sulfur, the special danger for aerobic aquatic ecosystems represents by water-soluble substances that effectively interact with hydrogen peroxide and at the same time are stable against oxidizing action of oxygen.

At excess receipt of similar substances reduced sulfur there can arise quasi-reduced conditions in the water when molecular oxygen is present at the water in "norm"(in a thermodynamic sense the water environment is in an oxidizing state), but instead of hydrogen peroxide there are detected substances which are effectively interacting with H_2O_2. Similar transition from a normal oxidizing state to *quasi*-reduced is followed by emergence in the water environment of toxic conditions for aerobic water organisms with intensive water exchange with the external environment, in particular, for larvae of fishes before their transition to branchiate breath and also creation the conditions in water favorable for mass development ("blossoming") of blue-green seaweed (cyanobacteria) and of accompanying them pathogenic microflora.

6.1   INTRODUCTION

The work purpose is to draw attention of water ecologists, developers of criteria of quality of water and programs of protection of water ecosystems, experts in the field of fishery to current situation when large-scale discharge into reservoirs and rivers of allegedly purified sewage (without control of water-soluble reduced sulfur compounds (RSH) and without toxicological control by methods of biotesting) leads to catastrophic consequences for ecosystems of the water objects accepting this sewage.

Biosphere as a whole and separate parts of it has been characterized normally by a high level of balance of biogeochemical cycles of carbon, oxygen, nitrogen, sulfur, and other biogenic elements [1]. The most important role in those productive-destructive processes (photosynthesis – respiration) is connected with the biogeochemical cycle of oxygen.

In the global biological cycle, molecular oxygen (O_2) formed during the oxidation of water molecules in the photosynthesis process using visible sun-light (hv), and it is reduced back to H_2O during the oxidative destruction of organic substances with the release of the stored solar energy (kT):

$$H_2O + CO_2 + hv \rightarrow \{CH_2O\} + O_2 \tag{1}$$

$$O_2 + \{CH_2O\} \rightarrow H_2O + CO_2 + kT \tag{2}$$

In an aerobic aquatic medium, these redox-processes are normally balanced: the content of O_2 in water varies in a relatively narrow range, which is physiologically acceptable for aerobic aquatic organisms.

6.2 MOLECULAR OXYGEN AS STORAGE OF OXIDATIVE EQUIVALENTS

The concentration of molecular oxygen dissolved in natural waters normally determined by its solubility and varies in a narrow range. On the other hand, the concentrations of active intermediate species of oxygen in natural waters depend on many factors, both natural and anthropogenic origin.

Despite its high oxidation potential, under normal conditions, molecular oxygen is chemical inert oxidant due to the peculiarity of its electronic structure (triplet basic state) and low value of one-electronic reduction potential [2]. Its content in water characterizes the total supply of oxidative equivalents, which are released only under certain conditions, namely, in the presence of catalysts, sensitizers, or free radicals in the aqueous medium. Concerning aquatic ecosystems, the participation of the intermediate species of molecular oxygen reduction, namely hydrogen peroxide and free radicals OH, $HO_2,(O_2^-)$ in abiotic oxidation processes [3], have more significance.

The catalysts of processes involving molecular oxygen as an oxidizer are generally mixed-valence metal ions [4] (M^+, M^{2+}, and M^{3+} are the conventional designations for the reduced, oxidized, and super-oxidizing states of the metal ion, respectively). As an oxidizer, O_2 participates mainly either in oxidase-type processes oxidations of organic donors H (DH_2) accompanied by the formation of hydrogen peroxide.

$$O_2 + DH_2 \xrightarrow{M^+/M^{2+}} D + H_2O_2 \qquad (3)$$

or in oxygenase-type processes accompanied by the introduction of the O atom at the C–H or C– bond, which increases the solubility of hydrophobic organic compounds:

$$O_2 + RH + DH_2 \xrightarrow{M^+/M^{3+}} ROH + D + H_2O \qquad (4)$$

In natural waters, molecular oxygen became chemical active either passing on singlet state under photochemical (sensitized) exiting [5–8] or forming complexes with partial one- or two-electronic innerspheric charge transfer at interaction with transition metal ions in the reductive state

[9, 10]. The molecular product of two-electronic reduction of dioxygen is hydrogen peroxide possessing more expressive reactivity as oxidant. Opposite to dissolved dioxygen the hydrogen peroxide content in natural waters varies in the large scale [11], undergo to daily and seasonal changes. Reaction ability of hydrogen peroxide is been determined by relatively low energy of O–O-bond and, respectively, by easiness formation of OH-radicals or radical-similar species, when splitting of O–O-bond occurs in the coordination sphere of transition metal ion without free radical withdrawal outside [12].

So, the oxidative processes involving O_2 are inevitably accompanied by the formation of hydrogen peroxide and OH-radicals in aqueous media.

6.3 KEY ROLE OF HYDROGEN PEROXIDE AN AEROBIC AQUATIC ECOSYSTEMS

Back in the 1960s, it was found that hydrogen peroxide is contained in both marine and surface freshwaters [11]. According to V.E. Sinelnikov [13, 14], hydrogen peroxide is one of the compounds that are constantly present in natural waters, and the content of H_2O_2 at a level of 10^{-6} M is typical for unpolluted areas of water. At the same time, the H_2O_2 concentration in the natural aquatic medium has seasonal and diurnal dynamics; it depends on the depth, weather conditions, and the presence of algae in the water.

Hydrogen peroxide has been formed in natural waters by several ways, mainly:

- The dismutation of superoxide radicals (in living cells catalyzed by superoxide dismutase, SOD)[15].

$$O_2^{-\cdot} + O_2^{-\cdot} \xrightarrow{\;2H^+,\; SOD\;} O_2 + H_2O_2 \tag{5}$$

- Two-electronic singlet oxygen reduction [16, 17] is given by:

$$^1O_2 + DH_2 \rightarrow D + H_2O_2 \tag{6}$$

- Secretion by algae as a by-product of photosynthesis process [18–20] or due to other photochemical reactions in algae's cells with the participation of O_2 [21].

Usually, superoxide radicals O_2^- are the predecessor of H_2O_2 in algae's cells too [22]. Formation of these radicals as a result of photochemical processes takes important play in aquatic cycling of hydrogen peroxide. Moreover, in the presence of H_2O_2 superoxide radicals can initiate of hydroxyl radicals formation by interaction with ions and complexes of copper (II) usually presented in natural waters in trace amounts.

Hydroxyl radicals are the most reactive oxidants in the aquatic environment. With rare exception, OH-radicals interact with organic substances with the rate constants closed to diffusion limit [23].

In excessive cell formation of OH-radicals can result in the destruction of physiologically important molecules. Similarly, in natural waters at high rates of OH-radicals formation, the oxidation of ecological metabolites can be possible, for example of unsaturated fatty acids that participate in aquatic biological cycling of organic substances [24]. Respectively, both in cell and in natural aquatic medium antioxidants are present – "scavengers" of free radicals. The rate of OH-radical formation and content of "scavengers" determine the steady-state of radicals normally in the range admissible for biological (or ecological) systems.

Anthropogenic impact on aquatic cycling of active intermediate species of oxygen can be manifested by several pathways, but the main of its are reduced to changing of hydrogen peroxide content and freeOH-radicals steady state concentration.

Hydrogen peroxide takes an important role in cells [25] and in aquatic environmental redox-processes [26]. H2O2 contains in cells of mammalians and alga varies mainly in range 10^{-7}–10^{-6} M [27], but sometimes it can reach 10^{-5} M [25]. Similar ranges of H_2O_2 contain changing are been also observed for natural water objects [28]. The increasing of hydrogen peroxide concentration in the medium (bothinnercells and aquatic environment) will result in the end to increase of formation rate of active intermediates that can cause destruction of important for biota molecules. At the sufficient great concentration of hydrogen peroxide can participate in direct, noncatalytic reactions (nucleophilic substitution) that also can cause to oxidation of biologically significant molecules. However, the effects of toxicity from those reactions will be observed only at the concentrations of H_2O_2 in the medium much more than typical values for physiological or aquatic ecological systems. Short-term increasing of hydrogen peroxide content in the medium for value by order 10^{-5} M is safe for the most aquatic organisms. The effects of toxicity have been observed at long-term maintaining of hydrogen peroxide on the levels more than 3×10^{-5} M. As test-organisms we used in these experiments infusoria *Tetrahymenapyriformis*, *Daphnia magna*, and larva sturgeon fish. The most sensitive presence of hydrogen peroxide in the aquatic medium is blue-green algae. Hydrogen peroxide suppresses the photosynthetic activity of these algae (semibacteria) at the concentration 10^{-5} M and less, especially in the presence of copper ions or under the impact of UV-B irradiation [29]. At the same time hydrogen peroxide at a concentration up to 10^{-4} M has practically no effects on the photosynthesis of green, diatomic, and other species of alga. On the contrary decrease of hydrogen peroxide concentration in the aquatic environment lower than 3×10^{-7} at the temperature upper 18 °C promotes the domination of blue-green alga's growth. So ecological danger represents not so much short-term increasing of hydrogen peroxide concentration but how much long-term deficiency of it in an aquatic medium.

When dissolved dioxygen is present in the aquatic medium the lack of hydrogen peroxide self is dangerous for the organisms that exchange intensively by water with the aquatic environment and evolutionarily adapted for consumption reactive oxidative equivalents from water in the form of hydrogen peroxide (for the moment when own enzymes of the respiratory system must be formed). For example, lack of hydrogen peroxide in the aquatic environment leads to the death of larvae of fishes at the early stages of their development – before the transition to branchiate breath [30, 31].

Thus, both in the case of dissolved dioxygen and in the case of hydrogen peroxide, there is certain ecological "corridor" for these oxidants admissible concentrations in the aquatic environment. Studies conducted in different countries and on different water objects showed that for fresh waters the optimal concentrations of hydrogen peroxide are 3×10^{-7}–3×10^{-6} M. At that, a decrease of hydrogen peroxide content in the water below minimal admissible level is dangerous for the stability of aquatic ecosystems. Hydrogen peroxide mainly forms in photochemical processes involving microalgae (including photosynthesis). Algae and the accompanying bacteria also make the main contribution to the decomposition of H_2O_2.

Decreasing of hydrogen peroxide contains in the medium below some admissible level consists in lowering of peculiar kind immunity of aquatic ecosystemrelatively to biological and chemical contamination of medium. When hydrogen peroxide is present in the water, the susceptibility of fish to nitrite-ions pollution decreases [32]. Lack of hydrogen peroxide promotes reproduction in the medium of pathogenic microorganisms, decreasing the ability of the aquatic environment to self-clearing from pollutants.

It is worth to note that in plant cells, special organelles (peroxydazo-somes) are responsible for the immunity system and use hydrogen peroxide as chemical "weapon" against strange substances [33].

In the living cell and in natural water, there are two types of H_2O_2 decomposition: catalase and peroxidase processes [34]. In the case of catalase process, one H_2O_2 molecule acts as an oxidizer and another as a donor of the H atom:

$$H_2O_2 + H_2O_2 \xrightarrow{\text{cat}} O_2 + 2H_2O \tag{7}$$

The rate of the catalase decomposition of $H_2O_2(W_{cat})$ is proportional to the hydrogen peroxide concentration over the whole range of concentration.

$$W_{cat} = k_d[H_2O_2] \tag{8}$$

In the peroxidase process, hydrogen peroxide participates in the oxidation of other H donors:

$$H_2O_2 + DH_2 \xrightarrow{\text{Per}} D + 2H_2O \tag{9}$$

The contains of hydrogen peroxide in natural waters depends in much from the presence of reactive substances – donors of electrons and/or H-atoms, DH_2. In the presence of transition metal ions and enzymes type of peroxidase, these reductive substances can interact with hydrogen peroxide rather fast.

In an open type system (both natural water and living cell with active water exchange with the external medium), the rate of hydrogen peroxide decomposition in the peroxidase process W_{per} may be independent of the initial or current H_2O_2 concentration; i.e., it may be determined by the rate of arrival of reducing substances that effectively interact with H_2O_2 in the medium or cell W_{red}:

$$W_{per} = W_{red} \tag{10}$$

In the frame of the dynamic character of aquatic environment as an opened type redox-catalytical system [35], it is significant not what concentrations of hydrogen peroxide as the ratio between rates of entering into system hydrogen peroxide and reductive substances interacting fast with it. Normally the entrance flux of $H_2O_2(W_{ox})$ exceeds the rate of reductive substances coming in (W_{red}) and the current concentration of H_2O_2 (when the rate of coming up is equal to the rate of loss) is determined by equation

$$[H_2O_2] = (W_{ox} - W_{red})/k_d \qquad (11)$$

where k_d is an effective constant rate of H_2O_2 decomposition in the aquatic medium as a result of catalase and other first order relatively to hydrogen peroxide concentration catalytic processes.

The danger of disappearance of hydrogen peroxide in natural waters originates not from the increase of catalase processes of it decomposition efficiency but from the intensification of the processes reductive substances formation, that interact with hydrogen peroxide effectively on pseudo-zero order relatively to H_2O_2 concentration. The zero order of the reaction is characteristic for water objects with redox-toxic conditions formed in the water medium during the period of mass development of blue-green algae. In the natural aqueous medium, substances that efficiently interact with hydrogen peroxide are titrated instead of H_2O_2 [31].

When $W_{red} > W_{ox}$, the aqueous medium passes from the normal oxidation state into the *quasi*-reductive state, which is toxic for aerobic organisms with intensive water exchange with the external environment [30]. The prefix *quasi* reflects the fact that the oxygen content in water may remain unchanged in this transition redox-state; in the thermodynamic sense, water remains in the oxidative state.

The presence of H_2O_2 in natural water is a sign of the ecological well-being of the aquatic ecosystem and provides a balance between the internal water oxygen-dependent redox processes. In contrast, the presence of reducing substances that effectively interact with H_2O_2 in water is a sign of intoxication of the aqueous medium in respect of hydrobionts with active water exchange.

In particular, for the reproduction of fishery resources, especially dangerous is the situation when instead of hydrogen peroxide, reductive substances are registered, which are effectively interacting with H_2O_2 [31].

The question arises what factors of human activity can cause an increase of reductant's flux in the aquatic environment.

6.4   ROLE OF BOTTOM SEDIMENTS AND BLUE-GREEN ALGAE IN THE CHANGING OF DYNAMIC REDOX-STATE OF NATURAL AQUATIC MEDIUM

As a result of many field studies, we have concluded that reductive substances that interact with hydrogen peroxide very effectively are produced mainly blue-green algae and probably accompanying them bacteria. The rate of reductant's production varies in large scale and sometimes reach 10^{-8}Msec^{-1} and even more (in the dark after concentrating on natural algae-cyanosis [31]. It depends on water temperature, from the rate of water exchange in water object, from species structure of phytoplankton and also from some unknown processes of a competitive fight between different representatives of algae-cyanosis.

The most favorable conditions for growing of reductive substances producers are formed in the reservoirs of regulated plain rivers, at low flowing, worming up, receipt into aquatic media of nutrients (compounds of nitrogen and phosphate), "flowering"('blooming") of blue-green algae, coming of reductants out bottom sediments [36].

The last factor is deserved by special attention. At slow flowing, the favorable conditions arise for bottom sediment's formation saturated by organic matters, out-cell's enzymes, transition metal ions. In the body of bottom sediments, reductive decomposition of organic compounds takes place, anaerobic conditions form. Carriers of reductive equivalents are ferrous sulfide in the solid phase and intermediate forms of sulfur reduction in water phase [37]. On the boundary of the system "water–bottom sediments" there are two opposite directions fluxes: on the one hand molecular oxygen and hydrogen peroxide come into sediments as oxidants from the aquatic medium, on the other hand, reductive species of sulfur come into aquatic media from sediment's body. Independence from the predominance of one or another from these fluxes on the boundary, different redox-situations can arise. If the flux of hydrogen peroxide prevails on the flux of reductive equivalents the oxidative layer forms on the sediment surface preventing to cross of phosphate-ions from the body of sediments to the body of water. In this case, unfavorable conditions for upper-bottom propagation of blue-green algae form.

Opposite in the case of the flux of water-soluble reactive sulfur compounds exceed the flux of hydrogen peroxide the *quasi*-reductive state of aquatic media is formed on the boundary "bottom sediments – water"(oxygen is else presence; hydrogen sulfide is not come out). At that: ferric (III) and manganese (IV) compounds are reduced, manganese (II), ferrous (II) and phosphates bounded with ferric (III) ions become free and come out to water body, the favorable conditions for massive growing of blue-green algae arisen, copper-ions in water form non-reactive inaccessible for biota firm complexes with thiolic(sulfides) compounds [38] that secreted from bottom sediments [36] and by blue-green algae [31]. As a result, the water became toxic relation to aerobic aquatic organisms. Also, dredging can

result in changing of dynamic redox-state of the aquatic medium in consequence of increasing of reductive equivalent flux into water under sediment
resuspension.

6.5 MUNICIPAL WASTE WATERS AS A POTENTIAL SOURCE OF DANGER FOR CHANGING OF DYNAMIC REDOX-STATE OF THE AQUATIC ENVIRONMENT

At the time of urbanization, the volume of municipal wastewaters increase.
As we have shown in Ref. [39], after biological aerobic treatment wastewaters contain reductive substances that interact with hydrogen peroxide
very effectively and does not interact with dioxygen. Without additional
oxidative treatment, wastewaters carrying reactive reductive substances
come to receiving aquatic ecosystems. As a result of it, natural hydrogen
peroxide in the aquatic medium is "eaten," the favorable conditions can
arise for pathogenic microorganisms growing and for redox-toxification
of the medium. A demonstrative example is Volga-river. The biological
treatment plants for municipal wastewater purification have built in the
1976 year for the most of big towns in Volga-river basin. And at the same
time (since 1976 and then annually) the massive death of fish larva has
begun on the south part of Volga-river, where many plants for artificial
reproduction of sturgeon fishes are located. In this case, sturgeon larvas
were served as involuntary bio-test organisms responding on changes of
dynamic redox-state of the aquatic medium in Volga-river. But the cost of
this "biotesting" was found very high because since last half of the 80s, the
sharp reduction of catches of sturgeon fishes in Volga-Caspian basin have
observed [40](10–12 years is the period of reproducibility of the most kind
of sturgeons).

The additions of hydrogen peroxide to wastewaters or natural waters
that contain reductive substances result in detoxification of water. This is
manifested when infusoria *Tetrahymenapyriformis* [41] or sturgeon larva
on the earliest steps of their life (after hatching from eggs before conversion to gill breath) [42] are used as bio-testing organisms. After titration of
reductive substances contained in initially acute toxic water by additives,
hydrogen peroxide water become non-toxic. In the case of sturgeon larva
toxicity of natural water be founded in the *quasi*-reductive state is accompanied by disturbance of lipid-carbohydrate exchange and death of larva
when all parental supply of glycogen is finished before conversion of them
to active nutrition. If hydrogen peroxide is added periodically into that

natural water before larva of fish would be inserted the lipid-carbohydrate exchange in larva have been normalized, and no disturbance was observed in enzymatic systems function, blood characteristics or physiological development of fish on the further steps their life. Using periodical additions of hydrogen peroxide to natural waters that used in fish plant for artificial reproduction of sturgeon more than 7 million of young fish (2–3 grams of weight each) were grown and let out to Volga-river during three seasons (June – August) when almost 100% death of larva in control is observed (without H_2O_2 additives).

Redox-detoxification of wastewaters before this biological treatment with using H_2O_2 additives or UV-B light is accompanied by improving the physiological state of active sludge, improving the efficiency of biological treatment and increasing of passing ability of water treatment station [39]. Application of UV-B light in combination with aeration of wastewater promotes the hydrogen peroxide formation and thereby the oxidation of toxic reductive substances.

6.6 REDUCED SULFUR COMPOUNDS (RSH) AS REDUCING EQUIVALENT CARRIERS

Among reductive substances, an RSH plays a special role. Many publications are devoted to the role of RSH in the natural aqueous medium and bottom sediments [43]. Main attention has been paid to studies of seawater [44]. For freshwater reservoirs, the most detailed studies were performed in the region of the Baikal Pulp and Paper Mill [45]. It was reported that reduced sulfur forms were discovered in the surface river waters, namely, in the region of wastewater discharge of the Arkhangelsk and Solombal Pulp and Paper Mills on the Northern Dvina River [46].

The presence of RSH in aerobic waters environment is a thermodynamically inexplicable phenomenon, which suggests the dynamic nature of its formation and oxidation. Among the reducing agents in the aquatic environment, there may be thiol compounds evolved by microalgae and related bacteria.

In the water body, the RSH interact with copper ions. We found for glutathione [47] and some other aminothiols[48] that Cu^+ ions interact with RSH to form very stable(1: 1)*mono*-complexes that is stable against the action of O_2, but interact effectively with H_2O_2.

Dioxygen reacts only with *bis*-thiolate Cu^+ complexes formed in appreciable amounts in the presence of a large excess of the thiol ligand. In this case, O_2 is reduced to water in a four-electron transfer process under the cooperative action of two these complexes. When the univalent copper ions are bound in the mono-thiolate complexes, copper as a microelement becomes biologically inaccessible, which leads to the toxicity of the aqueous medium for organisms at the stage of formation of respiratory enzymes.

Hydrogen sulfide (H_2S) has the greatest influence on aerobic processes in natural waters. The main source of H_2S supply in the natural aqueous medium is dissimilar sulfate reduction, which involves sulfate-reducing microorganisms since the sulfate ions cannot be reduced to sulfides in an abiotic way:

$$2CH_2O + SO_4^{2-} \xrightarrow{\quad 2H^+(H^+) \quad} 2CO_2 + H_2S(HS^-) + 2H_2O \quad (12)$$

On the water-bottom sediments interface, processes involving hydrogen sulfide and the product of its acid-base dissociation ($pK_a \sim 7$) hydrosulfide (HS⁻)play the major role in the formation of anaerobic conditions in the natural water medium. The contents of H_2S and HS⁻ in a neutral aqueous medium are approximately the same.

The mechanism of oxidation of H_2S/HS⁻ in water is not fully clear: H_2S does not react directly with O_2, whereas the oxidation of HS⁻ proceeds only after a long induction period. In order to determine the role of RSH in the redox-processes involving O_2/H_2O_2, we studied the kinetics of the oxidation of H_2S/HS⁻ by oxygen and hydrogen peroxide.

It was found that additions of ethylenediaminetetraacetate (10^{-5} M) to the aerobic solution of sodium sulfide (10^{-4} M, pH = 7, phosphate buffer with a concentration of 2×10^{-3} M) inhibit the oxidation of sulfide by molecular oxygen. This suggests the catalytic role of metal ions in the mechanism of the autocatalytic process. In contrast to the opinion expressed in Ref. [49], small additions of hydrogen peroxide to aerated water (10^{-5} M) did not lead to an acceleration of sulfide auto-oxidation.

Unlike the reaction with O_2, the interaction of HS⁻ with H_2O_2 proceeds without an induction period, and the presence of oxygen in the reaction medium has almost no effect on the reaction kinetics (Figure 6.1). Thus, neither H_2O_2 nor the products of hydrosulfide oxidation by hydrogen peroxide are involved in the autocatalysis of the oxidation of HS⁻ by molecular oxygen.

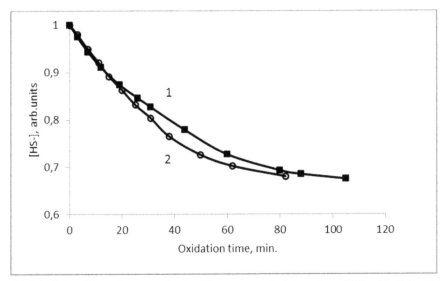

FIGURE 6.1 Effect of oxygen dissolved in water on the interaction of HS⁻ with H_2O_2: (1) without O_2 and *(2)* with O_2. Initial concentrations: $Na_2S - 10^{-4}$ M; phosphate buffer $- 2 \times 10^{-3}$ M; $H_2O_2 - 6 \times 10^{-4}$M.

An insignificant effect on the kinetics of hydrosulfide oxidation with hydrogen peroxide was produced by small additions of tert-butyl alcohol (acceptor of OH-radicals) to the reaction solution (Figure 6.2), which points to the two-electron mechanism of the process:

$$HS^- + H_2O_2 \rightarrow S + H_2O + OH^- \tag{13}$$

The main complexing agents of hydrogen sulfide in natural waters are ferrous ions, which form cluster complexes with HS⁻ [50]. At a ratio of 1:1at concentrations of, the interaction of HS⁻ with Fe^{2+} proceeds at the moment of mixing of the reagents with the formation of microcolloid(nano-) iron sulfide particles [FeS]<10^{-4} M:

$$HS^- + Fe^{2+} \rightarrow FeS + H^+ \tag{14}$$

The rate of oxidation of iron sulfide nanoparticles (clusters) by molecular oxygen depends quadratically on the FeS concentration [51], so that at low, natural concentrations, the process occurs slowly. In contrast, FeS effectively reacts with H_2O_2 at any concentration. The absence of any pronounced effect of small additions of ferrous ions on the kinetics of the interaction of HS⁻ with H_2O_2(Figure 6.2) indicates that there is no FeS regeneration stage in the mechanism. Evidently, when the FeS cluster interacts with H_2O_2 in a neutral aqueous medium in the presence of phosphate ions, the Fe(III) complexes form, which is not reactive with HS⁻.

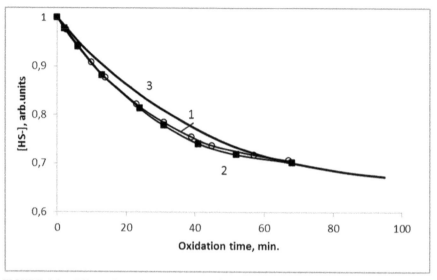

FIGURE 6.2   Effect of small additions of Fe^{2+} and *tert*-butanol on the rate of oxidation of hydrosulfide with hydrogen peroxide: (1) $Fe^{2+} + H_2O_2$; (2) $Fe^{2+} + H_2O_2$ in the presence of *tert*-butanol; and (3)H_2O_2 in the presence of *tert*-butanol. Initial concentrations: sulfide 10^{-4} M; phosphate buffer 2×10^{-3} M; H_2O_2 6×10^{-4} M; Fe^{2+} 10^{-5} M; *tert*-butanol 10^{-4} M.

Nevertheless, the bottom sediments can become a source of reducing equivalents in the aqueous medium not only in the form of H_2S/HS^-, but also as part of iron sulfide, which effectively interacts with hydrogen peroxide. The oxidative or quasi-reducing conditions can form at the interface between water and bottom sediments depending on the ratio of the streams of reactive oxidative equivalents in the form of H_2O_2 and reducing equivalents in the form of FeS.

Among the FeS oxidation products with both molecular oxygen and hydrogen peroxide is elemental sulfur, up to 50% of which(based on the amount of changed sulfide) is extracted with hexane. Elemental sulfur is relatively stable against further oxidation; in aqueous solutions, it can exist in various forms, including atomic sulfur S, hydrate SH_2O, thioperoxide HSOH, and various allotropic forms $S_n(n = 28)$, of which S_8 is thermodynamically most stable. In the presence of sulfide ions, elemental sulfur forms polysulfidescomposition $S^0(S_n^-)(n = 2–5)$. In the range pH 7.8, the steadiest are polysulfides of structure S_5^- and S_6^-.

Identification of the intermediate products of sulfide oxidation is difficult because of their low resistance to redox and associative interconversions. For example, in the presence of elemental sulfur, the sulfite ions rapidly transform into thiosulfate:

$$SO_3^- + S^0 \rightarrow S_2O_3^- \tag{15}$$

As a result of subsequent transformations of the thiosulfate ion in the aqueous medium, polythionates $S_mO_6^-(m = 2–6)$ can form, which typically interact with sulfide, sulfite, and thiosulfate.

To analyze the toxic properties of the intermediate products of FeS oxidation, we used bioassay methods. In the absence of iron ions, the products of the reaction of Na_2S (6.5×10^{-5} M) with H_2O_2 (7×10^{-5} M) proved low-toxic.

The freshly prepared iron sulfide FeS has no noticeable toxic effect on luminous bacteria and infusoria, while the primary products of the interaction of FeS with hydrogen peroxide have toxic properties, and toxicity persists in the aquatic medium for a long time after the reaction is completed. The removal of elemental sulfur from the reaction products by extraction with hexane does not lead to any significant reduction of toxicity.

6.7 CONCLUSIONS

There are two approaches to assess the biological usefulness and sustainability of aquatic ecosystems stability and health. These approaches can be characterized as hydrochemical and hydrobiological. The hydrochemical approach assumes measurements some sets both individual and group characteristics of a contaminant in an aquatic medium, bottom sediments, and biota. At that, the list of controlled parameters and ingredients is permanently supplemented and cannot be full in principal on obvious reasons. On the base of hydrochemical data, it is possible to conclude about the presence or lack of substances in an aquatic ecosystem that could be a danger for biota and human.

However, there is no guarantee that substances whose threshold limit concentration in water are lower than limits of sensitivity of control methods are absent, or that the water-soluble substances are presentwhich are not giving into neither concentrating, nor control in native samples of water, but making direct or indirect toxic impact on a water ecosystem(e.g., as in case of mass death of larvae of sturgeon fishes in the Volga water.

In the frame of hydrobiological approach, controlled parameters are primary production of phytoplankton, biological oxygen demand, biological diversity, biomass, andabundance of life in water object, acute, and/or chronic toxicity and so on. On the base of these parameters, it is possible to conclude about the favorable or unfavorable state of the aquatic ecosystem. However, in this case, it is impossible to get answer neither to identify a danger source, nor to take adequate measures for prevention of intoxication of the aquatic environment.

Both these approaches reciprocally complete each other, so it is the most productive to use them in parallel, without being limited only to the chemical analysis or ascertaining of the changes happening in a water ecosystem. Usually, the existing methods of quality control of the water environment and "measures" of water treatment are not going beyond confirmation of water pollution by heavy metals, pesticides, chlorine-organic substances, radionuclides, and other persistent contaminants. Uncontrollable is most toxic substances and unstable chemically reactive substances participating in inner-water chemical and biological processes. On the basis of the standard approaches, it is impossible to explain seasonal phenomena massive death of larva and adult fish in Volga-river and other water objects, to the subject of an impact.

At the same time in our opinion these phenomena have found logical explanation and connected with anthropogenic impacts on natural, evolutionary balanced processes of formation and decomposition of hydrogen peroxide as a natural participant of circulation of O_2 in aquatic ecosystems on the one hand and both formation and oxidizing transformation of RSH, predominantly H_2S/HS^- as products of sulfate reduction with another hand. The major negative factor of anthropogenic impact is an increase of inflow to the natural water environment of the RSH, which are effectively interacting with hydrogen peroxide. Correspondingly, at the elimination of an imbalance of oxidation-reduction processes the detoxification of the water environment won't be full, the danger of pollution of water by toxic substances, especially stable against oxidation, remains.

Heavy metals and low-soluble organic substances accumulated in sediments and biotic tissues and can cause long-term toxic effects type of genotoxic or mutagenic toxicity. Factors of changing of dynamical redox-state of the aquatic medium also can cause effects of long-term toxicity. For example, the mentioned before larva of sturgeon fish corresponded to active nutrition (not died) in the conditions of the quasi-reductive state of aquatic medium carry in the self mark of disturbance of fatty – carbohydrate exchange. It is manifested in the pre-spawning period (10 and more years after redox-toxification of larva) in the form of fatty stratification of muscular tissues, in resorption of eggs and at long last in stoppage of natural reproduction of sturgeon and other valuable types of fish.

KEYWORDS

- **free radicals**
- **hydrobionts**
- **hydrogen peroxide**
- **redox-processes**
- **wastewater**

REFERENCES

1. Gorshkov, V. G., (1990). *Energetics of Biosphere and Stability of Environment*(p. 238). Results of scientific and technique. Ser. theoretical and general questions of geography, 7. Moscow. All-Russian Institute for Scientific and Technical Information (VINITI RAS)(in Russian).
2. Duca, G. G., Scurlatov, Y. I., & Sychev, A. Y., (2002). *Redox Catalysis and Ecological Chemistry*(p. 316). Publishing Centre M.S.U., Chisinau (in Russian).
3. Skurlatov, Y. I., (1991). Important role of oxidative-reductive processes for formation of natural water quality. *Russian Chemical Reviews*, *60*(3), 575–580 (in Russian).
4. Sytchev, A. Y., Duka, G. G., Travin, S. O., & Skurlatov, Y. I., (1983). *Catalytic Reactions and Environment Protection*(p. 272). Kishinev: Shtiintsa(in Russian).
5. Haag, W. R., Hoigne, J., Grassman, E., & Braun, A. M., (1984). Singlet oxygen in surface waters. Part I: Furfuryl alcohol as a trapping agent. *Chemosphere*, *13*(5/6), 631–640.
6. Haag, W. R., Hoigne, J., Grassman, E., & Braun, A. M., (1984). Singlet oxygen in surface waters – Part II: Quantum yields of its production by some natural humic materials as a function of wavelength. *Chemosphere*, *13*(5/6), 641–650.
7. Haag, W. R., & Hoigne, J., (1986). Singlet oxygen in surface waters – Part III: Photochemical formation and steady-state concentrations in various types of waters. *Environ. Sci. Technol.*, *120*, 341–348.
8. Zepp, R. G., Wolfe, N. L., Baughman, G. L., & Hollis, R. C., (1977). Singlet oxygen in natural waters. *Nature*, *267*(5610), 421–423.
9. Skurlatov, Y. I., (1986). Partial charge transfer as the main factor that determines the reaction ability of coordinated O_2, H_2O_2, and others molecules. Advances of chemistry

of charge transfer complexes and ion-radical salts (KOMIS-5, October 1981). *Chernogolovka: Inst. Chem. Phys.*, 164–187 (in Russian).

10. Skurlatov, Y. I., Duka, G. G., Batyr, D. G., &Travin, S. O., (1989). Coordination complexes with partial charge transfer in oxidative-reductive processes. *Coordination Chemistry*, *15*(3), 291–307 (in Russian).

11. Shtamm, E. V., Purmal, A. P., & Skurlatov, Y. I., (1991). The role of hydrogen peroxide in natural aquatic media. *Russian Chemical Reviews*, *60*(11), 1228–1248 (in Russian).

12. Shtamm, E. V., Purmal, A. P., &Skurlatov, Y. I., (1986). Redox catalysis and problems of water quality. In: Shilov, A. E., (ed.), *Fundamental Research in Homogeneous Catalysis*(Vol. 3, pp. 1219–1233). NY: Gordon @ Breach Sci. Publ.

13. Sinelnicov, V. E., (1971). Content of hydrogen peroxide in river water and a method of its determination. *Hydrobiological Journal*, *7*(1), pp. 115–120 (in Russian).

14. Sinelnicov, V. E., & Demina, V. I., (1974). About origin of the hydrogen peroxide which is contained in water of superficial reservoirs. *Hydrochem. Materials,60*, 30–40 (in Russian).

15. Ernestova, L. S., & Skurlatov, Y. I., (1995). The formation and transformation of free radicals OH and O_2^- in natural waters. *J. Phys. Chem.*, *69*, 1157–1164 (in Russian).

16. Skurlatov, Y. I., Ernestova, L. S., &Vichutinskaya, E. V., (1997). Photochemical conversion of polychlorinated phenols in aquatic medium, Part 2. Sensitized transformations. *Chem. Phys. Reports*, *16*(11), 1943–1951.

17. Ehrhardt, M., &Petrick, G., (1983). *On the Photooxidation of Alkylbenzenes in Seawater.* Photochemistry of natural waters (extended abstracts of NATO-ARI workshop, woods hole, *Massachusetts*, USA).

18. Patterson, C. O., & Meyers, J., (1973). Photosynthetic production of hydrogen peroxide by *Anacystisnidulans.Plant Physiol.*, *51*, 104–109.

19. Stevens, S. E., Pat Patterson, C. O., & Meyers, J., (1973). The production of hydrogen peroxide by blue-green algae: A survey. *Journal of Phycology*, *9*, 427–430.

20. Palenic, B., Zafiriou, O. C., & Morel, F. M. M., (1987). Hydrogen peroxide production by marine phytoplankton. *Limnol. Oceanogr.*, *32*, 1365–1369.

21. Zepp, R. G., Skurlatov, Y. I., & Pierce, J. T., (1987). Algae-induced decay and formation of hydrogen peroxide in water: It's possible role in oxidation of anilines by algae. In: Zika, R. G., & Cooper, W. J., (eds.), *Photochemistry of Environmental Aquatic Systems*(Vol. 327, pp. 215–224). ACS Symposium. Series.

22. Batovskaya, L. O., Kozlova, N. B., Shtamm, E. V., &Skurlatov, Y. I., (1988). The role of microalgae in the governing of the hydrogen peroxide content in natural waters. *Reports of Academy of Sciences of the USSR*, *301*, 1513–1516 (in Russian).

23. Buxton, B. V., Greenstock, C. L., Helman, W. P., & Ross, A. B., (1988). Critical review of rate constants for reactions of hydrated electrons, hydrogen atoms and hydroxyl radicals (OH/O') in aquatic solution. *J. Phys. Chem. Ref. Data*, *17*, 513–886.

24. Telitchenko, M. M., (1988). Energy basis of hydrobiocenoses. In: Skurlatov, Y. I., (ed.), *Proceedings of the First All-Union Workshop on Ecological Chemistry of Natural Waters(Kishinev)*(pp. 165–182). Moscow: Centre for International Projects, GKNT – UNEP (in Russian).

25. Lukyanova, L. D., Balmukhanov, B. S., & Ugolev, A. T., (1982). *Oxygen-Depended Processes in Cell and its Functional State*(p. 301). Moscow: Nauka("Science," in Russian), (in Russian).

26. Skurlatov, Y. I., Ernestova, L. S., &Travin, S. O., (1985). Oxidative-reductive processes in natural waters. *Water Resources,5*, 66–72 (in Russian).

27. Oshino, N., Jamieson, D., Sugano, T., & Chance, B., (1975). Optical measurement of catalase-hydrogen peroxide intermediate (Compound I) in the liver of anesthetized rats and its implication to hydrogen peroxide production *in city. Biochem. J., 146*, 67–77.

28. Shtamm, E. V., Batovskaya, L. O., Skurlatov, Y. I., & Zepp, R. G., (1987). The role of algae in the hydrogen peroxide content regulation in natural waters. In:Novitskii, M. Y., (ed.), *Behavior of Pesticides in Environment (Proceedings of Soviet-American Symp., October Aiowa-City, USA)*(pp. 280–292). Leningrad: Gidrometeoizdat(in Russian).

29. Shtamm, E. V., &Batovskaya, L. O., (1988). Biotic and abiotic factors of natural water medium redox-state formation. In: Skurlatov, Y. I., (ed.), *Proceedings of the Second All-Union Workshop on the Ecological Chemistry of Aquatic Environment (Erevan, Armeniya)*(pp. 125–137). Moscow: Inst. Chem. Phys., (in Russian).

30. Skurlatov, Y. I., Ernestova, L. S., & Shtamm, E. V., (1984). Redox-state and seasonal toxicity of natural waters. *Reports of Academy of Sciences of the USSR, 276*, 1014–1016 (in Russian).

31. Shtamm, E. V., (1988). Redox state of water and fish reproduction. In: Skurlatov, Y. I., (ed.), *Proceedings of the First All-Union Workshop on Ecological Chemistry of Natural Waters (Chisinau)*(pp. 279–295). Moscow: Centre for International Projects, UNEP.

32. Vereschagin, G. V., (1987). Biological aspects of peroxide reactions studying in artificial aquatic ecosystems. *"PhD Thesis in Biology*(p. 20)." Moscow State University, Moscow. (in Russian).

33. Rogovin, V. V., Muravev, R. A., & Piruzan, L. A., (1983). Peroxidazosomes. *News of Academy of Sciences of the USSR, Ser. Biol., 4*, 510–529 (in Russian).

34. Skurlatov, Y. I., &Shtamm, E. V., (1995). Aquatic cycling of hydrogen peroxide as a principal way of natural waters self-purification. *Self-Purification Processes in Natural Waters*(pp. 27–37).Bulat Art Glob, Chisinau.

35. Skurlatov, Y. I., Duka, G. G., & Ernestova, L. S., (1983). Processes of toxication and mechanisms of self-clearing of aquatic environment under anthropogenic impacts. *Moldavian Academy of Sciences Bulletin, Ser. Biol. Chem. Sci.,5*, 3–20 (in Russian).

36. Pirumyan, G. P., (1988). Redox-processes in system water – bottom sediments and their role in transformation of pollutants. In: Skurlatov, Y. I., (ed.), *Ecological Chemistry of Aqueous Media, Proceedings of the Second All-Union Workshop on the Ecological Chemistry of Aquatic Environment (Erevan, Armeniya)*(pp. 179–193), Moscow: Inst. Chem. Phys., (in Russian).

37. Rodko, I. Y., Shtamm, E. V., & Skurlatov, Y. I., (1993). Redox-processes in freshwater sediments. *Proceedings of the 12th Annual Aquatic Toxicity Workshop*(p. 276). Quebec City, Quebec, Canada.

38. Semenyak, L. V., Skurlatov, Y. I., & Borodulin, R. R., (1992). Mechanism of copper-ions with glutathione interaction. *Chem. Phys., 11*, 1248–1252 (in Russian).

39. Duka, G. G., (1988). Ecological chemistry of wastewaters. In: Skurlatov, Y. I., (ed)., *Proceedings of the Second All-Union Workshop on the Ecological Chemistry of Aquatic Environment(Erevan, Armeniya)*(pp. 271–290), Moscow: Inst. Chem. Phys., (in Russian).

40. Zilanov, V. K., & Spivakova, T. I., (1993). New regulations need for fishery in Caspian Sea. *Fish Economy, 2*, 3–7 (in Russian).

41. Boldyreva, N. M., (1988). Complex bio-toxicological characteristic and redox-state of natural and wastewaters. *PhD Thesis in Biology*(p. 20). Moscow, State University, (in Russian).

42. Shtamm, E. V., & Skurlatov, Y. I., (1993). Toxicity caused by disbalance of Redox processes in natural waters. *Proceedings of the 12th Annual Aquatic Toxicity Workshop*(p. 282). Quebec City, Quebec, Canada.

43. Volkov, I. I., (1984). *Geochemistry of Sulfur in Oceanic Sediments*(p. 272). "Nauka"("Science," in Russian) Publishing House, Moscow (in Russian).

44. Makarov, S. V., (2001). The new directions in chemistry of sulfur-containing reductants. *Russ. Chem. Rev.*, *70*(10), 996–1007 (in Russian).

45. Beim, A. M., & Osharov, A. B., (1984). Ecological and toxicological criteria for the regulation of methylsulfur compounds in wastewaters of sulfate-cellulose production, *Review VNIPIEM –Lesprom*(p. 36). Moscow (in Russian)

46. Volkov, I. I., & Kokryatskaya, N. M., (2004). Compounds of the reduced inorganic sulfur in waters of the White Sea and the mouth of Northern Dvina. *Water Resour.*, *31*(4), 461–468 (in Russian)

47. Dyatchina, O. V., Semenyak, L. V., Skurlatov, Y. I., &Travin, S. O., (1992). Glutathione oxidation by oxygen and hydrogen peroxide in the presence of copper ions. *Chem. Phys.*, *11*(9), 1255–1259 (in Russian).

48. Bagiyan, G. A., Koroleva, I. K., Ufimtsev, A. V., Skurlatov, Y. I., & Semenyak, L. V., (2005). Ion-molecular mechanisms of catalytic oxidation of the thiol compounds in the presence of copper ions. *Chem. Phys.*, *24*(6), 51–62 (in Russian).

49. Resh, P., Field, R. J., Schneider, F. W., & Burger, M., (1989). Reduction of methylene blue by sulfide ion in the presence and absence of oxygen: Simulation of the methylene blue-O_2-HS⁻ CSTR oscillations. *J. Phys. Chem.*, *93*, 8181–8186.

50. Luther, G. W., & Rickard, D. T., (2005). Metal sulfide cluster complexes and their biogeochemical importance in the environment. *J. Nanopart. Res.*, *7*, 389–407.

51. Baykova, I. S., Shtamm, E. V., Vichutinskaya, E. V., Skurlatov, Y. I., (2009). The mechanism of oxidation of nanoparticles of FeS by molecular oxygen and hydrogen peroxide in the diluted water solutions. *J. Phys. Chem. B.*, *3*(2), 251–256.

PART II

Biological Activity of Antioxidants and Possibilities of Their Application

CHAPTER 7

Screening Eye Pigments as Natural Antioxidants

ALEXANDER E. DONTSOV and MIKHAIL A. OSTROVSKY

Emanuel Institute of Biochemical Physics of Russian Academy of Sciences, 4, Kosygin St., Moscow 119334, Russia, E-mail: adontsovnick@yahoo.com

ABSTRACT

The chapter is devoted to the screening pigments eyes of vertebrates and invertebrates animals and human eyes – melanosomes and ommochromes. A sum of literature and our own experimental data is presented, from which it follows that these screening pigments have a pronounced antioxidant activity. The role of these pigments in the physiology of the vision of vertebrates and invertebrate animals and humans is considered, as well as the role of melanosomes in aging and pathology of the human eye.

7.1 INTRODUCTION

Light for vision is the main source of information. However, light quanta, especially in the violet and blue regions of the spectrum, are highly energetic and therefore are a potentially dangerous damaging factor for eye structures, especially for retina photoreceptor cells (so-called "blue light hazards"). The damaging effect of light is mainly due to the occurrence of free radical oxidation processes. Their effectiveness is determined by three main factors: the presence in the eye structures of photosensitizers, high concentrations of oxygen, and substrates of oxidation. In the course of evolution, an effective system of protecting cellular structures from the potential danger of damaging effects of various pro-oxidant factors has been formed in the vertebrate and invertebrate eyes.

This protection system includes:

- Permanent renewal of photoreceptors, which may contain damaged and oxidized proteins and lipids;
- Complex of endogenous fat- and water-soluble antioxidants;
- Mechanism of rapid neutralization of active free aldehyde – all-*trans*-retinal, which can participate in the processes of photodamage and in the processes of modification of proteins and lipids;
- System of optical filters of the eye, which includes the cornea, iris, lens, vitreous, carotenoids, and screening pigments.

Optical systems of the eye of vertebrates and humans, in which the lens plays a key role, protect the photoreceptors from the penetration of short-wave light, mainly of ultraviolet light and to some extent from the visible light of the blue range. In many invertebrate animals, optical media of the complex eye are transparent, including for ultraviolet. Therefore, in this system of protecting the eye from the damaging effect of light, screening pigments play an important role.

The screening pigments of the eyes of vertebrate animals and humans are melanin-containing granules of melanosomes, and the screening pigments of the complex eye of invertebrates are granules containing ommochromes. Melanins and ommochromes are the main pigments of the eye, protecting the photoreceptor cells from prooxidant factors, both by optical screening and by chemical neutralization of free radical products.

An interesting fact of evolution: although practically all species of invertebrate animals can synthesize and accumulate melanins, their organs of vision contain ommochromes exclusively.

7.2 MELANOSOMES AND OMMOCHROMES STUDIES— A BRIEF HISTORY

Studies of melanin include a nearly 400-year period. However, the modern history of the study of melanins is usually begun from the moment of the first mention of the natural black pigment as melanin (from the Greek word *melas*, meaning black), which occurs in the late XIX to early XX centuries. In 1895, tyrosinase in mushrooms was found, which turns tyrosine into a black pigment [1]. In 1901, it was suggested that the formation of melanin occurs as a result of the interaction of intracellular oxidase with aromatic groups of proteins [2]. The first mention of the oxidation-reduction properties of melanin refers to 1938, when it was suggested that melanin could be a natural redox system comparable to glutathione and riboflavin [3]. Later it was shown that melanin oxidizes NADH and reduces ferricyanide, 2,6-dichlorophenolindophenol, and cytochrome c [4–6]. These results showed that melanins could participate in conjugated oxidation-reduction reactions.

An important event in the history of melanin research was the detection of its free radical properties by the electron spin resonance (ESR) technique. The work of Barry Commoner et al., which first mentions the free radical properties of melanin, was published in 1954 [7]. Subsequently, the free radical properties of melanin and the effects of light on them and inhibitors of free radical processes were investigated by Howard Mason and co-authors [8], Sever et al. [9], and also in a series of works by Mikhail Ostrovsky and co-authors [10–12]. The detection of a stable singlet ESR signal of melanins with a very high concentration of spins, which is almost two orders of magnitude higher than the usual concentration of paramagnetic centers in biological tissues, suggested the antiradical activity of melanin [13]. Consequently, melanin was considered as a buffer of free radicals or as a "scavenger" of excited molecules and free radicals, which the chemical energy of the excited bonds transfers and dissipates in the form of heat.

In 1978, it was shown by T. Sarna et al. [14] that photoexcitation of synthetic melanin led to the reduction of oxygen in superoxide and hydrogen peroxide. This finding allowed suggesting the phototoxicity of melanin. However, in 1981, Goodchild et al. [15] found that melanins are also capable of causing the dismutation of superoxide radicals. In 1979, in our laboratory, in a series of works by A. Dontsov, M. Ostrovsky and co-authors [16–18] was observed the high ability of natural melanins to suppress the processes of peroxide oxidation of lipids, induced by different prooxidant systems [16–18].

In 1984, we showed that both synthetic and natural melanins are able to reduce oxygen to superoxide [20]. In subsequent studies, we showed that melanins significantly accelerate the process of superoxide radicals dismutation. It was estimated that the rate constant of this process being more than five orders of magnitude is greater than the rate constant of the superoxide radicals generation by the same melanins [21, 22]. The results of these studies allowed us to assume the existence of an important way of protecting pigmented cells containing melanin. This refers to the protection associated with the ability of melanin to neutralize oxidative radicals and other active forms of oxygen that are generated in the course of chemical and photochemical reactions.

The history of the study of ommochromes, apparently, has not yet reached the 100-year period. The discovery of ommochromes is associated with genetic studies of a variety of eye mutations of flies, mainly *Drosophila*. Possibly, A. Johannsen first revealed the ommochromes in 1924 in the primary and secondary pigment cells of the *Drosophila* complex eye ommatidia, which contained two different types of pigment granules: a purplish-red color and an ocher-yellow color [23]. Among the mutant forms, A. Johannsen [23] observed the presence of pigments of the most diverse color: brown, violet, ruby, pink, etc. Probably, the very term "ommochromes" first appeared in the works of the German researcher E. Becker [24]. He proposed for the pigments contained in the eyes of the flies *Drosophila* and *Calliphora* the general name "Ommochromes," dividing them into two large groups of "Ommatines" and "Ommines." A more detailed history of ommochromes discovery can be found in B. Ephrussi and J. Herold from "The Johns Hopkins University," Baltimore, Maryland [25].

Then, in the fifties of the twentieth century, the molecular structure and basic physical and chemical characteristics of ommochromes from the most diverse classes of invertebrate animals were studied in detail. German chemists, Nobel laureate A. Butenandt and co-authors, performed these numerous works [26, 27].

The ability of ommochromes in the eye of shrimp *Pandalus latorostris* and the eye of insects *Calopteryx splendens* and *Pieris brassicae* to inhibit the lipid peroxidation reaction was first demonstrated by us in 1981 and in 1984 [28, 29]. In 1985, we also found that ommochromes have a stable ESR signal with a high concentration of paramagnetic centers, increasing with the action of ultraviolet and visible light [30]. Thus, the screening pigments of invertebrates show, like the vertebrate melanins, the paramagnetic properties. The nature of the dark and photo-induced paramagnetic centers of ommochromes is not yet known.

7.3   THE ROLE OF SCREENING PIGMENTS AS OPTICAL FILTERS

In the eye structures melanin, which is part of intracellular organelles melanosomes, was considered mainly as a screening pigment, carrying out the function of optical protection of photosensitive cells from excess light by its absorption. Melanosomes – melanin-containing black or brown pigment granules – are organelles that are the most characteristic of retinal pigment epithelium (RPE) cells (Figure 7.1). Melanosomes appear in the early stages of ontogenesis, developing structurally from the Golgi complex. Melanosomes of RPE cells have either an ovoid or a spherical shape. The formers are localized mainly in the apical part of the RPE cell and in the microvilli, and the latter, as a rule, are in the middle part of the cytoplasm. In the basal region of the cell, melanosome cells practically do not occur. The size of the melanosome is 2–3 µm in length and 1 µm in diameter. Melanosome is surrounded by a membrane, which, apparently, is in direct contact with melanoprotein, which fills almost all the internal space of the granule. Polymer melanin is found only in mature melanosomes, and it is always bound to a protein [31].

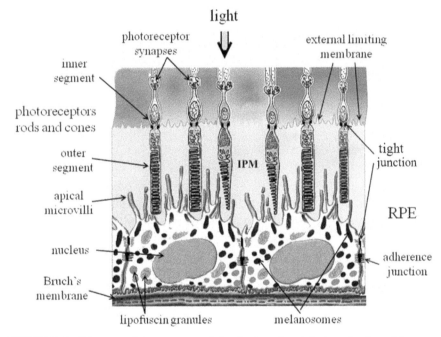

FIGURE 7.1 Diagram of the structure of the fundus of vertebrate animals and humans. (*Note:* RPE – retinal pigment epithelium, IPM – interphotoreceptor matrix).

Melanosomes of human RPE contain mainly a eumelanin type pigment. Eumelanins are highly heterogeneous polymers consisting of 5,6-dihydroxyindole and 5,5-dihydroxyindole-carboxylic units in the reduced or oxidized state and pyrrole units arising from their peroxide cleavage. The hypothetical structure of the eumelanin polymer is shown in Figure 7.2.

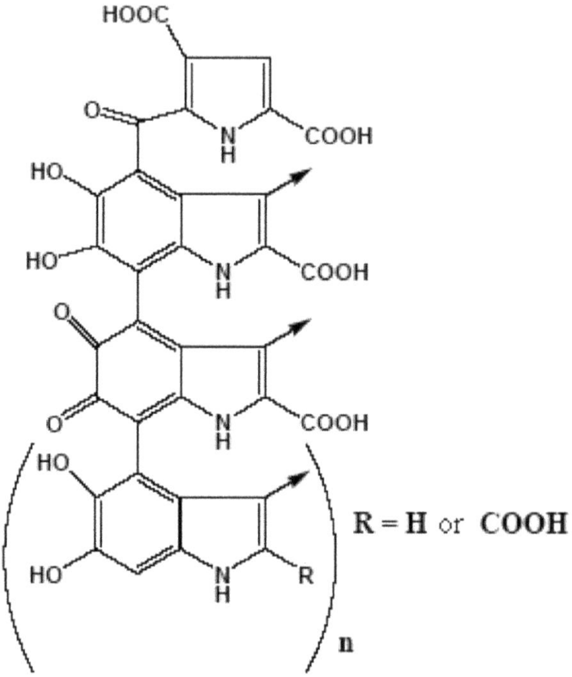

FIGURE 7.2 Hypothetical chemical structure of eumelanin pigment contained in melanosomes.

Absorption of light is one of the most important functions of melanin in the eye of vertebrates and humans. Melanins, like black or dark brown substances, have a broadband absorption spectrum that monotonically increases from the near infrared to the near ultraviolet. In this case, as a rule, there are no absorption maxima or minima in either the visible or the ultraviolet regions of the eumelanin spectrum. Most of the light energy absorbed by melanin quickly turns into heat by the internal conversion mechanism. Consequently, the risk of developing potentially harmful photochemical reactions mediated by melanin is significantly reduced. Melanosomes in the apical part of RPE cells absorb scattered light passing through the photoreceptors. Due to this, they significantly reduce the light scattering and light reflection processes, which leads to better image quality. It is estimated that the RPE melanosomes absorbs about 34–60% of the incident light in the foveal region of the retina and about 21–40% in its equatorial region [32].

Ommochromes are natural organic pigments wide spreading among invertebrates and are especially characteristic for arthropods (*Arthropoda)* [27]. Ommochromes are synthesized from tryptophan through the intermediate formation of kynurenine [33]. Screening pigments of invertebrate animals, like melanins, are localized in specific ommochromes granules.

Granules containing a screening pigment are inherent in almost all cells that make up the ommatidium. In the eyes of invertebrates, ommochromes are localized both in pigmentary and in photoreceptor reticular cells. The diameter of the granules can reach 2 microns, and in pigmentary cells, the diameter is usually much larger than in the photoreceptor cells [34].

Figure 7.3 shows a microscopic picture of the Baltic shrimp *Mysis relicta* [35]. Dark pigmentary granules filled with ommochromes are well observable, which in this case are located inside the retinular cells themselves.

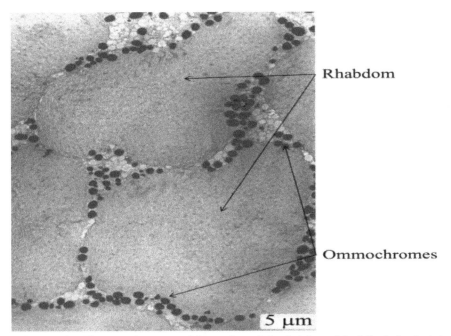

FIGURE 7.3 Electron microscopy of the cut through the rhabdom of the fully dark-adapted eye of *Mysis relicta* shrimp.

In addition to ommochromes in the eyes of invertebrates, there are other screening pigments, for example, pteridines [36]. In addition to eye structures, ommochromes have also been found in the cuticle, in feces and even in the cells of the nervous system of insects [37].

Ommochromes, as a rule, are divided into two types – ommatines and ommines. The difference between them is due to the fact that ommines have a higher molecular mass and contain sulfur in their composition. Of ommines, the most common is ommine A. Of ommatines, the most common are xanthommatines having a characteristic phenoxazine moiety in their structure (Figure 7.4).

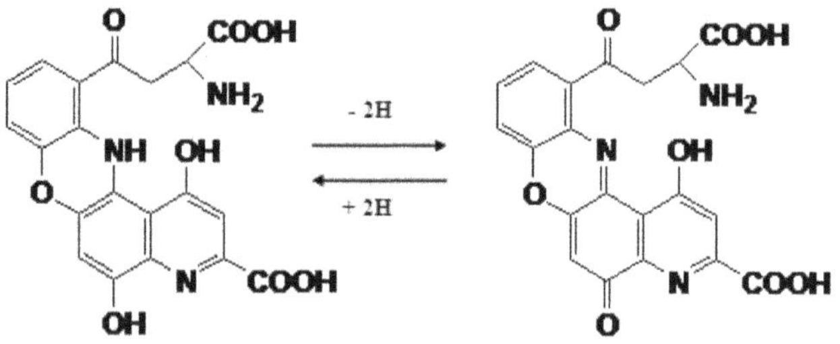

FIGURE 7.4 The chemical structure of one of the common ommochromes – xanthommatine in oxidized and reduced forms.

In the eyes of crustaceans and most insect species, there are ommines that have a maximum absorption in the visible region of the spectrum at 520–540 nm. In the eyes of some insects, for example, flies contains xanthomatine, which has a maximum absorption at 440–475 nm, depending on the solvent [33, 36, 38]. Up to now, it has been assumed that the main functions of ommochromes are:

a) the optical screening of photosensitive elements of retinal cells;
b) the screening of individual ommatidia from each other;
c) the increase in the resolution of the eyes;
d) the adjustment of the spectral sensitivity of the photoreceptors; and
e) the coloring of the skin and the detoxification of excess tryptophan.

We obtained experimental results demonstrating that ommochromes play an important role in the inhibition of photo-induced oxidation in ommatidia of the complex eye of invertebrate animals – insects and crustaceans.

7.4 ANTIOXIDANT ACTIVITY OF SCREENING PIGMENTS

In our experiments we used melanins of the eyes of man, cattle, rabbit, frog (*Rana temporaria*), an ink bag of cuttlefish (*Sepia officinalis*), insect cuticles (*Hermetia illucens* fly and *beetle Tenebrionidae latreille*), synthetic DOPA (dihydroxyphenylalanine) melanin and ommochrome eyes crustaceans (shrimps *Mysis relicta* and *Pandalus latirostris*) and insects (*Hermetia illucens* fly, *Pieris brassicae* butterfly, and dragonfly *Calopteryx splendens*). All experiments were carried out in accordance with international recommendations for guiding biomedical studies using experimental animals [39].

Antioxidant activity was evaluated by quenching the chemiluminescence of luminol initiated by hydrogen peroxide in the presence of hemoglobin, as well as inhibiting the lipid peroxidation reaction induced by various pro-oxidant factors.

Figure 7.5 shows the kinetics of the chemiluminescence of luminol in the presence of various concentrations of bovine melanosomes. The addition of melanosomes leads both to a decrease in the intensity (amplitude) of chemiluminescence, and to an increase in the latency period required to achieve maximum fluorescence. The processing of the obtained results in the Stern-Volmer coordinates [40] makes it possible to estimate the degree of antioxidant (antiradical) activity of the screening pigments of the eye.

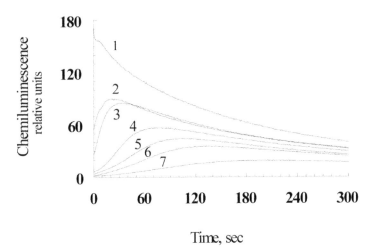

FIGURE 7.5 Kinetics of chemiluminescence of luminol in the presence of different amounts of bovine melanosomes. (*Note:* 1 – control without the addition of melanosomes; 2–7 – melanosomes were added in the amount from 2×10^7 to 9×10^7 granules/ml. The reaction was initiated by the addition of hydrogen peroxide).

Figure 7.6 shows the dependence of the decrease in the luminol luminescence intensity on the concentration of native melanosomes from the RPE
of a bull, a frog, and a human. It can be clearly seen that all melanosomes
effectively inhibit the process of chemiluminescence.

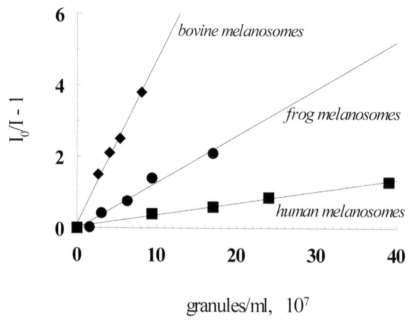

FIGURE 7.6   Comparison of antiradical activity of RPE melanosomes from different sources.
The y-axis is the decrease in the luminescence amplitude. (*Note:* I_0 is the chemiluminescence
intensity in control, I is the chemiluminescence in the presence of melanosomes. The abscissa
is the number of granules added).

The melanosomes isolated from the bovine and frog eyes were much
more active in relation to their ability to quench the chemiluminescence
of luminol than human melanosomes. This means that melanosomes from
human RPE have less antioxidant activity. The less antioxidant activity of
human melanosomes probably can be explained by two main reasons: either
by a higher packing density of the melanin pigment in the bovine and frog
melanosomes, or by the molecular structure peculiarities of melanin itself as
a polymer.

Similar experiments carried out with ommochromes isolated from the
eyes of various species of invertebrates showed that the eye screening
pigments also show a pronounced antioxidant activity. Figure 7.7 shows

the kinetics of chemiluminescence of luminol in the presence of various concentrations of ommochromes of *Pandalus latirostris* shrimp.

It is clearly seen that ommochromes of the complex eye, as well as melanosomes, cause a decrease in the amplitude (intensity) of chemiluminescence and increase the latent period of reaching the maximum of luminescence. The ability to effectively quench the chemiluminescence of luminol showed ommochromes from all sources used by us.

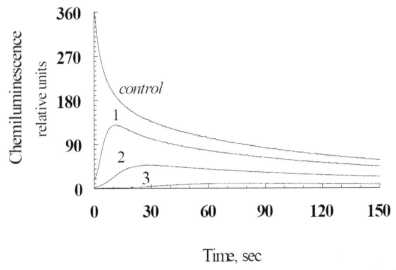

FIGURE 7.7 Kinetics of chemiluminescence of luminol in the presence of various concentrations of ommochromes of the eye of shrimp *Pandalus latirostris*. (*Note:* control – without the addition of ommochromes, (1–3) – added 40, 60 and 100 µg/ml ommochromes of the eye shrimp, respectively).

The amount of antioxidant activity of the screening pigments of invertebrates could be estimated for some ommochromes with a known molecular weight. In Figure 7.8, the constant of quenching of the chemiluminescence of luminol by ommochromes, extracted from the eyes of the black soldier fly (*Hermetia illucens*) was estimated.

The screening pigments of this insect are represented by xanthommatine, the molecular mass of which is known. Comparison of the latent period of induction of chemiluminescence in the presence of ascorbate and ommochromes made it possible to determine the constant of quenching of chemiluminescence by ommochromes of the fly's eye. It was $(0.65 \pm 0.15) \times 10^4 \text{ M}^{-1}$.

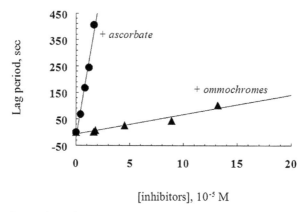

[inhibitors], 10^{-5} M

FIGURE 7.8   Comparison of antiradical activity of ommochromes of the fly's eye (*Hermetia illucens*) with ascorbic acid. The *y*-axis is the latent period of chemiluminescence in seconds; on the abscissa – the concentration of the inhibitor.

The screening pigments of vertebrates and invertebrates eyes effectively inhibit lipid peroxidation processes induced by a variety of prooxidant systems [16–18, 21, 41–44]. The main factors of oxidative stress (prooxidant factors) are the following factors:

a)   Oxygen and its active forms (superoxide radical, singlet oxygen, hydrogen peroxide, hydroxyl radical, lipid hydroperoxides and their radicals, nitrogen oxide radicals, etc.)

b)   Ions of metals of transition valency (especially divalent iron ions, which catalyze free radical processes);

c)   Carbonyl compounds, accumulating during oxidation of lipids and in the Maillard reaction;

d)   As well as various types of electromagnetic radiation (visible light, ultraviolet, γ-radiation) and photosensitizers.

Endogenous oxidants mainly formed due to the functioning of the mitochondrial electron-transport chain, NO oxidase reaction, Fenton's reaction, metabolism of arginine, microsomal detoxification, peroxisomal β-oxidation, in the processes of phagocytosis, inflammation, glycation, etc. Exogenous sources of oxidants are UV irradiation, *x*-ray, and γ-irradiation, visible light in the presence of photosensitizers, ultrasound, nanoparticles, pesticides, herbicides, ions of metals of variable valency, ozone, xenobiotics. The following Pro-oxidant factors have been applied in our experiments: (a) hydroperoxides, (b) superoxide radicals, (c) singlet oxygen, (d) ferrous ions,

(e) the irradiation of visible light in the presence of sensitizers, (f) hyperoxia, and (g) ultraviolet and *x*-ray irradiation.

In Figure 7.9 (A and B) shows the inhibitory effect of melanins and ommochromes on the peroxidation of lipids induced by ions of bivalent iron in the presence of ascorbic acid.

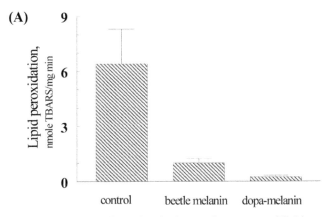

FIGURE 7.9A Inhibitory effect of melanins on the process of lipid peroxidation. *Note:* Inhibitory effect of natural melanin (beetle, *Tenebrionidae latreille*) and synthetic DOPA-melanin on the peroxidation process of the photoreceptor cells outer segments of the bovine eye. Control is the average rate of lipid peroxidation without melanin. The ordinate is the rate of peroxidation of lipids in *n* moles of TBA-reactive products (TBARS – "Thiobarbituric acid reactive substances") per minute per 1 mg of protein.

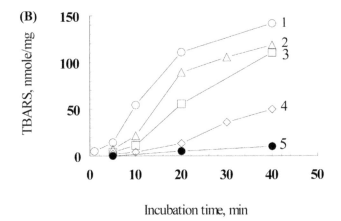

FIGURE 7.9B Inhibitory effect of ommochromes on the process of lipid peroxidation. *Note:* Kinetics of brain lipid peroxidation induced by ferrous ions in the presence of *Pandalus latirostris* shrimp ommochromes [21]. The ordinate is the concentration of TBARS; the abscissa is the incubation time. 1 –control, 2–5 – added 36, 73, 147, and 220 µg/ml ommochromes, respectively.

The substances of oxidation we have used were the membranes of photoreceptor cells, outer segments, or homogenates of the brain tissues containing high amounts of polyunsaturated fatty acid residues. The results show the pronounced inhibitory activity of both melanin (Figure 7.9A) and ommochromes (Figure 7.9B) with respect to the lipid peroxidation process. Even at a melanin concentration of 30 µg/ml and an ommochromes concentration equal to 150 µg/ml, the lipid peroxidation rate slows down approximately 6–10 times.

7.5   MAIN MECHANISMS OF ANTIOXIDANT ACTION OF SCREENING PIGMENTS

The ability of melanins and ommochromes to interact with superoxide radicals can be one of the mechanisms of antioxidant action of these screening pigments. Indeed, as we have shown, synthetic, and natural melanins, melanosomes, and ommochromes of the eye interact with the superoxide radicals, substantially accelerating their decease, both in aqueous and in anhydrous media [20, 21, 29].

Direct interaction of pigments with superoxide radicals is easy to observe in anhydrous media, for example, in a solution of superoxide in dimethylformamide. Superoxide radicals were obtained by electrolytic reduction of oxygen at a mercury cathode in a 0.01 M solution of tetrabutylammonium bromide in dimethylformamide. The superoxide concentration in the reaction medium was measured by the reduction of *nitro blue tetrazolium* to a *formazan* dye at an absorption wavelength of 560 nm. The molar extinction coefficient of formazan was assumed to be 2.3×10^3 M^{-1} cm^{-1}. The results showed that both melanins and ommochromes sharply accelerate the decay of superoxide in an anhydrous medium (Figure 7.10 A, B).

Since protein-containing melanosomes are unstable in dimethylformamide and therefore it is impossible to observe their interaction under these conditions, we also investigated the interaction of melanosomes, DOPA-melanin, and ommochromes with superoxide radicals in aqueous media by competitive inhibition method (Figure 7.10B). In these experiments, the superoxide radicals were generated during the xanthine oxidase reaction, and their concentration was measured either with *nitro blue tetrazolium* or with adrenaline (Figure 7.10 B); oxidation of adrenaline in adrenochrome, λ_{max} 485 nm), or cytochrome *c* (reduction of cytochrome *c* (Fe^{3+}) in cytochrome *c* (Fe^{2+}), λ_{max} 550 nm). The following data were used to calculate the interaction constants of pigments with superoxide radicals:

the interaction constant of superoxide with adrenaline is 4.0×10^4 $M^{-1}c^{-1}$ [46]; the interaction constant of superoxide with cytochrome c is 2.5×10^5 $M^{-1}c^{-1}$ [47]; the molecular weight of the monomeric unit of DOPA-melanin and melanosomes melanin is 170 Da; the approximate molecular weight of DOPA-melanin is 400 kDa [48], and the approximate molecular weight of shrimp ommochromes is 800 Da [20]. The obtained values of the constants are given in Table 7.1

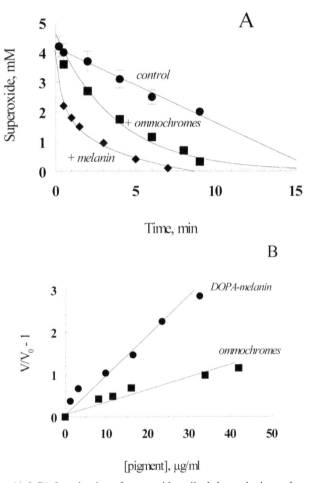

FIGURE 7.10 (A & B). Inactivation of superoxide radicals by melanins and ommochromes. (A: kinetics of superoxide radicals decay in an anhydrous medium in the presence of melanin and ommochromes [45], B: determination of the reaction constants of superoxide radicals with DOPA-melanin and shrimp ommochromes in aqueous media [20]. V_0 is the oxidation rate of adrenaline by superoxide in the absence of pigments; V is the oxidation rate of adrenaline by superoxide in the presence of various concentrations of pigments).

TABLE 7.1 The Interaction Constants of Screening Pigments with Superoxide Radicals

Pigment	The constant of interaction with superoxide radicals, Kp	
	$M^{-1} s^{-1}$	(M monomeric unit)$^{-1} s^{-1}$
DOPA-melanin	6.0×10^8	2.5×10^5
Melanosomes	—	9.5×10^4
Ommochromes	3.0×10^5	—

The results show that the melanins of the vertebrates' eyes and the ommochromes of invertebrates' eyes easily react with superoxide radicals and can, therefore, function as a sufficiently effective trap of this cytotoxic radical. The mechanisms of interaction of ommochromes with superoxide radicals are practically not studied.

At the same time, it is well known that the interaction of melanin melanosomes with superoxide radicals can be due to their reaction with the monomer unit of the polymer chain of the pigment, indole-5,6-quinone. Presumably, melanin reacts with superoxide in the form of hydroquinone with an intermediate formation of semiquinone according to the scheme given in Figure 7.11 [49].

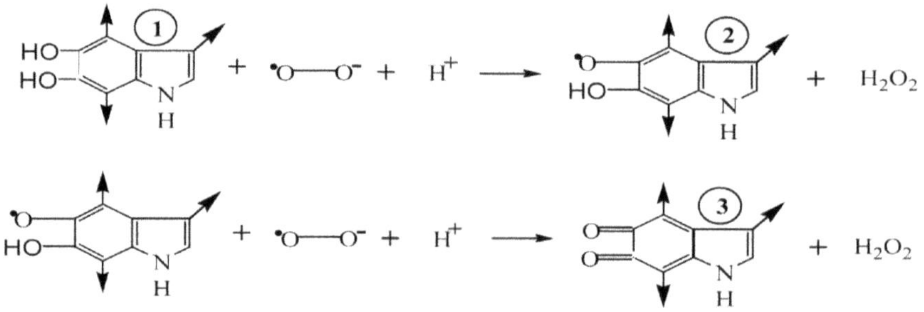

FIGURE 7.11 Scheme of the reaction of melanin in the reduced form with superoxide radical. *Note:* 1 – hydroquinone form of monomeric bond of melanin, 2 – semiquinone, 3 – quinone form (indole-5,6-quinone).

As a result, the superoxide turns into hydrogen peroxide, which can be inactivated by catalase and peroxidase. Melanin then turns into an oxidized form, which can be further reduced to its original form by various reducing agents, for example, NADH or NADPH.

However, the most interesting process is the reverse reduction of the oxidized form of melanin by the same superoxide radicals. It is well known that superoxide radicals can act both as oxidizing agents and as reducing agents. Therefore, the possibility of interaction of melanin in the form of indolequinone with superoxide was suggested [50], because of which a melanin radical and free oxygen are formed. The reaction+ can proceed according to the scheme shown in Figure 7.12 [51].

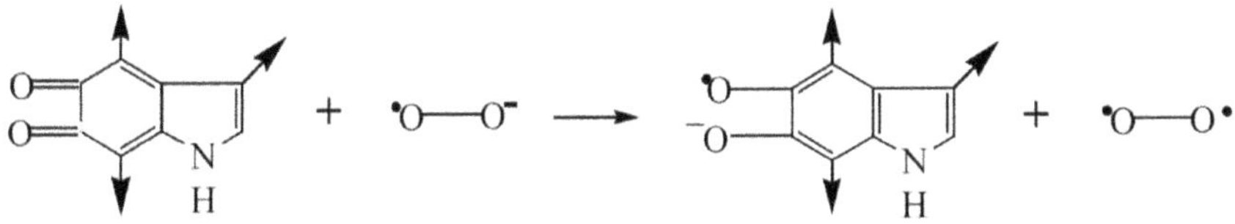

FIGURE 7.12   Scheme of the possible reaction of melanin in oxidized form with superoxide radical.

It was calculated that about 30% of the reaction between superoxide anions and DOPA melanin occurs along the way of reducing superoxide to hydrogen peroxide, and approximately 70% of the reaction goes through the oxidation of superoxide to molecular oxygen [52]. In this case, melanin will catalyze the decomposition of superoxide, without changing chemically, according to the scheme:

$$2\,O_2 + 2\,H^+ \rightarrow H_2O_2 + O_2$$

This indicates that melanin, both in oxidized and in reduced form, is able to interact with radicals of superoxide, converting them into non-radical products. However, at relatively high concentrations of superoxide, the melanin polymer breaks down (see below).

Another possible mechanism of antioxidant action of screening pigments can be their ability to bind prooxidant iron ions to inactive complexes. It is well known that melanins contained in melanosomes are an excellent chelator of metal ions [53]. We have carried out a study of the ability of melanosomes and ommochromes to bind iron ions by gamma-resonance spectroscopy technique [54, 55]. For the experiments, either solutions of $FeCl_2$ and $FeCl_3$ enriched in the isotope ^{57}Fe up to 90% or a solution of $^{57}FeSO_4$ obtained by dissolving of metallic α-^{57}Fe in dilute (0.1–0.2 M) sulfuric acid was used. Complexes of iron with melanosomes or ommochromes were separated from the excess reagent by centrifugation at 5000 g for 20 min. Figure 7.13 shows the gamma-resonance spectra of the $^{57}FeSO4$ complexes with the bovine eyes RPE melanosomes (A) and the shrimp eyes ommochromes (B).

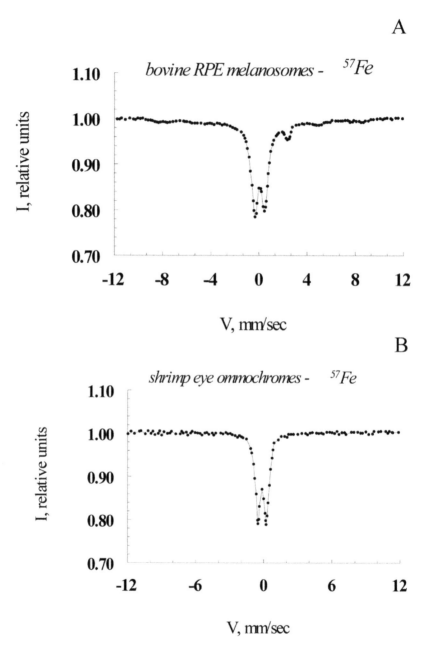

FIGURE 7.13   Formation of complexes of melanins and ommochromes with ferrous ions. (A – gamma-resonance spectrum of the bovine melanosome complex with $^{57}FeSO_4$ at 80K, B – gamma-resonance spectrum of the shrimp ommochromes complex with $^{57}FeSO_4$ at 80K).

7.6 PIGMENTED CELLS ARE MUCH MORE RESISTANT TO THE ACTION OF PROOXIDANTS

It is well known that albino animals, in which there are no screening pigments in the eye cells, exhibit increased photosensitivity (photophobia) and are poorly seen in bright daylight (day blindness). So, for example, for the development of degenerative processes in the retina of albino rats, 60 times less light intensity is sufficient than for pigmented animals of the same species [56, 57]. This is because the eye structures are extremely sensitive to various pro-oxidant factors. The high sensitivity of visual cells to oxidative stress is due to three main reasons. First, it is light; secondly, a high partial pressure of oxygen and, thirdly, high concentrations of easily oxidized lipids. Indeed, the combination of light and oxygen – a necessary circumstance for the realization of a normal photoreceptor process – simultaneously contributes to the emergence and development in the eye structures of destructive chemical reactions.

The sensitivity of photoreceptor cells to light damage is associated with the presence in them of all the factors necessary for a free radical reaction of photooxidation. This is, firstly, effectively light absorbing chromophores (mainly retinal and its derivatives, for example, A2E), secondly, the high partial pressure of oxygen and, thirdly, the presence of substrates in these cells, capable of oxidizing with great ease. The molecular components of the photoreceptor membrane are ideal substrates for the free radical oxidation reaction.

The three major phospholipids of the photoreceptor membrane are phosphatidylcholine, phosphatidylethanolamine, and phosphatidylserine. The composition of these phospholipids is unique: more than 60% are readily oxidable polyunsaturated fatty acids, with about 75% of all fatty acids accounting for the proportion of docosahexaenoic acid containing 22 carbon atoms and 6 double bonds (C 22:6).

All these factors contribute to the high sensitivity of photoreceptor cells to oxidative damage. However, all these factors are essential mainly for the cells of the albino animals' eye or for weakly pigmented eyes. In contrast, pigmented animals are extremely resistant to the action of various prooxidants [16, 21, 45, 58–68].

In our experiments, the resistance of eye structures of pigmented and unpigmented or slightly pigmented animals was compared to the action of various prooxidants. Figure 7.14 shows the kinetics of the peroxidation of the retina tissue of two subspecies of rabbits – albino and dark pigmented.

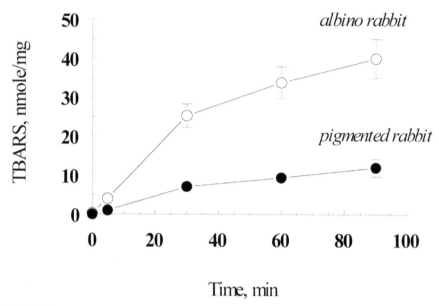

FIGURE 7.14   Comparative kinetics of lipid peroxidation induced by the Fe^{2+}-ascorbate in a mixture of the neural retina and RPE tissues of albino rabbits and normally pigmented rabbits. (*Note:* the y-axis is the concentration of TBARS in moles per mg of protein).

The rate of lipid peroxidation induced by ferrous ions was significantly higher in the retina of albino rabbits than in darkly pigmented rabbits. Additional experiments have shown that the activity of other antioxidant systems in RPE cells of pigmented animals and albinos does not differ significantly. Therefore, it can be assumed that if pigmented screen pigments are removed from pigmented tissue, then it will become more sensitive to the action of pro-oxidants.

A similar experiment was performed on the strongly pigmented RPE tissue of the frog *Rana temporaria* (Figure 7.15).

In this case, superoxide radicals generated in the course of the xanthine oxidase reaction were used as the prooxidant system. The results showed that this oxidation system does not cause any noticeable accumulation of TBARS, i.e., in the pigmented tissue, the process of lipid peroxidation was completely absent. Removal of melanosomes from the homogenate of pigmented RPE resulted in a significant increase in the peroxidation process (Figure 7.15). This result confirms our assumption that the main contribution to the inhibiting the peroxidation process belongs to screening pigments, in this case to melanosomes.

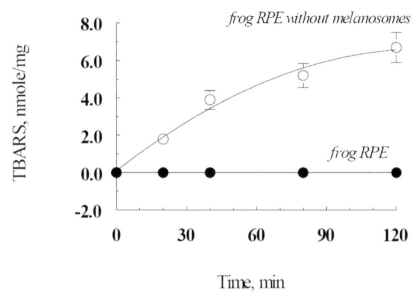

FIGURE 7.15 Kinetics of lipid peroxidation in the frog RPE tissue induced by the xanthine-xanthine oxidase system. (*Note:* the *y*-axis is the concentration of TBA-reactive products in moles per mg of protein).

Pigmented animals also exhibit a higher resistance to the action of ionizing radiation. The development of the process of lipid peroxidation in the retina and RPE of mice under X-ray irradiation and the role of melanin in these processes were investigated. The experiments were carried out on darkly pigmented mice (C57B1/6) and white SNK mice of weighing 18–20 g. X-ray irradiation of the animal was carried out on a RUT-200 apparatus at a voltage of 200 kV, the current strength of 10 mA, a Cu filter of 0.5 cm, a dose rate of 48.5 mC/kg. The animals were irradiated once in doses of 6 and 12 Gy or fractionated in a dose of 1.5 Gyx4 once a week (total dose of 6 Gy). The peroxidation products were determined in the homogenate of the neural retina and RPE mixture.

Figure 7.16 shows the accumulation of TBA-active products with a single total irradiation with a dose of 12 Gy. With such intensive irradiation leading to the death of all mice on the 7th–8th day, a sharp increase in the level of TBA active products in the eye structures of albino mice occurs 24 h after irradiation (Figure 7.16).

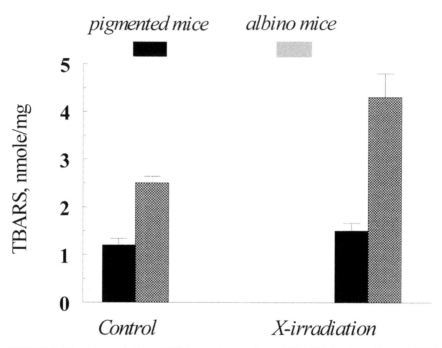

FIGURE 7.16   Accumulation of TBA-reactive products (TBARS) in the retina and RPE with a single total irradiation of pigmented and unpigmented mice at a dose of 12 Gy.

The maximum increase in the level of TBA-active products reached 65–75%, and in individual experiments, 100% compared with that of non-irradiated animals. At the same time, pigmented mice were found to be much less sensitive to irradiation in these doses by this criterion – the increase in the level of peroxidation products did not exceed 20–30%.

A similar pattern with respect to light damage is also observed for invertebrates: animals containing large amounts of screening pigments are less sensitive to various pro-oxidant factors [63, 67, 68]. We investigated two genetically similar populations of the Baltic shrimp of the *Mysis relicta* species. These shrimp at the end of the last glacial period (about 9000 years ago) were divided into several populations with different habitats. One of them (the marine population) lives in the sea at a shallow depth, where the illumination is quite high, and the other (lake population) in the lakes (in particular in the Paajarvi lake, whose maximum depth is about 90 m) at a very considerable depth, where the illumination extremely low. It turned out that the lake population is much more sensitive to light exposure compared to the marine population: even moderate illumination of animals, for example, on the surface of the water in the evening in the absence of the sun, leads to complete loss of vision [35]. An analysis of the content of screening pigments in the eyes of these animals showed that the amount of ommo-chromes in the marine population eyes is almost three times larger than in the lake population shrimps [63]. We compared the antioxidant activity of pigment and reticular cells in the eyes of two populations of shrimp *Mysis relicta*. Experiments have shown that the homogenates of the eyes of the marine population of shrimp proved to be much more resistant to peroxidation induced by ferrous ions in the presence of ascorbate.

Figure 7.17 shows the comparative kinetics of accumulation of lipid hydroperoxides during oxidation of eye homogenates of both types of shrimp populations.

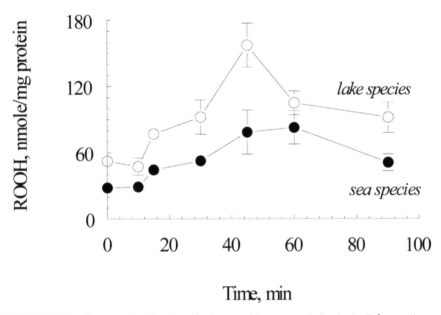

FIGURE 7.17 Comparative kinetics of hydroperoxides accumulation in the Fe^{2+}-ascorbate-induced peroxidation of the homogenate of the *Mysis relicta* shrimp eyes of the lake and marine populations [68].

It can be seen that the eye homogenate of the marine population is much more resistant to peroxidation than the homogenate of the lake population. We established [68] that the effect of the low sensitivity of the cells of the complex eye of the marine population shrimp to the action of pro-oxidants is not associated with a higher content of low-molecular antioxidants and antioxidant enzymes in them, but is due precisely to the presence of a large number of ommochrome granules.

In general, the results indicate that the rate of peroxide oxidation of lipids induced by such prooxidant systems as ionizing radiation, ultraviolet, intense visible light, divalent iron ions, hydroperoxides, hyperoxia, super-oxide oxygen radicals is much lower in the eye tissues of strongly pigmented animals than in poorly pigmented tissues or in albino animals. This allows us to conclude about the antioxidant role of screening pigments in the pigmented tissues of the eyes of vertebrates and invertebrates.

The protective action of melanosomes of vertebrates and ommochromes of invertebrates can be carried out in at least two ways. The first way is to regulate the intracellular concentration of pro-oxidants. Screening pigments fulfill their physiological role, being a particle, a granule. Molecules of pro-oxidants, such as ions of polyvalent metals (especially bivalent iron), exogenous, and endogenous photosensitizers, xenobiotics, can interact with the surface of granules of screening pigments, sorbing on them. Such heterogeneous reactions are characteristic of particles having pores. It is well known that melanin polymers have pores [69–71]. It is important, as it was shown in experiments on the binding of many exogenous photosensitizers by melanins, that their adsorption is physical in nature and that the chemical properties of bound molecules change radically [72].

Melanosomes and possibly ommochromes granules can regulate both the dark lipid peroxidation processes (induced, for example, by metal ions) and photo-induced lipid peroxidation processes sensitized by endogenous and exogenous photosensitizers. The intracellular granule containing the screening pigments practically acts as an intracellular sorbent, binding prooxidants to inactive complexes.

The second way of protective action of melanosomes of vertebrates and ommochromes of invertebrates eyes is the fusion of granules containing screening pigments both in pigment cells and also in reticular cells of the invertebrates complex eye with "dangerous" organelles in which the prob-ability of development of free radical processes and the spread of them to the whole cell is great. Such "dangerous" organelles, for example, in RPE cells of vertebrates are phagosomes containing a huge amount of material easily exposed to peroxidation in both darkness and illumination, as well as containing lipofuscin granules having toxic retinol fluorophores.

The fusion of melanosomes with lipofuscin granules and the formation of complex melanolipofuscin granules is a long-established fact [73, 74]. Being in direct contact with the object in which free radical oxidation reactions take place, the screening pigments can easily inhibit them. For example, they can inhibit them by interacting with radicals of superoxide and hydroperoxides, as well as with non-radical active products of the peroxidation process (for example, with aldehydes), inhibiting the process in the particles themselves and preventing its development in other compartments of the cell.

7.7 THE ROLE OF SCREENING PIGMENTS MELANOSOMES IN THE AGING PROCESSES

The aging process leads to significant changes in the cells of the human RPE [75]. In addition to various biochemical and morphological changes, in these non-dividing, post-mitotic cells, "age pigment" lipofuscin and complex pigment granules, such as melanolipofuscin, accumulate with age; while the number of melanin-containing melanomas decreases in these cells [76].

It has been shown that with age (after 40 years) a significant decrease in melanin concentration occurs in RPE cells [74, 77]. Morphological and ESR measurements showed that in the human RPE in the age group of 20–30 years, 36% more melanin is present than in the 60–90 age group [78]. In this case, there is a significant decrease in the number of melanosomes in the RPE cell. If in the age group up to 20 years melanosomes occupy approximately 8% of the RPE cells volume, then in the age group 41–90 years this volume gradually decreases to 3.5% [74]. In the age group 90–101, the melanosomes are almost completely replaced by mixed melanolipofuscin granules [79]. This indicates the processes of age-related biodegradation of melanosomes in RPE cells.

Certain mechanisms leading to age-related biodegradation of melanosomes in RPE cells have not yet been known. Analysis of literature data suggests that degradation of melanin is not associated with enzymatic reactions. Most likely, during aging, the chemical degradation of melanin occurs, caused either by light, or products of oxidation or photooxidation. Indeed, it has been shown that oxidative disintegration of melanin in melanosomes occurs with age, which is revealed by the increase in fluorescence and by the increase in oxygen uptake by melanosomes in elderly and older people [80, 81].

It is known that during the destruction of melanin, caused by the action of ultraviolet irradiation and/or hydrogen peroxide, fluorescent decay products are formed [82–84]. As shown by experiments [85–87], irradiation of melanin with intense visible light and ultraviolet leads to the destruction of the pigment. However, to achieve this effect, long-term irradiation of melanin is necessary.

Figure 7.18 shows the change in the antioxidant activity of DOPA-melanin, ascorbate, and glutathione when irradiated with UV-visible light.

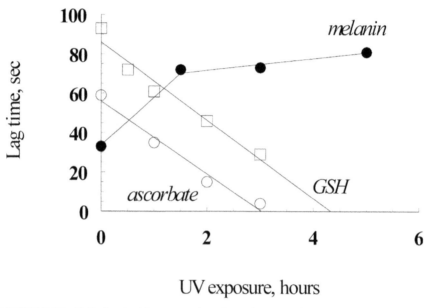

UV exposure, hours

FIGURE 7.18   Fall of antioxidant properties of ascorbate and glutathione (GSH), but not of melanin under the action of ultraviolet irradiation. (*Note:* the ordinate shows the latent period of luminal chemiluminescence retention; on the abscissa – the time of irradiation by an arc mercury-quartz lamp).

It can be seen that the antioxidant activity of ascorbate and glutathione drops sharply, while the ability of melanin to quench the chemiluminescence of luminol even slightly increases. These results indicate that the UV irradiation of melanin, which is not very long (up to 5 h), does not lead to any significant destruction of the polymer.

This is confirmed by the results of experiments on the comparative measurement of the antioxidant activity of native and UV-irradiated melanins. Figure 7.19 shows the kinetics of accumulation of TBARS with Fe^{2+}-induced peroxidation of cardiolipin liposomes (control). Melanosomes isolated from the bovine eye RPE effectively inhibited this process.

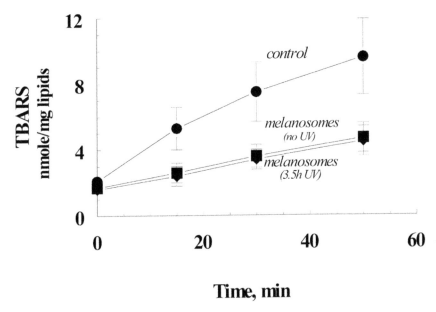

FIGURE 7.19 Effect of melanosomes from the bovine eye RPE cells on the kinetics of peroxidation products accumulation of cardiolipin liposomes. (*Note:* irradiated (3.5h UV) and non-irradiated (no UV) melanosomes were added at equal concentrations – 3 × 10^7 granules ml^{-1}. Control – accumulation of TBARS in the absence of melanosomes).

At the same time, melanosomes exposed to UV irradiation for 3.5 h proved to be equally effective inhibitors as native, non-irradiated melanosomes (see curves "melanosomes 3.5 h UV"). Only much longer irradiation of DOPA-melanin and melanosomes with UV-visible light leads to a decrease in their antiradical activity, determined by the degree of quenching of the chemiluminescence of luminol (Figure 7.20).

It is clearly seen that if 7-h irradiation causes only a slight decrease in the ability to quench the chemiluminescence of luminol, then a 32-h irradiation resulted in a significant, but not catastrophic, drop in the antiradical activity of DOPA-melanin. These experiments, therefore, confirm the high resistance of melanins to the destructive effect of light.

Even severe and prolonged UV irradiation (wavelengths from 190 nm and above) of DOPA-melanin did not lead to a significant decrease in the pigment's ability to exert antioxidant effects. It is seen from Figure 7.21 that UV irradiation of the pigment for 32 h resulted in only about a 1.5-fold decrease in the inhibitory capacity of DOPA-melanin against the peroxidation of cardiolipin liposomes.

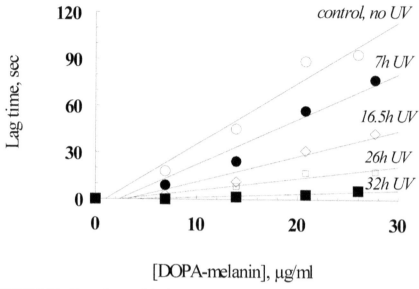

FIGURE 7.20 Dependence of the latent period duration of luminol chemiluminescence initiation on the concentration of non-irradiated (control) and exposed to UV irradiation for 7 to 32 h DOPA-melanin.

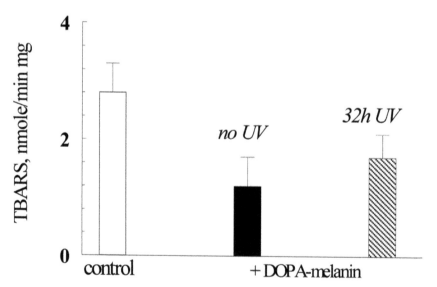

FIGURE 7.21 Effect of non-irradiated and UV-irradiated DOPA melanin on the rate of UV-induced peroxidation of cardiolipin liposomes. (*Note:* control is the rate of accumulation of TBARS with UV irradiation in the absence of melanin. In the test samples, 25 µg/ml of the non-irradiated and irradiated DOPA-melanin were added).

Thus, the destruction of melanin by ultraviolet or intense visible light requires too high irradiation energies and long exposures. It should be noted that the structure of the eye – the cornea, the lens, and the vitreous humor – practically do not pass the shortwave ultraviolet to the RPE cells containing melanin. It is also unlikely that the visible light of such a large intensity and duration, which is used to destroy melanin in the experiment, has ever acted on the retina in vivo.

Therefore, the destruction of melanin within the RPE melanosomes for these reasons is generally excluded for the eye, but may occur for melanin of hair and skin exposed to direct sunlight.

In contrast, strong oxidants, such as hydrogen peroxide or superoxide radicals, quite easily lead to the destruction of melanin and the loss of its antioxidant properties [87, 88]. We have shown, for example, that superoxide radicals lead simultaneously both to a decrease in the number of melanosomes and to a drop in the concentration of the paramagnetic centers of the melanin. Complete degradation of the melanosomes leads to the formation of a transparent solution, the disappearance of the ESR signal, and the complete loss of the antioxidant properties of the pigment [88].

Figure 7.22 shows the dependence of the decrease in the concentration of melanosomes from human RPE, depending on the concentration of potassium superoxide. At a superoxide concentration of 100 mM after two hours of incubation, an almost threefold decrease in the number of melanosomes in the sample is observed. Increasing the concentration of superoxide up to 170–180 mM can achieve complete degradation of all the melanosomes initially present in the suspension.

The suspension becomes transparent and microscopic control shows a complete absence of visible granules. Under these conditions, the disappearance of the initial ESR signal of melanin is also observed, which indicates complete destruction of the polymer.

The obtained data on the dependence of the decrease for human RPE melanosomes on the superoxide concentration, allow us to estimate approximately the amount of superoxide molecules necessary for complete degradation of one melanosome.

In our experiments, an average of 175±25 micromoles of superoxide was required to destroy 2.7×10^8 melanosomes. Thus, it is necessary to 650±100 femtomoles of superoxide for complete degradation of one melanosome. It is known [89] that in the human RPE cell containing approximately 700 lipofuscin granules. With such a number of lipofuscin granules, 0.56 femtomoles of superoxide can be formed in the RPE cell at one minute with relatively weak blue light irradiation (1.5 mW/cm²). It follows that it takes less than 24 h (about 19 h) to generate 650 femtomoles of superoxide, which are necessary for the complete destruction of one melanosome.

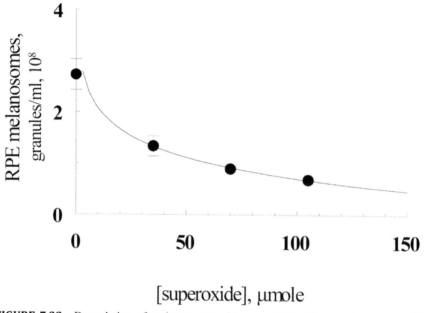

[superoxide], μmole

FIGURE 7.22 Degradation of melanosomes of human RPE with potassium superoxide [88]. (*Note:* incubation of melanosomes for 2 h with superoxide leads to a decrease in their number. The initial concentration of melanosomes was 2.7×10^8 granules/ml).

The above calculations are based, as already mentioned, on the results of our in vitro experiments. However, in a situation in vivo, melanosomes do not quickly break down. First, melanosomes, up to the rather old age, mostly exist separately. They are separated from lipofuscin granules, and a significant accumulation of mixed melanolipofuscin granules is observed already in senile age. Melanosomes themselves are extremely resistant to irradiation. Secondly, there is a powerful antioxidant system in the RPE cell, a key role in which is played by superoxide dismutase, which destroys the superoxide radicals. Therefore, it is unlikely that before melanosomes fuse with lipofuscin granules superoxide radicals could significantly destroy melanin in melanosomes. However, in the melanolipofuscin granule, the situation becomes different. Superoxide radicals, which are generated by the action of light on the lipofuscin part of this mixed granule, can affect melanin "in situ," without leaving to the cytoplasm. In this case, these radicals, as we have shown, actively destroy melanin. Based on these considerations, it can be assumed that the destruction of melanosomes in the RPE cell is determined not so much by light as by the intensity of melanosome fusion with lipofuscin granules. Consequently, the more lipofuscin granules accumulate in the RPE cells, the greater the probability of their fusion with melanosomes and the formation of mixed melanolipofuscin granules. The process of lipofuscin granules accumulation in the RPE cell directly depends on age and is associated with the progression of a degenerative retina disease, a vivid example of which can be Stargard's disease.

It is well known that in the process of aging melanosomes fuse with lipofuscin granules, and the formation of melanolipofuscin granules take place; as a result, the number of melanolipofuscin granules increases with age [74, 76]. In the age group of 90–101 years, the melanosomes are almost completely replaced by mixed melanolipofuscin granules [79]. We suggested that a possible cause of an age-related decrease in melanosome content in the human RPE cells is their fusion with lipofuscin granules and further photo-induced destruction of melanin pigment within the melanosome by superoxide radicals formed upon absorption of light by lipofuscin fluorophores.

The results obtained to support this assumption. These results suggest that the age-related disappearance of melanin in the RPE cell is due to the destruction of the pigment in the melanolipofuscin granule and that this destruction is caused by the superoxide radicals generated by lipofuscin fluorophores under the action of visible light [88].

The fall in the RPE cells with the age of melanosomes amount performing both screening and antioxidant functions, and simultaneous accumulation in these cells of melanolipofuscin and lipofuscin granules can lead to photo-oxidative stress. This stress is directly related to the development of such serious eye pathologies as age-related macular degeneration and retinitis pigmentosa.

7.8 CONCLUSIONS

Thus, we can conclude that intracellular organelles – melanosomes and ommochromes as screening pigments of the vertebrates and invertebrates eye and the human eye – perform two important physiological functions simultaneously. First, they serve as cutoff filters that protect photosensitive cells from the hazard of damage by too bright light and increase the resolving power of the eye.

Indeed, since the outer segments of the vertebrate and human retina photoreceptor cells are surrounded by the outgrowths of RPE cells containing a mass of melanosomes, these processes, effectively absorbing the scattered light inside the eye, act as shielding "pads." Thus, they absorb the scattered light and increase the resolution of the eye.

Secondly, screening pigments, with a pronounced antioxidant activity, effectively protect the cell in which they are located, from the hazard of oxidative stress, including above all photo-oxidative stress.

Excess accumulation of lipofuscin granules in retinal pigment epithelial cells is considered in eye pathology (degenerative diseases of the retina, for example, Stargardt's macular dystrophy) as an essential pathogenesis factor. This accumulation of lipofuscin granules can lead to the accelerated disappearance of melanosomes due to their fusion with lipofuscin granules and degradation in the composition of mixed melanolipofuscin granules. The disappearance of melanosomes as light filters and antioxidants creates, as already mentioned above, the danger of oxidative, primarily photo-oxidative stress in the structures of the eye.

KEYWORDS

- aging
- albino animals
- melanin
- melanosomes
- ommochromes
- oxidative stress
- retinal pigment epithelium
- superoxide

REFERENCES

1. Bertrand, G., (1896). On a new oxidase is oxidative soluble enzyme of plant origin. *Comp. Rend. H. S. Acad. Sci.*, *122*, 1215–1217 (in French).
2. Furth, V., Schneider, O., & Schneider, H., (1902). On animal tyrosinases and their relationship to pigment formation. *Beitr. Chem. Physiol. Pathol.*, *1*, 229–242 (in German).

3. Figge, F. H. J., (1939). Melanin: A natural reversible oxidation-reduction system and indicator. *Proceedings of the Society of Experimental Biology and Medicine, 41,* 127–127.
4. Van Woert, M. N., (1967). Oxidation of reduced nicotinamide adenine dinucleotide by melanin. *Life Sciences, 6,* 2605–2612.
5. Gan, E. V., Haberman, H. F., & Menon, I. A., (1976). Electron transfer properties of melanin. *Archives of Biochemistry and Biophysics, 173,* 666–669.
6. Gan, E. V., Lam, K. M., Haberman, H. F., & Menon, I. A., (1977). Electron transfer properties of melanins. *British Journal of Dermatology, 96,* 25–28.
7. Commoner, B., Townsend, J., & Pake, G., (1954). Free radicals in biological materials. *Nature, 174,* 689–691.
8. Mason, H. S., Ingram, D. J. E., & Allen, B., (1960). The free radical property of melanins. *Archives of Biochemistry, 86,* 225–230.
9. Sever, R. J., Cope, F. W., & Polis, B. D., (1962). Generation by visible light of labile free radicals in the melanin granules of the eye. *Science, 137,* 128–129.
10. Ostrovsky, M. A., & Kayushin, L. P., (1963). A study of electron paramagnetic resonance in the retina under the action of light. *Reports of Academy of Sciences of the USSR, 151*(4), 986–988 (in Russian).
11. Ostrovsky, M. A., (1966). Reversible change in ESR signals of the pigment epithelium of the eye under the action of visible light. *Free-Radical Processes in Biological Systems: Proceedings of the MSN (Moscow Society of Naturalists), Science* (pp. 275–279). Moscow, (in Russian).
12. Sakina, N. L., Pulatova, M. K., Ostrovskii, M. A., & Smirnov, L. D., (1979). Effect of inhibitors of free radical processes on the "dark" paramagnetic centers of the melanoprotein granules of the pigment epithelium of the eye. *Biophysics, 24*(4), 646–650 (in Russian).
13. Lukiewicz, S., (1972). The biological role of melanin. I. New concepts and methodical approaches. *Folia Histochem. et Cytochem., 10,* 93–108.
14. Felix, C. C., Hyde, J. S., Sarna, T., & Sealy, R. C., (1978). Melanin photoreactions in aerated media: Electron spin resonance evidence for production of superoxide and hydrogen peroxide. *Biochem. Biophys. Res. Com., 84,* 335–341.
15. Goodchild, N. T., Kwock, L., & Lin, P. S., (1981). Melanin: A possible cellular superoxide scavenger. In: Rodgers, M. A. J., & Powers, E. L., (eds.), *Oxygen and Oxy-Radicals in Chemistry and Biology* (pp. 645–648). N.Y/San Francisco/ London, Acad. Press.
16. Dontsov, A. E., Sakina, N. L., & Ostrovsky, M. A., (1980). Comparative study of lipid peroxidation in the eye pigment epithelium of pigmented and albino animals. *Biochemistry, 45*(5), 923–928 (in Russian).
17. Sakina, N. L., Dontsov, A. E., Kuznetsova, G. P., Ostrovsky, M. A., & Archakov, A. I., (1980). Inhibition of lipid peroxidation by melanoprotein granules. *Biochemistry, 45*(8), 1476–1480 (in Russian).
18. Ostrovsky, M. A., Sakina, N. L., & Dontsov, A. E., (1980). Antioxidative functions of screening pigments of the eye. *Reports of Academy of Sciences of the USSR, 255*(3), 749–751 (in Russian).
19. Lapina, V. A., Dontsov, A. E., & Ostrovsky, M. A., (1984). Superoxide generation via melanin interaction with oxygen. *Biochemistry, 49*(10), 1712–1718 (in Russian).

20. Lapina, V. A., Dontsov, A. E., Ostrovsky, M. A., & Emanuel, N. M., (1984). Interaction of oxygen anion radicals with eye melanins and ommochromes. *Reports of Academy of Sciences of the USSR, 280*(6), 1463–1465 (in Russian).

21. Ostrovsky, M. A., Sakina, N. L., & Dontsov, A. E., (1987). An antioxidative role of ocular screening pigments, *Vision Res., 27*, 893–899.

22. Dontsov, A. E., Sakina, N. L., & Ostrovsky, M. A., (1999). Lipofuscin granules and melanosomes from human retinal pigment epithelium play opposite roles in photooxidation of cardiolipin. *Biophysics, 44*(5), 853–858 (in Russian).

23. Johannsen, A. O., (1924). Eye structure in normal and eye-mutant Drosophila. *J. Morph., 39*, 337–350.

24. Becker, E., (1941). The pigments of the ommine and ommatine group, a new class of natural dyes. *Naturwissenschaften,* 237–238 (in German).

25. Ephrussi, B., & Herold, J. L., (1944). Studies of eye pigments of drosophila. Methods of extraction and quantitative estimation of the pigment components. *Genetics, 29*, 148–175.

26. Butenandt, A., Schiedt, U., & Biekert, E., (1954). About ommochrome. I. Communication: Isolation of xanthommatine, rhodommatine and ommatine *C* from the excreta of *Vanessa urticae. European J. Organic Chem., 586*, 217–228 (in German).

27. Butenandt, A., & Schafer, W., (1962). Ommochromes. Gore, T. S., et al., (eds.), *Recent Progress in the Chemistry of Natural and Synthetic Coloring Matters and Related Fields* (pp. 13–62). N.Y., Acad. Press.

28. Dontsov, A. E., (1981). Antioxidative function of shrimp ommochromes *Pandalus latirostris. J. Evolut. Biochem. Physiol., 17*, 53–56 (in Russian).

29. Dontsov, A. E., Lapina, V. A., & Ostrovsky, M. A., (1984). Superoxide photogeneration by ommochromes and their role in the system of antioxidative protection of invertebrate eye cells. *Biophysics, 29*(5), 878–882 (in Russian).

30. Dontsov, A. E., Mordvintsev, P. I., & Lapina, V. A., (1985). Dark and light induced ESR signals of the ommochromes of the invertebrate eye. *Biophysics, 30*(1), 6–8 (in Russian).

31. Duchon, J., Borovanski, J., & Hach, P., (1973). Chemical composition of ten kinds of various melanosomes. In: McGover, V. J., & Rassel, P., (eds.), *Mechanisms of Pigmentation* (Vol. 1, pp. 165–170). Basel, S. Karger.

32. Rozanowska, M., (2011). Properties and functions of ocular melanins and melanosomes. In: Borovansky, J., & Riley, P. A., (eds.), *Melanins and Melanosomes: Biosynthesis, Biogenesis, Physiological, and Pathological Functions* (pp. 187–224). Wileg-VCH Verlag Gmbh & Co, Weinheim Germany.

33. Linzen, B., (1974). The tryptophan ommochrome pathway in insects. In: Treherne, J. E., et al., (eds.), *Advances in Insect Physiology* (Vol. 10, pp. 117–246). London and N-Y, Acad. Press.

34. Gribakin, F. G., & Chesnokova, E. G., (1984). A study of physiology insect vision with using of eye mutants. *Advances in Modern Biology, 97*, 69–82 (in Russian).

35. Lindstrom, M., & Nilsson, H. L., (1988). Eye function of *Mysis relicta* Loven (Crustacea) from two photic environments. Spectral sensitivity and light tolerance. *J. Exp. Mar. Biol. Ecol., 120*, 23–37.

36. Langer, H., (1975). Properties and functions of screening pigments in insect eyes. *Photoreceptor Optics* (pp. 429–455). Berlin – N.Y., Springer.

37. Sawada, H., Nakagoshi, M., Mase, K., & Yamamoto, T., (2000). Occurrence of ommochrome-containing pigment granules in the central nervous system of silkworm, *Bombyx mori. Comp. Biochem. Physiol. B., 125,* 421–428.

38. Strother, G. K., (1966). Absorption of Musca domestica screening pigment. *J. Gen. Physiol., 49,* 1087–1088.

39. *International Guiding Principles for Biomedical Research Involving Animals,* (1985). CIOMS & ICLAS, Altern. Lab. Anim. (ATLA), *12*(4).

40. Smirnov, L. D., Kuznetsov, I. V., Proskuriakov, S. I., Skvortsov, V. G., Nosko, T. N., & Dontsov, A. E., (2011). Antiradical and NO-inhibiting activities of beta-hydroxy(ethoxy) derivatives of nitrous heterocycles. *Biophysics, 56*(2), 316–321 (in Russian).

41. Sakina, N. L., Dontsov, A. E., & Ostrovsky, M. A., (1986). Inhibition by melanin of lipid photo-oxidation. *Biochemistry, 51*(5), 864–868 (in Russian).

42. Ostrovsky, M. A., Sakina, N. L., & Dontsov, A. E., (1987). The system of protection of the eye structures from photodamage. Screening pigments of vertebrates – melanosomes as inhibitors of photooxidation processes I. *J. Evolutionary Biochem. Physiol., 23*(5), 575–581 (in Russian).

43. Sakina, N. L., Dontsov, A. E., Lapina, V. A., & Ostrovsky, M. A., (1987). The system of protection of the eye structures from photodamage. Screening pigments of arthropods – ommochromes as inhibitors of photo-oxidation processes. *J. Evolutionary Biochem. Physiol., 23*(6), 702–706 (in Russian).

44. Pustynnikov, M. G., & Dontsov, A. E., (1988). Inhibition of UV-induced accumulation of lipid peroxides by melanins and ommochromes. *Biochemistry, 53*(7), 1117–1120 (in Russian).

45. Dontsov, A. E., & Ostrovsky, M. A., (2005). The antioxidant role of shielding eye pigments – melanins and ommochromes, and physicochemical mechanism of their action. In: Varfolomeev, S. D., & Burlakova, E. B., (eds.), *Chemical and Biological Kinetics: New Horizons* (Vol. 2, pp. 133–150). Danvers, MA 0193, USA.

46. Asada, K., Takahashi, M., & Nagate, M., (1976). Reactivity of tiols with superoxide radicals. *Agr. Biol. Chem., 40,* 1891–1892.

47. Yamashita, T., (1982). Selective inhibition by the sulfhydryl reagent maleimide of zymosan particle phagocytosis by neutrophils. *FEBS Lett., 141,* 68–73.

48. Bull, A. T., (1970). Kinetics of cellulase inactivation by melanin. *Enzymologia, 39,* 333–347.

49. Dontsov, A. E., (2014). *Protective Effect of Melanins in Oxidative Stress* (p. 165). Lambert Acad. Publ., Saarbrucken, Deutschland, (in Russian).

50. Korytowski, W., Kalyanaraman, I. A., Menon, I. A., Sarna, T., & Sealy, R. C., (1986). Reaction of superoxide anions with melanins: Electron spin resonance and spin trapping studies. *Biochim. Biophys. Acta, 882,* 145–153.

51. Dontsov, A. E., Glickman, R. D., & Ostrovsky, M. A., (1999). Retinal pigment epithelium pigment granules stimulate the photo-oxidation of unsaturated fatty acids. *Free Radic. Biol. Med., 26*(11/12), 1436–1446.

52. Korytowski, W., Pilas, B., Sarna, T., & Kalyanaraman, I. A., (1987). Photoinduced generation of hydrogen peroxide and hydroxyl radicals in melanins. *Photochem. Photobiol., 45,* 185–190.

53. Potts, A. M., & Au, P. C., (1976). The affinity of melanin for inorganic ions. *Exp. Eye Res., 22.* 487–491.

54. Bagirov, R. M., Stukan, R. A., Dontsov, A. E., Ostrovsky, M. A., & Lapina, V. A., (1986). Study of the binding of iron ions by melanoprotein granules of the eye pigment epithelium using gamma-resonance spectroscopy. *Biophysics, 31*(3), 469–474 (in Russian).

55. Bagirov, R. M., Stukan, R. A., Dontsov, A. E., Ostrovsky, M. A., & Lapina, V. A., (1986). Gamma-resonance spectroscopic study of iron ion binding by ommochromes of the invertebrate eye. *Biophysics, 31*(6), 1017–1022 (in Russian).

56. Lavail, M. M., Sidman, R. L., & Gerhardt, C. O., (1975). Congenic strains of RCS rats with inherited retinal dystrophy. *J. Hered., 66*, 242–244.

57. Rapp, L. M., & Williams, T. P., (1980). The role of ocular pigmentation in protecting against retinal light damage. *Vision Res., 20*, 1127–1131.

58. Sakina, N. L., Dontsov, A. E., Afanasiev, G. G., Ostrovsky, M. A., & Pelevina, I. I., (1990). Accumulation of lipid peroxidation products in eye structures of mice subjected to whole-body X-irradiation. *Radiobiology, 30*(1), 28–31 (in Russian).

59. Scalia, M., Geremia, E., Corsaro, C., Santoro, C., Baratta, D., & Sichel, G., (1990). Lipid peroxidation in pigmented and unpigmented liver tissues: Protective role of melanin. *Pigment Cell Res., 3*, 115–119.

60. Bilgihan, A., Bilgihan, M. K., Akata, R. F., Aricioglu, A., & Hasanreisoglu, B., (1995). Antioxidative role of ocular melanin pigment in the model of Lenz induced uveitis. *Free Radic. Biol. Med., 19*(6), 883–885.

61. Zemel, E., Loewenstein, A., Lei, B., Lazar, M., & Periman, I., (1995). Ocular pigmentation protects the rabbit retina from gentamicin-induced toxicity. *Invest. Ophthalmol. Vis. Sci., 36*(9), 1875–1884.

62. Corsaro, C., Scalia, M., Blanco, A. R., Aiello, I., & Sichel, G., (1995). Melanins in physiological conditions protect against lipo-peroxidation. A study on albino and pigmented *Xenopus. Pigment Cell Res., 8*, 279–282.

63. Dontsov, A. E., Fedorovich, I. B., Lindström, M., & Ostrovsky, M. A., (1999). Comparative study of spectral and antioxidant properties of pigments from the eyes of two *Mysis relicta* (Crustacea, Mysidacea) populations, with the different light damage resistance. *J. Comp. Physiol. B., 169*, 157–164.

64. Safa, R., & Osborne, N. N., (2000). Retinas from albino rats are more susceptible to ischemic damage than age-matched pigmented animals. *Brain Res., 862*, 36–42.

65. Sundelin, S. P., Nilsson, S. E., & Brunk, U. T., (2001). Lipofuscin-formation in cultured retinal pigment epithelial cells is related to their melanin content. *Free Radic. Biol. Med., 30*, 74–81.

66. Wang, Z., Dillon, J., & Gaillard, E. R., (2006). Antioxidant properties of melanin in retinal pigment epithelial cells. *Photochem. Photobiol., 82*, 474–479.

67. Insausti, T. C., Le Gall, M., & Lazzari, C. R., (2013). Oxidative stress, photodamage and role of screening pigment in insect eyes. *J. Exp. Biol., 216*, 3200–3207.

68. Feldman, T. B., Dontsov, A. E., Yakovleva, M. A., Fedorovich, I. B., Lindstrom, M., Donner, K., & Ostrovsky, M. A., (2008). Comparison of antioxidant systems in the eyes of two *Mysis relicta* (Crustacea: Mysidacea) populations, with different light damage resistance. *Sensory Systems, 22*(4), 309–316 (in Russian).

69. Zeise, L., Addison, R. B., & Chedekel, M. R., (1992). Bio-analytical studies of eumelanins. I. Characterizations of melanin the particle. *Pigment Cell Res., 5*(2), 48–53.

70. Zeise, L., Murr, B. L., & Chedekel, M. R., (1992). Melanin standard method – particle description. *Pigment Cell Res., 5*, 132–142.

71. Bridelli, M. G., Ciati, A., & Crippa, P. R., (2006). Binding of chemicals to melanins reexamined: Adsorption of some drugs to the surface of melanin particles. *Biophys. Chem., 119*(2), 137–145.
72. Crippa, P. R., Fornes, J. A., & Ito, A. S., (2004). Photophysical properties of pyrene in interaction with the surface of melanin particles. *Colloids and Surfaces B.: Biointerfaces, 35*, 137–141.
73. Feeney-Burns, L., Berman, E. R., & Rothman, H., (1980). Lipofuscin of human retinal pigment epithelium. *Am. J. Ophthalmol., 90*(6), 783–791.
74. Feeney-Burns, L., Hilderbrand, E. S., & Eldridge, S., (1984). Aging human RPE: Morphometric analysis of macular, equatorial, and peripheral cells. *Invest. Ophthalmol. Vis. Sci., 25*(2), 195–200.
75. Boulton, M., (1998). Melanin and the retinal pigment epithelium. In: Marmor, M. F., & Wolfensberger, T. J., (eds.), *The Retinal Pigment Epithelium: Function and Disease* (pp. 65–85). Oxford University Press, New York.
76. Feeney-Burns, L., (1980). The pigments of the retinal pigment epithelium. *Curr. Top. Eye Res., 2* 119–178.
77. Weiter, J. J., Delori, F. C., Wing, G. L., & Fitch, K. A., (1986). Retinal pigment epithelial lipofuscin and melanin and choroidal melanin in human eyes. *Invest. Ophthalmol. Vis. Sci., 27*, 145–152.
78. Rozanowski, B., Cuenco, J., Davies, S., Shamsi, F. A., Zadlo, A., Dayhaw-Barker, P., Rozanowska, M., Sarna, T., & Boulton, M., (2008). The phototoxicity of aged human retinal melanosomes. *Photochem. Photobiol., 84*, 650–657.
79. Feeney-Burns, L., Burns, R. P., & Gao, C. L., (1990). Age-related macular changes in humans over 90 years old. *Am. J. Ophthalmol., 109*, 265–278.
80. Rozanowska, M., Korytowski, W., Rozanowski, B., Skumatz, C. M., Boulton, M., Burke, J. M., & Sarna, T., (2002). Photoreactivity of aged human RPE melanosomes: A comparison with lipofuscin. *Invest. Ophthalmol. Vis. Sci., 43*(7), 2088–2096.
81. Sarna, T., Burke, J. M., Korytowski, W., Rozanowska, M., Skumatz, C. M., Zareba, A., & Zareba, M., (2003). Loss of melanin from human RPE with aging: Possible role of melanin photooxidation. *Exp. Eye Res., 76*(1), 89–98.
82. Kayatz, P., Thumann, G., Luther, T. T., Jordan, J. F., Bartz-Schmidt, K. V., Esser, P. J., & Schraermeyer, U., (2001). Oxidation causes melanin fluorescence. *Invest. Ophthalmol. Vis. Sci., 42*, 241–246.
83. Elleder, M., & Borovansky, J., (2001). Autofluorescence of melanin induced by ultraviolet radiation and near ultraviolet light. A histochemical and biochemical study. *Histochem. J., 33*, 273–281.
84. Borovansky, J., & Elleder, M., (2003). Melanosomes degradation: Fact or fiction. *Pigment Cell Res., 16*, 280–286.
85. Korzhova, L. P., Frolova, E. V., Romanov, Y. A., & Kuznetsova, I. A., (1989). Photo-induced destruction of DOPA-melanin. *Biochemistry, 54*, 992–998 (in Russian).
86. Zadlo, A., Rozanowska, M. B., Burke, J. M., & Sarna, T. J., (2006). Photobleaching of retinal pigment epithelium melanosomes reduces their ability to inhibit iron-induced peroxidation of lipids. *Pigment Cell Res., 20*, 52–60.
87. Dontsov, A. E., Sakina, N. L., Koromyslova, A. D., & Ostrovsky, M. A., (2015). Effect of UV irradiation and hydrogen peroxide on the antiradical and antioxidant activity of DOPA-melanin and melanosomes from retinal pigment epithelial cells. *Russian Chemical Bulletin, 7*, 1623–1628 (in Russian).

88. Dontsov, A. E., Sakina, N. L., & Ostrovsky, M. A., (2017). Loss of melanin by eye retinal pigment epithelium cells is associated with its oxidative destruction in melanolipofuscin granules. *Biochemistry, 82*(8), 916–924 (in Russian).
89. Boulton, M., Dontsov, A., Jarvis-Evans, J., Ostrovsky, M., & Svistunenko, D., (1993). Lipofuscin is a photoinducible free radical generator. *J. Photochem. Photobiol. B. Biol., 19*, 201–204.

CHAPTER 8

Blue Light and Retinal Aging: Morphofunctional Study of Retinal Pigment Epithelium and Choroid on Japanese Quail as an Accelerated Aging Model

PAVEL P. ZAK,[1] ALEXANDER E. DONTSOV,[1] LARISA S. POGODINA,[2] NATALIA B. SEREZNIKOVA,[1,2] and MIKHAIL A. OSTROVSKY[1]

[1]*Emanuel Institute of Biochemical Physics of Russian Academy of Sciences, Kosygin Str., 4, Moscow 119334, Russia,*
E-mail: natalia.serj@yandex.ru

[2]*Biological Department of Lomonosov Moscow State University, Lenin Hills, 1, Page 12, Moscow 119234, Russia*

ABSTRACT

The review of our retinal studies performed on Japanese quail *Coturnix japonica* in 2010–2017 years is presented. There are data on: (a) age-related morphological changes in retinal pigment epithelium (RPE) and choroid; (b) the dependence of the age-related changes in RPE and choroid on the spectral composition of everyday illumination; (c) evaluation of retinal structures age resistance to the photodamaged effect of blue light; and (d) study of RPE cellular activity photomodulation by low-dose light emitting diodes radiation. Methods of light-optical and electron microscopy, biochemical studies with the use of vital fluorescent dyes were applied. It is shown that pronounced age-related disorders are manifested in deformations of the RPE cell nuclei and the structures of the hemato-retinal barrier. The presumed causes of senile retinal disorders are genetic aging, as well as the accumulation of lipofuscin granules in RPE. There were adaptive responses to compensate for age-related disorders at the level of RPE mitochondria and choriocapillaris. The study on the dependence of the age-related changes on the spectral composition of daily illumination was performed with the use of blue (440–470 nm) and yellow (500–650 nm) light in the physiological range. In the data we obtained, blue everyday illumination compared to yellow light had a notable activation effect in an increase in the number of RPE mitochondria and choriocapillaris at a young age. The overall picture of RPE and choroid cells aging under blue and yellow light was approximately the same. In studies on blue light damage (440–460 nm, 4 J/cm^2), it was found that blue irradiation induced RPE cells of young animals' changes similar to the usual age-related state of RPE (deformation of the cell nuclei and basal processes and adaptive changes in the mitochondria shapes). Studies on the photomodulatory action of weak blue light (0.001–1 J/cm^2) on the cellular activity were carried out on a suspension of the RPE cells. We revealed that such illumination led to a marked activation of cellular and mitochondrial activity (estimated by an increase in the fluorescence of resazurin and tetramethylrhodamine ethyl ester), as well as to the increase in the number of RPE mitochondria (detected by electron microscopy). All the obtained results correlate with the data known for human.

8.1 INTRODUCTION

The Japanese quail (*Coturnix japonica*), as a fast aging organism, is a popular experimental animal model in studies of age and pathological retina processes [1–9]. For retinal studies, using this as the object presents a number of advantages compared to traditional laboratory rodents. Thus, like a human, the quail has a color vision, its retina has a mixed cone-rod assembly of photoreceptors [10], a central region of acute vision [11], and oxycarotenoid protection of the cone photoreceptors against light damage by blue light, similar to a human yellow spot [5, 6, 12, 13], as well as its own exceptionally powerful own retinal system for regulating and producing the hormone melatonin [14–16]. The cellular structure of the Japanese quail's retina is quite typical for the retina of vertebrates. Figure 8.1 shows a light-optical micrograph of the transverse section of the Japanese quail's retina with the adjacent RPE (retinal pigment epithelium) and the CHR (choroid).

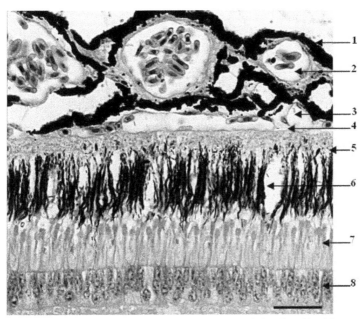

FIGURE 8.1 Cross-section of the outer retinal layers and the choroid of Japanese quail. Light microscopy (microscope Leica DM 1000, Leica Microsystems), staining with hematoxylin and eosin, scale bar is 20 μm (*Note:* 1 – melanocytes within the choroid; 2 – choroidal blood vessels filled with erythrocytes; 3 – choriocapillaris; 4 – Bruch's membrane; 5 – cell bodies of RPE; 6 – RPE melanin-containing apical processes surround outer segments of photoreceptors; 7 – inner segments of photoreceptors; 8 – outer nuclear layer (basal parts of photoreceptors with nuclei)).

In this case, the biological life of the quail is about one and a half years and, with a high degree of certainty, it can be assumed that both the general aging of the organism and retinal aging in the Japanese quail occur about 50 times faster than in humans. In this regard, the Japanese quail is a convenient experimental animal model for investigating the mechanisms and modeling of age-related retinal diseases [6–9, 12, 13, 17]. This review summarizes the results of our own studies, published in 2010–2017 by Russian journals [7–9, 18–24], regarding the interaction of retinal aging processes with the photobiological effect of visible light.

8.2 MATERIALS AND METHODOLOGY

The studies were performed on Japanese quail females beginning at 9 weeks of age for 78 weeks of age inclusive. The animals were kept in a daily light mode entailing 16 h of daylight and 9 h of night darkness. The following was carried out in four fields of study: (1) age-related morphological changes in RPE and CHR [18, 20]; (2) the dependence of the age-related changes in RPE and CHR on the spectral composition of everyday illumination [20, 21, 24]; (3) evaluation of retinal structures age resistance to the photodamaged effect of blue light [19]; and (4) study of RPE cellular activity photomodulation by low-dose LED (light emitting diode) radiation [22]. Methods of light-optical and electron microscopy were used. Methods of cellular biochemistry with the use of fluorescent dyes were applied in the section on LED photomodulation. Three age groups of animals were used: a juvenile group with an initial egg-laying capacity (age 9–25 weeks), middle-aged animals with maximum egg-laying capacity (30–40 weeks), and aged birds with a lowered egg-laying capacity (over 50 weeks). The subjects of the study were RPE and CHR females of the Japanese quail. RPE, in combination with the extracellular Bruch membrane, forms the hemato-retinal barrier and is a necessary life support system for the neural retina [25]. The stratified localization of subcellular structures with a parallel arrangement along the connective tissue of Bruch's membrane is characteristic for the quail RPE (Figure 8.2).

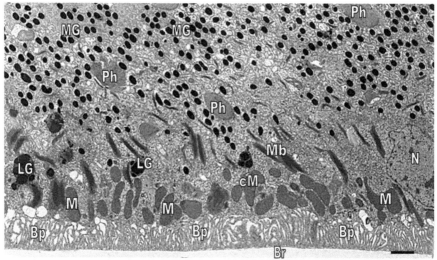

FIGURE 8.2 Ultrastructure of RPE cell of Japanese quail at 35 weeks. Transmission electron microscopy (microscope JEOL-1011, JEM), ultrathin section, scale bar is 1 μm (Modified from Ref. 19) (*Note:* N – nucleus, M – mitochondria, cM – changed forms of mitochondria (ring-shaped and dumbbell-like), Ph – phagosome, Mb – myeloid bodies, LG – lipofuscin granules, MG – melanin granules into apical cytoplasmic processes, Bp – basal processes, Br – Bruch's membrane).

The well-ordered basal processes of RPE cells directly adjoin the Bruch membrane, which multiplies the area of RPE contact with the extracellular space. Numerous basal processes of RPE carry out the active transport of metabolites between RPE and extracellular space, and the Bruch membrane performs the passive filtration of substances with respect to the circulatory capillaries of the choroid. As can be seen in the microphotography, there is a layer of numerous RPE mitochondria providing the energy costs of RPE in the region of the bases of the basal processes. Cell nuclei rounded at a young age, and deformed in older animals are located in the immediate vicinity of the basal membrane. Lipofuscin granules are located in the central thickness of the cytoplasm, being undigested metabolites and, at the same time, indicators of cellular aging. Phagosomes are closer to the photoreceptor side, the number of which makes it possible to evaluate the level of RPE phagocytic activity and shedding cells of the outer segments of photoreceptor cells. The photoreceptor side of RPE is filled with melanin granules located in the apical processes of RPE, which shields the bodies of RPE cells from excess light. Choriocapillaris of the circulatory system is located on the other side of the Bruch membrane, delivering oxygen and nutrients to the RPE and removing metabolic products.

8.3   RESULTS AND DISCUSSION

8.3.1   AGE-RELATED MORPHOLOGICAL CHANGES IN RETINAL PIGMENT EPITHELIUM (RPE) AND CHOROID (CHR)

The purpose of this section is to identify the most pronounced senile changes in the submicroscopic structures of the hematoretinal barrier. A common indicator of the aging of any animal cells is the "pigment of old age" – lipofuscin. Lipofuscin is a phototoxic "slag" of cellular metabolism [26, 27] and is one of the causes of cell aging. The specificity of RPE lipofuscin granules is their phototoxicity, based on the generation of free radicals contained in them by bisretinoids [27]. In human RPE, the accumulation of lipofuscin begins already at an early age, reaching a maximum level by about 50 years and does not change later [28]. According to our research [18], the process of lipofuscin granules' accumulation in Japanese quail RPE begins after 20 weeks of age; the amount of lipofuscin increases by 2–3 times by the age of 40 weeks, and, starting from 50 weeks of age, it reaches the saturation level, exceeding the original by 5–7 times (Figure 8.3).

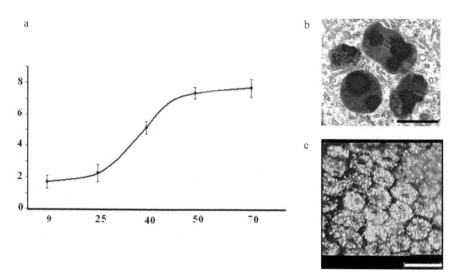

FIGURE 8.3   Lipofuscin granules in RPE cells of Japanese quail (*Note:* a – Accumulation of lipofuscin granules in RPE cells of Japanese quail during aging. The ordinate shows the number of lipofuscin granules per 100 μm^2 of the cytoplasm (numerical density), the abscissa shows the age of birds (in weeks). b –Lipofuscin granules in RPE cells of Japanese quail have different sizes and heterogeneous contents. Transmission electron microscopy (microscope JEOL-1011, JEM), ultrathin section, scale bar is 0.5 µm. c – Freshly isolated monolayer of RPE of Japanese quail was studied by the method of fluorescent microscopy (microscope Axiovert, Carl Zeiss), excitation wavelength – 450 nm. The hexagonal packing of cells filled with numerous luminous lipofuscin granules was visible. Scale bar is 50 µm).

The curve of lipofuscin accumulation shown in Figure 8.3 allows distinguishing three main age-related retinal processes: juvenile (less than 20 weeks old), middle-aged (30–40 weeks old) and senile (over 50 weeks old). Accordingly, our studies were conducted in a comparison of the results between these three age groups.

Figure 8.4 shows photomicrographs characterizing the state of RPE in young and old animals.

The most pronounced features of cell aging manifestation in Japanese quail RPE was the deformation of many cell nuclei with chromatin condensation, which, apparently, is associated with the genetic aging of DNA (deoxyribonucleic acid). The deformation of nuclei is an indirect sign of apoptosis [29]. Based on the results of TUNEL (terminal deoxynucleotidyl transferase dUTP nick end labeling) analysis with light microscopy, we found that about 4% of apoptotic cell nuclei are detected in the RPE of old birds, while in younger birds, such nuclei are not detected. In addition, destruction of the basal processes could be observed in the RPE cells of old animals, indicating a violation of the transport functions in RPE cell membrane. In a number of cases, the pattern of destruction of the basal processes resembled the process of druse formation, characteristic of the age-related macular degeneration of humans. Druses are pathological intercellular inclusions resulting from old age in the interval between the basal processes of RPE and the Bruch membrane. The process of druse formation is based on the development of age-related macular degeneration, the most serious eye disease. Earlier druse inclusions have also been described for RPE Japanese quail [3].

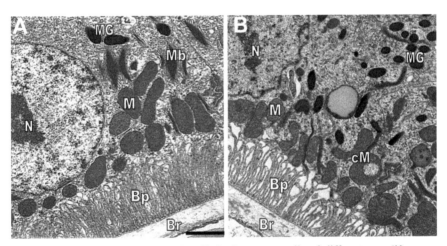

FIGURE 8.4 Ultrastructure of RPE cells in Japanese quails of different age (*Note:* a – young 9-week quails; b – old 55-week quails. Transmission electron microscopy (microscope JEOL-1011, JEM), ultrathin section, scale bar is 1 μm. The characteristic signs of aging are deformation of the cell nuclei (N), disorganization of the basal processes (Bp), the frequent presence of changed forms of mitochondria (cM), the appearance of the heterogeneous inclusions in the Bruch's membrane (Br). M – mitochondria, Mb – myeloid body, MG – melanin granules into apical cytoplasmic projections).

Along with the destruction of the basal processes in old birds, one could observe a violation in the homogeneity of the Bruch membrane with the appearance of fine-dispersed inclusions of the vesicular and granular character (Figure 8.5).

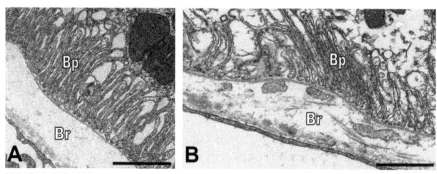

FIGURE 8.5 Ultrastructure of basal processes of RPE cells and Bruch membrane in Japanese quails of different age [Transmission electron microscopy (microscope JEOL-1011, JEM), ultrathin section, scale bar is 1 μm] (*Note:* a – young 9-week quails; b – old 55-week quails. Bp – basal processes, Br – Bruch's membrane).

Similar phenomena are described in the retina of mammals, including humans [30]. They are probably associated with age-related degradation of the extracellular matrix, the accumulation of lipids and fatty acids in it [31]. In general, senile changes in the basal processes of RPE and the Bruch membrane indicate a violation of the transport of substances between the RPE and the blood vessels of the chorus. The plate and vesicular inclusions in the Bruch membrane of the old birds observed are similar in structure to the components of "soft" human friends [30] and, possibly, can serve as precursors of drusiform deposits. It is believed that the formation of drusen results from the disruption in the normal functioning of RPE cells, in which the transfer of various material accumulating in RPE cells to the Bruch membrane is due to its insufficient elimination [32]. In humans, a correlation was found between the density of these deposits and the degree of RPE cells degeneration and photoreceptors [33].

Along with the age-related disorders of the hemato-retinal barrier in the eyes of Japanese quail, we found that aging is accompanied by a decrease in the phagocytosis level: the volume of phagocytic mass in RPE decreased in three times in old age compared to a young age with the phagosomal number decreasing by two times (Figure 8.6).

The reduction of phagosomal and lysosomal activities during aging is shown for RPE of rodents [34] and, in the authors' opinion, this is associated with the development of oxidative stress. In general, the observed age-related changes indicate a worsening of the metabolic state of RPE cells and fit into the idea that the aging of RPE involves genetic aging, in combination with the deterioration of the transporting structures of the hemato-retinal barrier.

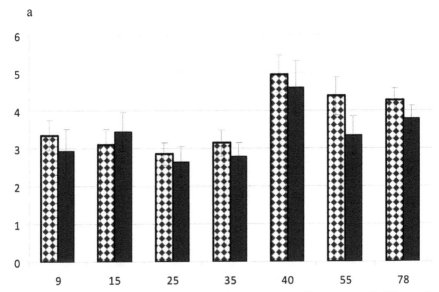

FIGURE 8.6 Age changes in the content of RPE phagosomes of Japanese quails. The results of the morphometric analysis of RPE cells ultrathin sections are shown. The ordinate shows the number of organelles per 100 μm² of the cytoplasm (numerical density, hatched columns) and the volume occupied by the organelles per unit volume of the cytoplasm (relative volume, in%, black columns); the abscissa shows the age of birds (in weeks). The data are presented as mean values, M ± m.

We have studied age-related changes in the mitochondrial apparatus in detail. As it turned out, the mitochondrial mass in the RPE cells of the Japanese quail remained at practically the same level throughout life. However, in old age, the specific fraction of changed mitochondria (ring-shaped and dumbbell-like) increased by approximately two times (Figure 8.7).

According to literature data on a variety of animal cells, such mitochondria are formed in conditions of metabolic stress [35] and provide the highest level of cellular metabolism [36], presumably due to an increase in the surface of contact with the cytoplasm compared to conventional round mitochondria. The age-related increase in the number of ring-shaped mitochondria was also observed in RPE chicken cells [37]. They are also described with regards to aging in the neurons of a monkey brain [38]. In human RPE, such mitochondria were observed with gyrate retinal atrophy [39]. By our assumption, these age-related mitochondrial changes in RPE cells are aimed at counteracting the processes of retinal aging.

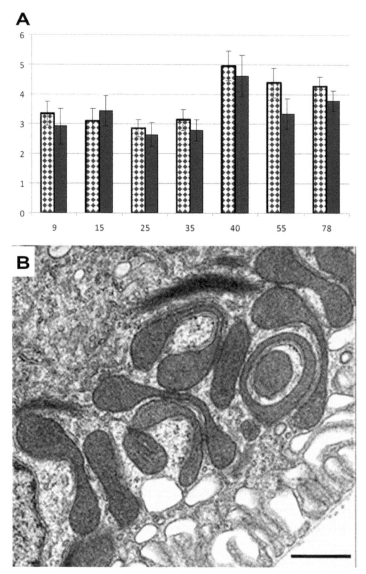

FIGURE 8.7 Age changes in the content of RPE mitochondria of Japanese quails. Transmission electron microscopy (microscope JEOL-1011, JEM), ultrathin section, scale bar is 1 μm. (A) The results of morphometric analysis of RPE cells ultrathin sections are shown; the hatched columns indicate numerical density of mitochondria while the black columns indicate relative volume of mitochondria. Age groups of birds are displayed on the abscissa (in weeks). The results are expressed as mean values, M ± m. (B) Ultrastructure of the basal part of the RPE cell of old 55-week Japanese quail, showing the appearance of changed forms of mitochondria (ring-shaped and dumbbell-like) and destruction of basal processes.

We have also found that changes in CHR vessels are added to the age-related disorders of the hemato-retinal barrier. According to light-optical microscopy, it was found that the thickness of the CHR in old birds was one-third less than in young birds [20]. This correlates with the literature data that the thickness of the human CHR also decreases with aging [40]. Detailed morphometric studies were carried out to measure the total number of the choriocapillaris and to reveal the proportion of active open capillaries (Figure 8.8), which showed that the total number of capillaries remained virtually unchanged throughout life. However, at a young age, about 30% of choriocapillaris are in a closed non-working state, but gradually become active with aging (Figure 8.8).

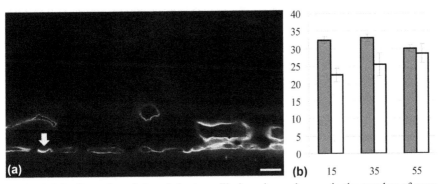

FIGURE 8.8 Detection of closed choriocapillaris and age changes in the number of open capillaries of Japanese quails. (*Note:* a – Detection of closed choriocapillaris in a common pool by staining the retinal sections with lectin *Maclura pomifera*, conjugated with FITC (fluorescein isothiocyanate), revealing specific luminescence of the endothelium of blood vessels and choriocapillaris. Vessels of the choroid of Japanese quail (15 weeks old). The arrow indicates a closed choriocapillary, and the rest choriocapillaris are open. Fluorescence microscopy (microscope Axiovert, Carl Zeiss), scale bar is 10 μm. b – Age changes in the number of the open choriocapillaris (involved in the bloodstream) in comparison with the constant total number of choriocapillaris in the choroid of Japanese quail. The ordinate shows the number of the choriocapillaris (gray columns – number of all choriocapillaris, white columns – number of open choriocapillaris); the abscissa shows the age of birds (in weeks)).

On the results of electron microscopy, we have established that the number of fenestrae and transendothelial channels (Figure 8.9) in choriocapillaris contacting the Bruch membrane, i.e., in the structures providing transport functions in relation to RPE, decreases nearly by two times as the quail ages.

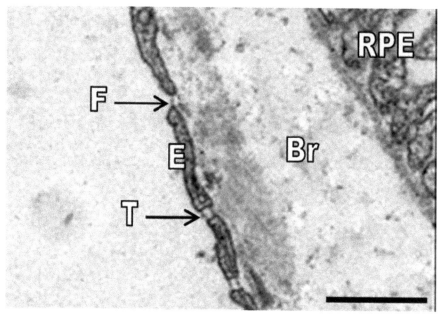

FIGURE 8.9 Ultrastructure of the endothelial wall of the choriocapillaris bordering Bruch's membrane [Transmission electron microscopy (microscope JEOL-1011, JEM), ultrathin section, scale bar is 0.5 µm. E – endothelium, F – fenestra, T – transepithelial channel, Br – Bruch's membrane, RPE – retinal pigment epithelium].

It can be assumed that these losses of transport functions are compensated, to some extent, by an increase in the number of open choriocapillaris. Our data on the age-related changes in the fenestration of the choriocapillaris are similar to the literature data known for human eyes about the decrease in the degree of choriocapillaris' fenestration during the development of senile macular degeneration [41].

In general, the retinal aging data obtained in Japanese quail correlate in many ways with those known for humans and other vertebrates, which supports the use of Japanese quail *Coturnix japonica* as an experimental animal model for the study of retinal aging according to an accelerated time regime. The data obtained make it possible to conclude that the most pronounced senile retinal disorders are related to the state of the RPE cell nuclei and the hemato-retinal barrier.

8.3.2 *DEPENDENCE OF AGE-RELATED CHANGES IN RETINAL PIGMENT EPITHELIUM (RPE) AND CHOROID (CHR) ON THE SPECTRAL COMPOSITION OF DAYLIGHT ILLUMINATION*

The purpose of the studies was to evaluate the possible danger of everyday LED lighting. Modern illuminating LEDs of cold and neutral white light with a correlated color temperature above 4500 K have two spectral emission bands in the blue band 440–460 nm and the yellow area 500–650 nm. Now, the issue of the possible danger of excessive blue component of LED lighting [42–45] is being widely discussed in scientific and technical literature. The main risk factors, according to literature data, are the toxicity of RPE lipofuscin granules [27], as well as the disturbance of the daily production of melatonin [15], which is controlled by the blue-sensitive melanopsin ganglion cells of the retina [46]. The groups of increased visual risk in relation to the blue component of lighting include young and old people [42, 44, 45]. For young people, the increased risk is based on the high transparency of the eyes in the blue region of the spectrum and, for the elderly, it is based on the high content of phototoxic lipofuscin in RPE cells. In this regard, the issue of introducing age-related adjustments to light safety standards for artificial light sources is being discussed [45, 47–49]. Estimates of age sensitivity to daylight are not amenable to any lighting calculations and require additional studies in experimental animals. At the same time, studies on primates are poorly acceptable for ethical and financial reasons, and also due to their long biological life span. The retinas of rats and mice lack a number of required components, such as the presence of cones, the central zone of acute vision, and the photoprotection of cones with oxycarotenoid antioxidant action. In this regard, the Japanese quail seems to be a more adequate animal model for research of this kind. Just like the human retina, the retina of Japanese quail, along with rods, contains cones of SWS (short wavelength sensitive), MWS (middle wavelength sensitive) and LWS (long wavelength sensitive) types [10]. The quail has a central zone of acute vision similar to the macular area (yellow spot) of the human retina [11]. Similar to the cones of the yellow spot of the human retina, the cones of Japanese quail contain photoprotective oxycarotenoids of antioxidant activity [5–9, 12, 13]. In this connection, we have conducted a study of RPE cells state and choroids of uneven-aged Japanese quails in two types of everyday LED lighting: blue LED lighting with a wavelength of 450 nm, and yellow illumination in the spectral band 500–650 nm, obtained with white LEDs combined with a cut-off filter for wavelengths shorter than 500 nm. Both light sources were aligned in energy:

0.002 W/cm² in the center of the bird's cage (50 cm) which corresponded to 200 lx illumination from a 50-watt incandescent lamp, i.e., in the physiological limits of illumination. The daylight regime consisted of 15 h of lighting and 9 h of darkness. The studies were conducted at three age points: 15, 35, and 55 weeks. Birds were kept for life for different types of lighting, starting at 6 weeks old.

Electron microscopic studies showed that both groups of birds do not differ in the state of RPE cell nucleus, basal processes, and Bruch's membrane, which indicated the absence of obvious signs of photodamage. Light microscopy by the method of TUNEL found that in both groups of animals' apoptotic nuclei are equally present (7% of all cell nuclei) only in old quails. However, morphofunctional differences in the content of phagosomes, lipofuscin granules, mitochondria in RPE cells, and also in the number of the choriocapillaris, were observed between the animals of the "blue" and "yellow" groups. These data are summarized in Figures 8.10 and 8.11.

Thus, in birds of the "blue" group, regardless of age, the content of phagosomes in RPE cells was significantly lower than in birds grown under yellow light (Figure 8.10a). A possible cause of these differences may be the minor portion of the blue cones [10] in the general photoreceptor retinal pool, and hence the formation of a phagosomal mass in blue illumination can be reduced.

The final stage of the phagosomal transformation is lipofuscin granules. The lipofuscin content in RPE is increased gradually with aging: by the age of 55 weeks, its specific volume increased by 6–7 times compared with the age of 15 weeks. In young and middle-aged birds, the "blue" and "yellow" groups did not differ in lipofuscin content. At the same time, in older animals (55 weeks) of the "blue" group, the lipofuscin content in RPE cells turned out to be one and a half times lower than under yellow light (see Figure 8.10b). This fact is rather unexpected and requires further study. Based on these data on the measurements of lipofuscin and phagosomes, it can be concluded that other things being equal, the blue component of the LED lighting creates a lesser load on the RPE intracellular digestion system.

A comparison of mitochondria content in RPE cells showed that the daily blue illumination in comparison with the yellow one, increased the number of all mitochondria (see Figure 8.10c), including changed mitochondria (see Figure 8.10d), most significantly in young birds. In our opinion, this indicates the activating effect of blue light on mitochondrial activity. For changed mitochondria, it is known that, in comparison with mitochondria of the usual form, they provide increased production of reactive oxygen species and higher levels of respiration and energy metabolism [36].

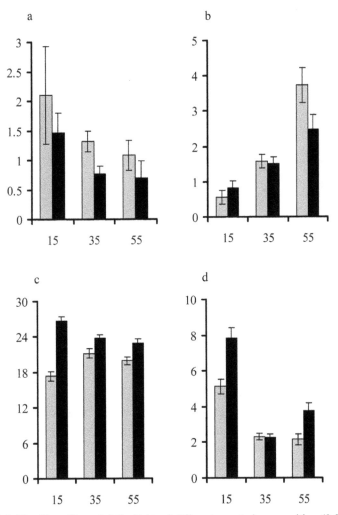

FIGURE 8.10 The effect of daily light of different spectral composition (0,002 W/cm²) on the age-related changes in the RPE of Japanese quail. The data of the morphometric analysis of RPE ultrathin sections are presented (*Note:* a – relative volume of phagosomes, b – the relative volume of lipofuscin granules, c – numerical density of all mitochondria, d – numerical density of changed mitochondria. The ordinate shows the volume occupied by the organelles per unit volume of the cytoplasm (relative volume, in%) or the number of organelles per 100 μm² of the cytoplasm (numerical density); the abscissa shows the age of birds (in weeks). Black columns – illuminated with blue light (λ = 450–470 nm), gray columns – illuminated with yellow light (λ = 500–650 nm)).

a
b

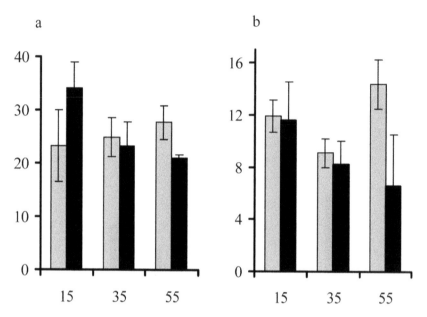

FIGURE 8.11 The effect of daily light of different spectral composition (0,002 W/cm²) on the age-related changes in the choroid of Japanese quail. The data of morphometric analysis of light microscopic retinal sections are presented (*Note:* a – numerical density of open choriocapillaris, b – numerical density of choroid vessels. The ordinate shows the number of structures per unit length of the Bruch membrane, and the abscissa shows the age of birds (in weeks). Black columns – illuminated with blue light ($\lambda = 450–470$ nm), gray columns – illuminated with yellow light ($\lambda = 500–650$ nm)).

Thus, we have shown that the blue component of LED lighting causes the pronounced adaptive response of RPE cells, mainly at a young age, manifested in an increase in the number of mitochondria, a change in their shape and structure. At the same time, older-age birds of the blue group show signs of a decrease in phagocytic activity and the formation of lipofuscin granules.

We also found age differences between the animals of the blue and yellow groups in the state of CHR vessels. Due to light microscopy, it was found that the total thickness of the CHR in 52-week-old birds kept under blue illumination was 30% higher than in the "yellow" group birds, while no differences were observed in young and middle-aged birds [20]. The increase in the thickness of CHR was due to an increase in the volume of the connective tissue part of CHR. In measurements of the number of the open choriocapillaris, it was found that the number of open capillaries increases slightly with aging in young birds of the yellow group, while, in young birds of the blue group the number of open choriocapillaris is maximal and gradually decreases with age to normal values (see Figure 8.11a). The number of hydrophilic channels estimated by electron microscopy (fenestrae and transendothelial channels) in the walls of choriocapillaris was the same for all ages of birds in the blue and yellow groups.

The number of CHR vessels at a young age was equally high in birds of the blue and yellow groups, but decreased significantly with aging in the blue group, while the number of vessels remained at a high level in the yellow group (see Figure 8.11b). In general, these results on the study of changes in CHR show that blue daylighting at a young age causes the intensification of blood flow in the choriocapillaris, whereas, through aging, the blood flow is apparently reduced due to a decrease in the number of vessels.

In the data we obtained, the most pronounced specific effect of blue illumination was an increase in the number of mitochondria in RPE and choriocapillary cells in CHR in the youngest group of animals. These effects do not entail photodamage and rather testify to the photoactivation of retinal processes. The RPE cell bodies and the CHR are well protected from light by melanin granules, and the illumination applied was very low power. Therefore, it is difficult to assume that the observed effects of photoactivation are based on the direct action of light on the structures studied. These processes most likely involve retinal melatonin, as a regulator of mitochondrial activity and growth factors of blood vessels. According to our unpublished data, the melatonin content in the neural retina of quail is 100 times higher than in the blood plasma and in the vitreous body, both at night and daytime. At the same time, the retinal content of melatonin appeared to be 30% higher in birds of the yellow group than in the birds of the blue group (Figure 8.12).

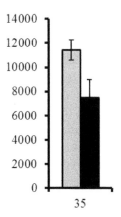

FIGURE 8.12 The effect of daily light of different spectral composition (0.002 W/cm^2) on the content of melatonin in the retina of Japanese quail.

The ordinate shows the content of melatonin (in pg/ml), the abscissa shows the age of birds (in weeks). Gray column – illuminated with yellow light (λ = 500–650 nm), black column – illuminated with blue light (λ = 450–470 nm).

As is known, melatonin is a universal regulator of cellular processes and is able to freely penetrate through cell membranes [50]. It is fairly well known that, at the level of the whole organism, blue illumination (450–470 nm), interfering with the development of melatonin, activates the general vital activity at the expense of maintaining the normal daily rhythm of metabolic processes. In experiments with mice knocked out by the melanopsin gene (photopigment of the so-called melanopsin ganglion cells, involved in the regulation of circadian rhythms and providing suppression of the epiphyseal melatonin), the key role of the melanopsin signaling pathway and melanopsin-containing retinal cells were demonstrated in the light-dependent enhancement of the choroidal blood flow [51]. Melanopsin receptors present both on RPE cells and on the endothelium of CHR [52] are most efficiently excited by visible light in the 420–450 nm region [46], which corresponds to the spectral composition of the light used in the experiments. Thus, both changes in RPE cells, as well as, the involvement of functional reserve choriocapillaris in the total blood flow with daily exposure to low-intensity blue light, can be the result of this system' activation.

In general, the obtained data specifically indicated the activating effect of blue daylight on age-related retinal processes. The most significant detected age-related disorders due to blue LED lighting includes a significant 30% increase in the choroid's thickness and a two-fold decrease in the number of blood vessels of the choroids.

8.3.3 DAMAGING EFFECT OF BLUE LIGHT ON RETINAL STRUCTURES OF JAPANESE QUAILS OF DIFFERENT AGES

The danger of blue light sources of illumination for vision is widely known and is taken into account in the standards for laser radiation safety [48]. In a variety of mammalian species, it has been shown that light in the blue visible range at doses exceeding 10 J/cm^2 of the surface of the cornea with exposures up to 1 h causes photochemical retinal photodamage [43, 53, 54], developing 1–2 days after the accumulation of toxic products [27, 55]. Retinal photochemical damage acceptors are phototoxic bisretinoids of RPE lipo-fuscin granules and mitochondrial cytochrome C oxidase of synaptic retinal layers and RPE [55–57]. Virtually the studies on photochemical damage by blue light are performed without taking the possible age specificity into account. The purpose of this section of the research was to assess the age resistance of subcellular retinal structures to the photodamaging effect of blue light. In these experiments, the eyes of whole animals were irradiated with a blue LED (λ = 450–470 nm) in sublethal photodamage doses of 4 J/cm^2 of the corneal surface at 40 min of exposure. The results of the irradiation were evaluated after 24 h, in comparison with the control non-irradiated eye. Comparisons were made between the data obtained in juvenile animals (9 and 25 weeks), animals of middle age (40 weeks) and old animals (52 weeks).

According to the data obtained with conventional light microscopy, the result of blue irradiation was an increase in the size of the RPE cell bodies, as compared to paired unirradiated eyes. Thus, in 25-week-old quails, the thickness of the RPE layer increased by 44%, and by 25% in 40-week-old quails and 13% in 52-week-old birds [19]. Changes in the RPE monolayer thickness as a result of light irradiation indicate swelling of the cells, and this effect was more pronounced in young than in older birds. With a high degree of probability, it can be assumed that the light-induced swelling of RPE cells is a consequence of damage to the hemato-retinal barrier. A similar chain of these events is described in the literature for rabbit eyes, where the examination of the fundus revealed the edema of the retina accompanying subcellular photo induced disturbances in RPE cells [58]. The electron microscopic study carried out showed that irradiation with a blue light at the level of the photoreceptors caused the destruction of mitochondrial cristes of stick ellipsoids, while the structure of the cone mitochondria, protected from light by oxycarotinoid oil droplets (OD), did not differ from the control object (Figure 8.13). According to the literature, the cones of the Japanese quail are protected from light damage by oxycarotenoid OD, while such droplets are absent [5].

At the RPE level, the response of cells to radiation was significantly different in different age groups. The most pronounced changes could be observed in the RPE cells of juvenile animals, while the overall picture of these photo induced changes was extremely similar to the structural changes observed in normal aging (Figure 8.14).

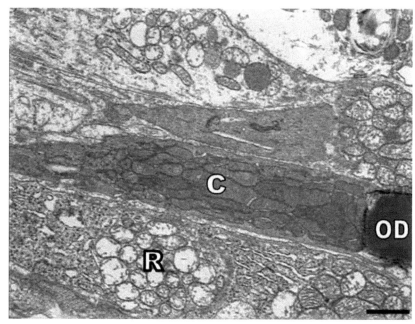

FIGURE 8.13   Blue light damage (λ=450–470 nm, 4 J/cm^2 of the cornea) to rods mitochondria (R) but not cones mitochondria (C) due to their oil droplets (OD). Transmission electron microscopy (microscope JEOL–1011, JEM), ultrathin section, scale bar is 1 μm (Modified from Ref. 19).

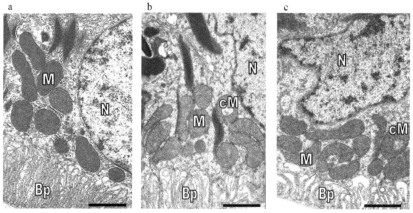

FIGURE 8.14   Blue light action (λ = 450–470 nm, 4 J/cm^2 of the cornea) on the RPE ultrastructure (*Note:* a – RPE of the unirradiated eye of young 9-week quail, b – RPE of the paired irradiated eye of young 9-week quail, c – RPE of the unirradiated eye of old 52-week quail. Transmission electron microscopy (microscope JEOL-1011, JEM), ultrathin section, scale bar is 1 μm. N – nucleus, M – mitochondria, cM – changed forms of mitochondria (ring-shaped and dumbbell-like), Bp – basal processes).

Thus, blue irradiation led to the appearance of deformed cell nuclei with increased condensation of chromatin, which can be evidence of the cells' apoptotic state. Indeed, our light-microscopic data of TUNEL analysis showed that about 4% of RPE cells in the irradiated eyes of young birds are in a state of apoptosis. The irradiation of the eyes of juvenile animals also resulted in the deformation of the basal processes, in combination with the appearance of heterogeneous micro inclusions in the Bruch's membrane. In general, the morphological picture of irradiated eyes' RPE cells of photo-induced disorders in young animals, visually, insignificantly differed from the picture of the usual senile state of RPE animals, without any irradiation. For older animals (over 40 weeks), the additional effects of light damage were poorly expressed against the background of age changes already taking place. The results of the morphometric analysis of RPE cells after blue irradiation are summarized in Figure 8.15.

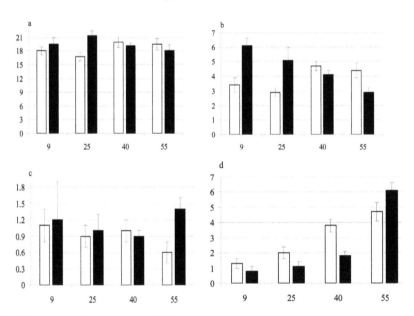

FIGURE 8.15 The reaction of RPE of Japanese quails of different ages a day after a single short-term blue light exposure (λ = 450–470 nm, 4 J/cm^2 of the cornea). The data of morphometric analysis of the RPE ultrathin sections are presented (*Note:* a – numerical density of all mitochondria, b – numerical density of changed mitochondria, c – numerical density of phagosomes, d – numerical density of lipofuscin granules. The ordinate shows the number of organelles or granules per 100 μm^2 of the cytoplasm (numerical density), the abscissa shows the age of birds (in weeks). White columns – control: paired unirradiated eyes, black columns – irradiated eyes.

The number of phagosomes and the total phagocytic mass in RPE cells of birds of different ages did not exert a noticeable influence on blue irradiation (see Figure 8.15c). At the same time, blue irradiation led to a 2-fold decrease in the number of lipofuscin granules (see Figure 8.15d) in young and middle-aged birds when the total lipofuscin mass was conserved. In other words, blue irradiation resulted in the aggregation of small granules into larger ones. The increase in the size of the granule is accompanied by a geometric decrease in its contact surface with the cytoplasm, which, in principle, is a positive factor limiting the rate of diffusion of free radicals from the granule into the cytoplasm.

The total number of mitochondria remained at approximately at the same level, regardless of age and irradiation. At the same time, the fraction of changed mitochondria of RPE irradiated eyes of young animals doubled in relation to paired unirradiated eyes. In middle-aged animals, these values were identical and, in older animals, the fraction of modified mitochondria in irradiated eyes decreased by one-third. Data on mitochondrial changes under the influence of light irradiation are shown in Figure 8.15a, b. As follows from the figure, the ability of RPE *C. japonica* cells to react to irradiation with an increase in the number of modified mitochondria linearly decreases with the age of birds.

In general, both the actual damaging effect which manifests itself in the death of photoreceptor mitochondria of rods and in deformation of nuclei and basal processes of RPE, and adaptive responses of RPE cells by way of an increase in the number of modified mitochondria and enlargement of lipofuscin granules, could be identified in the observed changes. These adaptive reactions were well expressed in young and middle-aged birds, but were practically not observed in the RPE of old animals.

The irradiation doses used in the 450–470 nm band corresponded to the known threshold doses of retina photo damage for a variety of mammals: from laboratory rodents to rhesus macaques [43]. Accordingly, it has shown that the Japanese quail *C. japonica* can be considered as an adequate test object for comparative evaluation of biosafety of artificial light sources in the blue region of the spectrum, i.e., in relation to the function B (λ) of the international requirements for the safety of artificial light sources [47].

8.3.4 EVALUATION OF RETINAL PROCESSES PHOTOMODULATION BY LOW-DOSE LIGHT-DIODES ILLUMINATION OF THE BLUE BAND

The photostimulating effect of light on regenerative processes is a popular trend in ophthalmic retinal therapy. According to existing data, low-dose illumination in the blue and far red region is able to activate the cellular activity of a wide variety of organs, and the light of the medium-wavelength green-yellow band does not exibit such an activating action [59–61]. It is assumed that the mechanism of this activation is based on the selective absorption of light in the blue and red regions by enzymes of the mitochondrial respiratory chain [59, 62, 63]. At high doses, these enzymes die [64], but at low doses, they are presumably activated [59, 65]. The activating effect of red light is widely used in therapeutic ophthalmology [66–68]. In the development of studies on the possible activating effect of blue light on retinal processes, short-term 15- to 30-minutes irradiation of RPE quail cells was analyzed with measurements of cellular and mitochondrial activity.

In our experiments [22], we tried to evaluate the possible activation effect of low-dose blue irradiation of RPE with the prospect of its use in therapeutic ophthalmology. The studies were carried out on a cell suspension of RPE Japanese quail, obtained from freshly insulated tissue by soft homogenization with the microscopic control of the homogenate composition. The light irradiation of RPE cells was carried out under blue (450 nm) and red (630 nm) illumination in comparison with the dark control, with stirring on a thermostable magnetic stirrer. The optical path from the surface of the solution to the bottom was 2 cm. We used irradiation doses from 1.0 to 0.001 J/cm^2 on the surface of the sample. The data obtained were normalized with respect to the amount of protein in the sample. The viability of RPE cells was evaluated by binding a specific resazurin dye (an indicator of the total metabolic activity), and the level of the mitochondrial potential was estimated by the potential-dependent mitochondrial marker TMRE (tetramethylrhodamine ethyl ester) fluorescence. In these experiments, it was found that the low-dose blue light used (below the known thresholds for photodamage of the retina) increased the mitochondrial potential and the viability of irradiated RPE cells by about 30% (Figure 8.16).

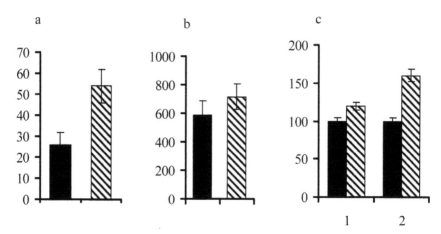

FIGURE 8.16 Influence of short-term low-dose blue light (λ = 450–470 nm, 0.001–1.0 J/ cm² of cornea) on the metabolic state of RPE mitochondria (*Note:* a – The results of fluorescent biochemical analysis of the RPE total metabolic activity. The ordinate shows fluorescence of resazurin (in relative units). b – The results of fluorescent biochemical analysis of the level of RPE mitochondrial potential. The ordinate shows fluorescence of TMRE (in relative units). c – The results of morphometric analysis of RPE mitochondria in birds after light exposure. The ordinate shows the number of mitochondria per 100 μm² of the cytoplasm or the volume occupied by mitochondria per unit volume of the cytoplasm. The abscissa shows: 1 – numerical density, 2 – specific volume. Hatched columns – illuminated with blue light (λ = 450–470 nm), black columns – dark control).

The effects of red irradiation were less pronounced. One of the possible causes of the photo-induced increase in mitochondrial activity may be an increase in the number of mitochondria. Indeed, according to our electron microscopic data, the illumination of the eyes of the Japanese quail under in vivo conditions at equally low doses resulted in a 30% increase in the mitochondrial content with respect to the dark control (Figure 8.16c). The electron microscopic literature data from other authors indicate that low-dose red irradiation does cause mitochondrial fission, leading to an increase in their numbers [59].

In general, the data obtained testify to the prospects of continuing research on the possible use of low-dose blue irradiation in ophthalmic therapy.

8.4 CONCLUSION

In general, our practice of research on Japanese quail in combination with the analysis of ophthalmologic and gerontological literature shows that this animal is a unique object for predicting the functional state of RPE and CHR cells during the aging process and in regards to light exposure.

ACKNOWLEDGMENTS

The authors thank DSc. Tamara S. Gurieva, DSc. Alexander E. Dontsov, Alina O. Sigaeva and the personnel of Laboratory of Electron Microscopy of Lomonosov Moscow State University and Clinic of New medical technologies "Arhi-Med" for technical assistance.

The Russian Foundation supported this study for Basic Research, grants No.14–04–01072, 15–29–03865, 17–04–00708.

KEYWORDS

- **age-related disorders**
- **choriocapillaris**
- **electron microscopy**
- **hemato-retinal barrier**
- **mitochondria**
- **photodamage**
- **photomodulation**
- **retina**
- **shortwave radiation**

REFERENCES

1. Fite, K. V., Bengston, L., & Donaghey, B., (1989). Aging and sex-related changes in the outer retina of Japanese quail. *Experimental Eye Research, 8*(10), 1039–1048.
2. Fite, K. V., Bengston, L., & Donaghey, B., (1993). Experimental light damage increases lipofuscin in the retinal pigment epithelium of Japanese quail (Coturnix japonica). *Experimental Eye Research, 57*(4), 448–453.
3. Fite, K. V., Bengston, L., & Cousins, F., (1994). Drusen-like deposits in the outer retina of Japanese quail. *Experimental Eye Research, 59*(4), 417–424.
4. Mizutani, M., (2002). Establishment of inbred strains of chicken and Japanese quail and their potential as animal models. *Experimental Animals, 51*(5), 417–429.
5. Thomson, L. R., Toyoda, Y., Langner, A., Delori, F. C., Garnett, K. M., Craft, N., Nichols, C. R., Cheng, K. M., & Dorey, C. K., (2002). Elevated retinal zeaxanthin and prevention

of light-induced photoreceptor cell death in quail. *Investigative Ophthalmology & Visual Science*, *43*(11), 3538–3549.

6. Toomey, M. B., & McGraw, K. J., (2007). Modified saponification and HPLC methods for analyzing carotenoids from the retina of quail: Implications for its use as a nonprimate model species. *Investigative Ophthalmology & Visual Science*, *48*(9), 3976–3982.

7. Zak, P. P., Zykova, A. V., Trofimova, N. N., Abu Khamidakh, A. E., Fokin, A. I., Eskina, E. N., & Ostrovsky, M. A., (2010). The experimental model for studying of human age retinal degeneration (Japanese quail *C. Japonica*). *Reports of Biological Sciences*, *434*(1), 297–299.

8. Zak, P. P., Zykova, A. V., Trofimova, N. N., & Ostrovsky, M. A., (2013). Japanese quail (*Coturnix japonica*) as a model of accelerated aging of human retinal processes. Report 1. Assessment of the retinal pigment epithelium lipofuscin accumulation depending on the retinal amount of oxycarotenoids. *Ophthalmosurgery*, *1*, 9–12.

9. Yakovleva, M. A., Zykova, A. V., Trofimova, N. N., Ostrovsky, M. A., Zak, P. P., Arbuhanova, P. M., & Borzenok, S. A., (2013). Japanese quail (*Coturnix japonica*) as a model of accelerated aging of human retinal processes. Report 2. Comparative analysis of the composition of retinoids isolated from retinal pigment epithelium of human and Japanese quail eyes. *Ophthalmosurgery*, *2*, 47–51.

10. Bowmaker, J. K., Kovach, J. K., Whitmore, A. V., & Loew, E. R., (1993). Visual pigments and oil droplets in genetically manipulated and carotenoid deprived quail: A microspectrophotometric study. *Vision Research*, *33*(5/6), 571–578.

11. Lee, J. Y., Holden, L. A., & Djamgoz, M. B., (1997). Effects of aging on spatial aspects of the pattern electroretinogram in male and female quail. *Vision Research*, *37*(5), 505–514.

12. Khachik, F., De Moura, F. F., Zhao, D. Y., Aebischer, C. P., & Bernstein, P. S., (2002). Transformations of selected carotenoids in plasma, liver, and ocular tissues of humans and in nonprimate animal models. *Investigative Ophthalmology & Visual Science*, *43*(11), 3383–3392.

13. Bhosale, P., Serban, B., & Bernstein, P. S., (2009). Retinal carotenoids can attenuate the formation of A2E in the retinal pigment epithelium. *Archives of Biochemistry and Biophysics*, *483*(2), 175–181.

14. Steele, C. T., Tosini, G., Siopes, T., & Underwood, H., (2006). Timekeeping by the quail's eye: Circadian regulation of melatonin production. *General and Comparative Endocrinology*, *145*(3), 232–236.

15. Tosini, G., Baba, K., Hwang, C. K., & Iuvone, P. M., (2012). Melatonin: An underappreciated player in retinal physiology and pathophysiology. *Experimental Eye Research*, *103*, 82–89.

16. Blasiak, J., Reiter, R. J., & Kaarniranta, K., (2016). Melatonin in retinal physiology and pathology: The case of age-related macular degeneration. *Oxidative Medicine and Cellular Longevity*, 1–12.

17. Toyoda, Y., Thomson, L. R., Langner, A., Craft, N. E., Garnett, K. M., Nichols, C. R., Cheng, K. M., & Dorey, C. K., (2002). Effect of dietary zeaxanthin on tissue distribution of zeaxanthin and lutein in quail. *Investigative Ophthalmology & Visual Science*, *43*(4), 1210–1221.

18. Seryoznikova, N. B., Zak, P. P., Pogodina, L. S., Lipina, T. V., Trofimova, N. N., & Ostrovsky, M. A., (2013). Subcellular aging markers of Japanese quail *Coturnix*

japonica retinal pigment epithelium (electron-microscopic investigation). *Moscow University Biological Sciences Bulletin*, *68*(4), 149–155 (in Russian).

19. Zak, P. P., Serezhnikova, N. B., Pogodina, L. S., Trofimova, N. N., & Ostrovsky, M. A., (2014). The estimation of age-related sensitivity of the retinal pigment epithelium of Japanese quail *C. japonica* to the light damage. *Russian Journal of Physiology*, *100*(7), 841–851 (in Russian).

20. Sigaeva, A. O., Seriozhnikova, N. B., Pogodina, L. S., Trofimova, N. N., Dadasheva, O. A., Gureva, T. S., & Zak, P. P., (2015). Changes of the choroid of different age groups of Japanese quails *Coturnix japonica* depending on the spectrum composition of illumination. *Sensory Systems*, *29*(4), 354–361 (in Russian).

21. Zak, P. P., Serezhnikova, N. B., Pogodina, L. S., Trofimova, N. N., Lipina, T. V., Gurieva, T. S., & Dadasheva, O. A., (2015). Photo-induced changes in subcellular structures of the retinal pigment epithelium of Japanese quail *Coturnix japonica*. *Biochemistry (Moscow)*, *80*(6), 785–789.

22. Dontsov, A. E., Vorobjev, I. A., Zolnikova, I. V., Pogodina, L. S., Potashnikova, D. M., Seryoznikova, N. B., & Zack, P. P., (2017). Photobiomodulating effect of low-dose LED blue range (450 nm) radiation on mitochondrial activity. *Sensory Systems*, *31*(4), 311–320 (in Russian).

23. Sereznikova, N. B., Pogodina, L. S., Lipina, T. V., Trofimova, N. N., Gurieva, T. S., Dadasheva, O. A., & Zak, P. P., (2017). Age-related changes in retinal pigment epithelium of Japanese quail *Coturnix japonica* under blue light during the daytime. *Clinical and Experimental Morphology*, *21*(1), 48–53 (in Russian).

24. Sereznikova, N. N., Pogodina, L. S., Lipina, T. V., Trofimova, N. N., Gurieva, T. S., & Zak, P. P., (2017). Age-related adaptive responses of mitochondria of the retinal pigment epithelium to the everyday blue LED lighting. *Reports of Biological Sciences*, *475*(2), 1–3.

25. Strauss, O., (2005). The retinal pigment epithelium in visual function. *Physiological Reviews*, *85*(3), 845–881.

26. Katz, M. L., Drea, C. M., Eldred, G. E., Hess, H. H., & Robison, W. G., (1986). Influence of early photoreceptor degeneration on lipofuscin in the retinal pigment epithelium. *Experimental Eye Research*, *43*(4), 561–573.

27. Boulton, M., Dontsov, A., Ostrovsky, M., Jarvis-Evans, J., & Svistunenko, D., (1993). Lipofuscin is a photoinducible free radical generator. *Journal of Photochemistry and Photobiology*, *19*(3), 201–204.

28. Feeney-Burns, L., Hilderbrand, E. S., & Eldridqe, S., (1984). Aging human RPE: Morphometric analysis of macular, equatorial, and peripheral cells. *Investigative Ophthalmology & Visual Science*, *25*(2), 195–200.

29. Kerr, J. F., Wyllie, A. H., & Currie, A. R., (1972). Apoptosis: A basic biological phenomenon with wide-ranging implications in tissue kinetics. *British Journal of Cancer*, *26*(4), 239–257.

30. Johnson, P. T., Lewis, G. P., Talada, K. C., Brown, M. N., Kappel, P. J., & Johnson, L. V., (2003). Drusen-associated degeneration in the retina. *Investigative Ophthalmology & Visual Science*, *44*(10), 4481–4488.

31. Bonilha, V. L., (2008). Age and disease-related structural changes in the retinal pigment epithelium. *Clinical Ophthalmology*, *2*(2), 413–424.

32. Birch, D. G., & Liang, F. Q., (2007). Age-related macular degeneration: A target for nanotechnology derived medicines. *International Journal of Nanomedicine*, *2*(1), 65–77.

33. Ehrlich, R., Harris, A., Kheradiya, N. S., Winston, D. M., Ciulla, T. A., & Wirostko, B., (2008). Age-related macular degeneration and the aging eye. *Clinical Interventions in Aging, 3*(3), 473–482.

34. Chen, H., Lukas, T. J., Du, N., Suyeoka, G., & Neufeld, A. H., (2009). Dysfunction of the retinal pigment epithelium with age: Increased iron decreases phagocytosis and lysosomal activity. *Investigative Ophthalmology & Visual Science, 50*(4), 1895–1902.

35. Liu, X., & Hajnoczky, G., (2011). Altered fusion dynamics underlie unique morphological changes in mitochondria during hypoxia-reoxygenation stress. *Cell Death and Differentiation, 18*(10), 1561–1572.

36. Ahmad, T., Aggarwa, K., Pattnaik, B., Mukherjee, S., Sethi, T., Tiwari, B. K., Kumar, M., Micheal, A., Mabalirajan, U., Ghosh, B., Sinha, R. S., & Agrawal, A., (2013). Computational classification of mitochondrial shapes reflects stress and redox state. *Cell Death & Disease, 4*, 1–10.

37. Lauber, J. K., (1982/1983). Retinal pigment epithelium: Ring mitochondria and lesions induced by continuous light. *Current Eye Research, 2*(12), 855–862.

38. Picard, M., & McEwen, B. S., (2014). Mitochondria impact brain function and cognition. *Proceedings of the National Academy of Sciences of the United States of America, 111*(1), 7–8.

39. Liang, H., Crewther, S. G., & Crewther, D. P., (1995). A model for the formation of ring mitochondria in retinal pigment epithelium. *Yan Ke Xue Bao, 11*(1), 9–15.

40. Ramrattan, R. S., Van der Schaft, T. L., Mooy, C. M., De Bruijn, W. C., Mulder, P. G., De Jong, P. T., (1994). Morphometric analysis of Bruch's membrane, the choriocapillaris, and the choroid in aging. *Investigative Ophtalmology & Visual Science, 35*(6), 2857–2864.

41. Biesemeier, A., Taubitz, T., Julien, S., Yoeruek, E., & Schraermeyer, U., (2014). Choriocapillaris breakdown precedes retinal degeneration in age-related macular degeneration. *Neurobiology of Aging, 35*(11), 2562–2573.

42. Behar-Cohen, F., Martinsons, C., Vienot, F., Zissis, G., Barlier-Salsi, A., Cesarini, J. P., Enouf, O., García, M., Picaud, S., & Attia, D., (2011). Light-emitting diods (LED) for domestic lighting: Any risks for the eye? *Progress in Retinal and Eye Research, 30*(4), 239–257.

43. Van Norren, D., & Gorgels, T. G. M. F., (2011). The action spectrum of photochemical to the retina: A review of monochromatical threshold data. *Photochemistry and Photobiology, 87*(4), 747–753.

44. Zak, P. P., & Ostrovsky, M. A., (2012). Potential danger of light emitting diode illumination to the eye, in children and teenagers. *Light & Engineering, 20*(3), 5–8.

45. State Standart SCENIHR, (2017). Preliminary opinion. *Potential Risks to Human Health of Light Emitting Diodes.* URL: https://ec.europa.eu/health/sites/health/files/scientific_committees/scheer/ docs/scheer_o_011.pdf (Accessed on 29 May 2019).

46. Newman, L. A., Walker, M. T., Brown, R. L., Cronin, T. W., & Robinson, P. R., (2003). Melanopsin forms a functional short-wavelength photopigment. *Biochemistry, 42*(44), 12734–12738.

47. State Standart. IEC 62471, (2006). *Photobiological Safety of Lamps and Lamp Systems.* URL: http://www.ledsmagazine.com/articles/2012/11/led-based-products-must-meet-photobiological-safety-standards-part-2-magazine.html (Accessed on 29 May 2019).

48. State Standart IEC/EN 60825, (2014). *Laser Safety.* URL: https://webstore.iec.ch/publication/3587.

49. State Standart ANSI Z80.3, (2015). *Sunglasses Standard.* URL: https://ansidotorg.blogspot.com/2016/05/ansi-z803-sunglasses-requirements.html#gref (Accessed on 29 May 2019).

50. Pautler, E. L., & Hall, F. L., (1987). Movement of melatonin across the retinal pigment epithelium. *Experiment Eye Research, 45*(2), 351–355.

51. Berkowitz, B. A., Schmidt, T., Podolsky. R. H., & Roberts, R., (2016). Melanopsin phototransduction contributes to light-evoked choroidal expansion and rod L-type calcium channel function in vivo. *Investigative Ophthalmology & Visual Science, 57*(13), 5314–5319.

52. Alarma-Estrany, P., & Pintor, J., (2007). Melatonin receptors in the eye: Location, second messengers and role in ocular physiology. *Pharmacology & Therapeutics, 113*(3), 507–522.

53. Ham, W., Mueller, H., & Sliney, D., (1976). Retinal sensitivity to damage from short wavelength light. *Nature, 260*(5547), 153–155.

54. Jaadane, I., Boulenguez, P., Chahory, S., Carré, S., Savoldelli, M., Jonet, L., Behar-Cohen, F., Martinsons, C., & Torriglia, A., (2015). Retinal damage induced by commercial light emitting diodes (LEDs). *Free Radical Biology & Medicine, 84*, 373–384.

55. Hunter, J. J., Morgan, J. I., Merigan, W. H., Sliney, D. H., Sparrow, J. R., & Williams, D. R., (2012). The susceptibility of the retina to photochemical damage from visible light. *Progress in Retinal and Eye Research, 31*(1), 28–42.

56. Chen, E., (1993). Inhibition of cytochrome oxidase and blue-light damage in rat retina. *Graefe's Archive for Clinical and Experimental Ophthalmology, 231*(7), 416–423.

57. Suter, M., Reme, C., Grimm, C., Wenzel, A., Jaattela, M., Esser, P., Kociok, N., Leist, M., & Richter, C., (2000). Age-related macular degeneration. The lipofusion component N-retinyl-N-retinylideneethanolamine detaches proapoptotic proteins from mitochondria and induces apoptosis in mammalian retinal pigment epithelial cells. *Journal of Biological Chemistry, 275*(50), 39625–39630.

58. Putting, B. J., Zweypfenning, R. C. V. J., Vrensen, G. F. J. M., Oosrerhuis, J. A., Van Best, J., (1992). Blood-retinal barrier dysfunction at the pigment epithelium induced by blue light. *Investigative Ophthalmology & Visual Science, 33*(12), 3385–3393.

59. Passarella, S., & Karu, T., (2014). Absorption of monochromatic and narrowband radiation in the visible and near IR by both mitochondrial and non-mitochondrial photoacceptors results in photobiomodulation. *Journal of Photochemistry and Photobiology B: Biology, 140*, 344–358.

60. Fuma, S., Murase, H., Kuse, Y., Tsuruma, K., Shimazawa, M., & Hara, H., (2015). Photobiomodulation with 670 nm light increased phagocytosis in human retinal pigment epithelial cells. *Molecular Vision, 21*, 883–892.

61. Meesters, Y., Winthorst, W. H., Duijzer, W. B., & Hommes, V., (2016). The effects of low-intensity narrow-band blue-light treatment compared to bright white-light treatment in sub-syndromal seasonal affective disorder. *Bio Med. Central Psychiatry, 16*(27), 1–10.

62. Vekshin, N. L., (1991). Light-dependent ATP synthesis in mitochondria. *Journal of Biochemistry International, 25*(4), 603–611.

63. Karu, T., (1999). Primary and secondary mechanisms of action of visible to near-IR radiation on cells. *Journal of Photochemistry and Photobiology B: Biology, 49*(1), 1–17.

64. Osborne, N. N., Núñez-Álvarez, C., & Del Olmo-Aguado, S., (2014). The effect of visual blue light on mitochondrial function associated with retinal ganglions cells. *Experimental Eye Research, 128,* 8–14.

65. Buravlev, E. A., Zhidkova, T. V., Osipov, A. N., & Vladimirov, Y. A., (2015). Are the mitochondrial respiratory complexes blocked by NO the targets for the laser and LED therapy? *Lasers in Medical Science, 30*(1), 173–180.

66. Begum, R., Powner, M. B., Hudson, N., Hogg, C., & Jeffery, G., (2013). Treatment with 670 nm light upregulates cytochrome C oxidase expression and reduces inflammation in an age-related macular degeneration model. *PLoS ONE, 8*(2), 1–11.

67. Tang, J., Du, Y., Lee, C. A., Talahalli, R., Eells, J. T., & Kern, T. S., (2013). Low-intensity far-red light inhibits early lesions that contribute to diabetic retinopathy: *In vivo* and in vitro. *Investigative Ophthalmology & Visual Science, 54*(5), 3681–3690.

68. Beirne, K., Rozanowska, M., & Votruba, M., (2017). Photostimulation of mitochondria as a treatment for retinal neurodegeneration. *Mitochondrion, 36,* 85–95.

CHAPTER 9

Effect of the Potassium Phenosan Salt and Radiation at Low Doses on the Functional Activity and State of the Brain Membranes of Mice

JULIA A. TRESHCHENKOVA, ALEXANDER N. GOLOSHCHAPOV,
LYUDMILA N. SHISHKINA, and ELENA B. BURLAKOVA

Emanuel Institute of Biochemical Physics of Russian Academy of Sciences, Kosygin St., 4, Moscow, 119334, Russia, E-mail: tresch@sky.chph.ras.ru

ABSTRACT

The influence of the synthetic antioxidant potassium salt of phenosan (potassium phenosan) in a wide range of concentrations, the exposure of gamma radiation at a dose of 15 cGy and their combined effect on the activity of aldolase and lactate dehydrogenase (LDH) in the cytoplasm, microsomes, and synaptosomes of the mice brain at different duration after the cessation of, action, as well as the structural state of the membranes and the composition of the mice, brain lipids have been studied. Maintaining the changes in the activity of enzymes and microviscosity of membranes of the brain structures after the termination of chronic γ-irradiation of mice at a dose of 1.2 cGy is shown for a long time.

Different sensitivity and ability to normalize of the studied parameters depending on the dose of both a substance and radiation, as well as on the duration after action have been revealed. In the early period after a combined action of the preparation and irradiation, an increase of the microviscosity of the probe localized in the surface layer of the lipid phase of the brain subcellular structures is accompanied by an increase in the ratio of the main fractions of phospholipids (PL) in the mice brain.

The set of the obtained data testifies to a high sensitivity of the brain structures to the effects of potassium phenosan and low-intensity radiation at the low doses, as well as to the absence of a linear dependence on the dose of the substance, which results in a complex nature of changes in the functional activity of the brain.

9.1   INTRODUCTION

Interest in studying the action of biologically active substances (BAS) and ionizing radiation at the low doses on living organisms at different levels of biological organization has not weakened for several decades. Studies of the dose dependences of BAS in a wide range of concentrations and the ionizing radiation at the low doses on biological objects help to reveal some general regularities, such as bimodal and nonlinear dependences. However, the mechanisms of action of BAS and physical factors at the low and ultra-low doses (ULD) have not yet been fully understood [1–4]. *In vitro* and in vivo experiments have established that antioxidants (AO) have an effect on the structural state of the cell membranes [5–10] and regulate the expression of genes of enzymes [11].

Environmental background radiation, the use of radiological methods in medicine, radiation disasters can affect living organisms to different extents [12–14]. So, ionizing radiation at the low doses causes a stimulating effect (hormesis), adaptive response, radioresistance, genomic instability, and epigenetic changes, which can be weakened by the AO action [15–18]. It has also been shown that ionizing radiation at the low doses induces a metabolic shift from oxidative phosphorylation to aerobic glycolysis, thus causing an increase in radioresistance [19]. Metabolic changes manifest themselves in the increased regulation of the gene encoding proteins glucose transporters, glycolytic enzymes, and the pentosophosphate pathway, as well as in the reduced regulation of the genes encoding mitochondrial enzymes. It has been shown that metabolic reprogramming depends on the HIFα transcriptional factor, which is specifically induced by the low doses of radiation [20].

AO can mitigate the damage caused by the radiation and regulate gene expression [21]. However, data on the study of the effect of SLD on the individual enzymes of metabolic pathways and the structural state of the cell membranes in vivo is insufficient. We used a synthetic AO – the potassium salt of phenosan (1-β-4-hydroxy-3,5-di-*tert*-butylphenyl-1-propionate potassium), which, along with antioxidant properties [22], has a broad spectrum of the biological activity.

Glycolysis is the main way of providing the brain with energy. The animal brain cells contain the basic forms of glycolytic enzymes: aldolase – muscle type A_4 and tissue-specific C_4, and lactate dehydrogenase (LDH) – M_4 or LDH5 – muscle type and H_4 or LDH1 – heart type. Aldolase (A_4 and C_4) and LDH (M_4 and H_4) have a four-subunit structure, are encoded by the different genes, and their hybrid forms (5 isoenzymes) differ in kinetic and immunological properties [23–25].

Despite the fact that glycolytic enzymes are predominantly contained in the cytoplasm of cells (soluble forms), aldolase, and LDH are reversibly associated with the membranes of the subcellular structures with the formation of complexes, which is one of the ways of regulating their activity in a cell [19, 26–28]. In this case, substrates of aldolase and LDH are important for the functioning of the subcellular structures and neurons [29–31]. The choice of these enzymes for research is attributed to the fact that aldolase is one of the key velocity-limiting enzymes of energy pathways of glycolysis and gluconeogenesis and LDH is the final enzyme of the metabolic pathway.

The aldolase and LDH activities are changed under the pathology state processes and have sensitive to various effects. For example, earlier we showed the influence of chronic γ-radiation in the dose range from 0.6 to 5.4 cGy on the change of the kinetic properties and the isozyme spectra of aldolase and LDH of cytoplasm [32]. We also revealed the role of the initial values of the kinetic parameters of LDH and the brain microsome microviscosity in the response reaction on potassium phenosan administration in a wide range of concentrations [33] and studied the effect of potassium phenosan at the low doses and low-intensity γ-radiation on the kinetic properties of aldolase and LDH fractions of myelin and nuclei, as well as the microviscosity of the lipid bilayer in the mice, brain nuclei [34].

The aim of this work is to study the effect of potassium phenosan and irradiation at the low doses and their combined action on the kinetic parameters (V_{max}, K_m, V_{max}/K_m) of aldolase and LDH in the cytoplasm, microsomes, and synaptosomes, as well as microviscosity of the membranes of the subcellular structures of the brain and the antioxidant properties and the lipid composition of the mice brain.

9.2 MATERIALS AND METHODOLOGY

We used white mice SHK (males) weighing 20–24 g. Animals were maintained on a standard diet. Solutions of potassium phenosan (1-β-4-hydroxy-3,5-di-*tert*-butylphenyl-1-propionate potassium) were prepared by the method of successive 10-fold dilution with the distilled water of the initial solution 10^{-3} M. Different doses of potassium phenosan (mol/kg) were administered intraperitoneally based on 0.2 mL per 20 g of the weight of the mouse. Control mice were injected a corresponding amount of the distilled water.

Several series of experiments were carried out, in each of which groups of animals consisting of 10 individuals were studied. In the first series of experiments, mice were exposed to γ-irradiation on the GUT-Co-400 apparatus at a dose of 15 cGy with a dose rate of 0.01 cGy/min. The animal cells were irradiated with preservation of feed and drink; the duration of irradiation exposure was 25 h. The following groups of mice were formed:

- Group 1 is the intact control;
- Group 2 is the administration of potassium phenosan at a dose of 10^{-5} mol/kg and/or 10^{-15} mol/kg; mice were decapitated during 2 days after the administration of the preparation;
- Group 3 is γ-irradiation of mice at a dose of 15 cGy; animals were decapitated during 1 day after the completion of the irradiation;
- Group 4 is the administration of potassium phenosan at a dose of 10^{-5} mol/kg and/or 10^{-15} mol/kg 30 min before the irradiation of mice at a dose of 15 cGy; mice were decapitated during 1 day after the irradiation.

The dose of 10^{-15} mol/kg was considered as ULD. After mice were decapitated, the group was divided into subgroups of 3–4 individuals in each. The control mice from the same group were decapitated simultaneously with the animals from the experimental groups. The total number of mice in this series of experiments was 80 individuals.

In another series of experiments, F_1 CBA/C57Bl mice (males) weighing 18–22 g were chronically exposed to γ-radiation from a source of ^{137}Cs at a total dose of 1.2 cGy with a dose rate of 0.6 cGy/day. The mice were decapitated within a month at different times after the termination of irradiation. The total number of mice in this series of experiments was 140 individuals.

After extraction, the brain was placed in the ice-cooled weighing bottles. 10% solution of homogenate was prepared in 0.32 mol/l sucrose, pH 7.4, at 5 °C. Subcellular fractions of the brain were obtained by the differential centrifugation at 105,000 g for 1 h on a Beckman centrifuge (Germany). Synaptosomes from the heavy mitochondrial fraction were purified using the gradient of the sucrose density (0.8 and 1.2 M) at 105,000 g for 1 h.

The activity of aldolase was determined spectrophotometrically at 540 nm [35]. Fructose-1,6-diphosphate (FDP) of the VDN company (Great Britain) in the range of concentrations $(0.1–1.0)×10^{-4}$ mol/L served as a substrate. The LDH activity was determined spectrophotometrically at 340 nm [36]. Sodium pyruvate (Merk, Germany) in the concentration range $(0.4–11.0) \times 10^{-4}$ mol/l served as a substrate. The kinetic parameters of enzymes, namely the maximum rate V_{max} (mmol of the substrate/min×mg of protein) and the apparent Michaelis constant (K_m), were calculated from the initial rates of the reactions catalyzed by aldolase and LDH using the Cornish-Bowden method [37], after which the V_{max}/K_m ratio characterizing the enzyme efficiency was calculated. Protein was determined by the Lowry method [38].

Microviscosity in different regions of the membrane lipid bilayer in microsomes and synaptosomes was estimated on the basis of the rotational correlation times τ_c of paramagnetic probes – the stable aminoxyl radicals: 2.2,6.6-tetramethyl-4-capriloiloxypiperidine-1-oxyl (probe 1) and 5,6-benzo-2,2,6,6-tetramethyl-1,2,3,4-tetrahydro-γ-carboline-3-oxyl (probe 2). The values of τ_c were calculated from the EPR spectra registered at room temperature on the EPR spectrometer by Brucker (Germany), model ER-200D. Probe 1 is localized in the surface layer of the membrane lipids, and probe 2 is mainly localized in the area of the near-protein lipids. The details of the work with the probes are given in [6].

Lipids were extracted using the Bligh and Dyer method in the Kates modification [39]. The value of the antioxidant activity (AOA) of lipids was determined on the methyl oleate oxidative model. The qualitative and quantitative composition of phospholipids (PL) was estimated by the method of thin-layer chromatography using glass plates of 9×12 cm, silica gel type G or H of the Sigma company (USA) and a mixture of chloroform/methanol/glacial acetic acid/distilled water in the ratio of 12.5/7.5/4/2 (v/v) as the mobile phase. All details of the above methods for the lipid research are given in Ref. [40].

The obtained data were processed with standard methods of variation statistics and using software packages Origin 6.1 and KINS [41]. The results are presented in the form of arithmetic means with the indication of their mean square errors (M±m). Differences at $p < 0.05$ were considered to be reliable.

9.3 RESULTS AND DISCUSSION

The first stage of our research was aimed at studying the response of the glycolytic enzymes in the early period after the action of potassium phenosan in a wide range of doses, low-intensity γ-radiation at a dose of 15 cGy and their combined action at the level of cytoplasm, microsomes, and synaptosomes of the mice SHK brain cells. As seen from the data presented in Figure 9.1, the maximum rate (V_{max}) of aldolase is increased in the cytoplasm during 2 days after administration of potassium phenosan at a dose of 10^{-5} mol/kg by approximately 15% and by 25% when the preparation is administered at a dose of 10^{-15} mol/kg.

The dose of the preparation had the opposite effect on the efficiency (V_{max}/K_m) of the enzyme in the cytoplasm: it was revealed a significant increase after administration of potassium phenosan at a dose of 10^{-5} mol/kg and a decrease after its administration at the ULD. The differences observed in the efficiency of aldolase are likely to be due to the changes in the affinity of the enzyme to the FDP substrate, because it increased with the administration of AO at the dose of 10^{-5} mol/kg (the K_m value grew smaller) and, on the contrary, decreased with its administration at the ULD.

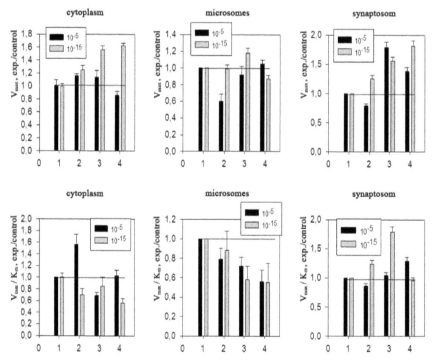

FIGURE 9.1   Relative changes in the maximum rate (V_{max}, mmol/min×mg of protein) and the efficiency (V_{max}/K_m) of aldolase in the cytoplasm, microsomes, and synaptosomes of the mice SHK brain during 2 days after administration of potassium phenosan at a dose of 10^{-5} mol/kg (•) or at a dose of 10^{-15} mol/kg (o) (column 2), during 1 day after γ-irradiation at a dose of 15 cGy (column 3), and during 1 day after termination of irradiation under a combined action of potassium phenosan and γ-radiation (column 4). Column 1 is the reference values.

A significant decrease in V_{max} of aldolase was observed in microsomes (Figure 9.1) after administration of potassium phenosan at a dose of 10^{-5} mol/kg, as well as the absence of AO influence after the administration at synaptosomes. The effectiveness of aldolase slightly decreased relative to the control. At the same time, a different direction of the influence of the AO dose is observed in synaptosomes (see Figure 9.1): with a unidirectional decrease in the V_{max} values and the efficiency of aldolase after administration of potassium phenosan at a dose of 10^{-5} mol/kg, the values of these parameters were found to be increased by 20–25% as compared with the value in the control group after administration of AO at ULD.

The direction of the changes in V_{max} of aldolase and LDH in response to exposure to γ-radiation at the low doses depends on the initial values of the parameter in the control group of mice [32]. So, it is not surprising that the scope of the changes in the maximum rate of aldolase after irradiation varies in the different experiments (see Figure 9.1, columns 3). Nevertheless, an increase in V_{max} of aldolase during 1 day after γ-irradiation of mice at a dose of 15 cGy was found in all subcellular structures of the brain, and it was most pronounced in synaptosomes. At the same time, the efficiency (V_{max}/K_m) of the enzyme decreased in the cytoplasm and microsomes and increased in synaptosomes.

Multidirectional changes in the kinetic parameters of aldolase were found upon a combined action of potassium phenosan and irradiation (Figure 9.1, columns 4). The prophylactic administration of the preparation at both doses had a modulating effect on the changes in V_{max} and V_{max}/K_m of aldolase, thus increasing or decreasing the parameters of the enzyme relative to their values in the groups of irradiated mice. At the same time, the administration of potassium phenosan at both doses did not significantly affect the efficiency (V_{max}/K_m) of microsomal aldolase in comparison with its value in the irradiated mice; it remained almost twice lower than the level of the enzyme parameter in control. The normalization of V_{max}/K_m is observed after a combined action of AO and irradiation with the administration of potassium phenosan at a dose of 10^{-5} mol/kg in the cytoplasm and at a dose of 10^{-15} mol/kg in synaptosomes.

Thus, the changes in the kinetic parameters of aldolase, revealed in the early period in response to the effects of potassium phenosan in a wide range of doses, γ-radiation at a low dose and their combined action, had multi-directional nature, and the scope and direction of their changes depended on the investigated subcellular structure of the brain. A significant positive correlation of the changes in V_{max} of aldolase was detected between the cytoplasm and synaptosomes ($p = 0.023$) after the administration of potassium phenosan at ULD.

Figure 9.2 shows the changes in the kinetic properties of LDH after administration of potassium phenosan at the different doses, the exposure to γ-radiation at a dose of 15 cGy and their combined action in the subcellular structures of the mice brain. It was found that the V_{max} value of LDH significantly varies depending on the dose of potassium phenosan only in the cytoplasm, increasing 1.5 times and decreasing by 25% during 2 days after the administration of AO at the doses of 10^{-5} and 10^{-15} mol/kg, respectively. In microsomes and synaptosomes, this parameter remains practically unchanged as compared to the same value in control (see Figure 9.2). The administration of potassium phenosan at various doses induces more significant changes in the efficiency (V_{max}/K_m) of LDH. While the administration of the preparation at a dose of 10^{-5} mol/kg caused an increase in V_{max}/K_m of LDH in all studied structures of the brain, which was the most pronounced in microsomes (by 2.5 times), the efficiency of LDH when AO administered at a dose of 10^{-15} mol/kg increased by 2.25 times only in microsomes and decreased significantly, especially in synaptosomes of the mice brain (see Figure 9.2).

The scope of changes in the studied parameters of LDH after gamma irradiation also depended on the time of the experiments; however, the analysis of the data shows that there is an increase in the value of V_{max} of LDH in microsomes and especially in synaptosomes, while the effect in the cytoplasm largely depended on the initial level of the parameter (see Figure 9.2, columns 3). The efficiency (V_{max}/K_m) of LDH increased in the cytoplasm and microsomes, but drastically decreased in synaptosomes.

Under a combined action of two factors, potassium phenosan administered before the irradiation exerted a modulating effect on the changes in the parameters of LDH in the subcellular structures of the mice brain. The prophylactic administration of potassium phenosan at a dose of 10^{-5} mol/kg induced a decrease in V_{max} of LDH at all subcellular structures and V_{max}/K_m of LDH in the cytoplasm and microsomes as compared to their values in the group of irradiated mice, while substantially reducing the values of both kinetic parameters of LDH in synaptosomes were found under the potassium phenosan administration at both doses, which had the same low level as in the group of irradiated mice (Figure 9.2, columns 3 and 4). A normalized value of V_{max} of LDH in the cytoplasm and microsomes, a significant increase in V_{max} in synaptosomes and effectiveness of LDH in microsomes were revealed upon administration of potassium phenosan at ULD (ultra-low doses) (Figure 9.2, columns 4).

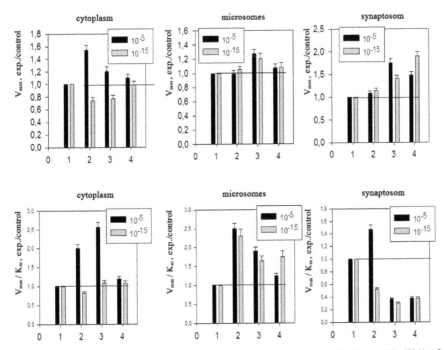

FIGURE 9.2   Relative changes in the maximum rate (V_{max}) and the efficiency (V_{max}/K_m)) of lactate dehydrogenase in the cytoplasm, microsomes, and synaptosomes of the mice SHK brain during 2 days after the administration of potassium phenosan at a dose of 10^{-5} mol/kg (•) or at a dose of 10^{-15} mol/kg (o) (column 2), during 1 day after γ-irradiation at a dose of 15 cGy (column 3), and during 1 day after the termination of the irradiation under a combined action of potassium phenosan and γ-radiation (column 4). Column 1-reference values.

Thus, the multidirectional nature of changes in the kinetic properties of LDH in the subcellular structures of the brain was found in the early period in response to the effects of AO in a wide range of doses, irradiation, and their combined action, as for aldolase.

It is pertinent to note both direct and reverse direction of changes in the parameters between aldolase and LDH in the subcellular structures of the brain caused by the potassium phenosan administration, the γ-irradiation and their combined action. Besides, the administration of AO at a dose of 10^{-5} mol/kg results in a positive relationship of the V_{max} changes between aldolase and LDH in synaptosomes ($p = 0.046$). Special consideration should be given to a stimulating effect of the irradiation at low dose on the activity of enzymes: an increase, to a greater or lesser extent, in the value of V_{max} of aldolase and LDH was observed in all subcellular fractions of the mice brain.

The brain lipids of animals are characterized by the pro-oxidant activity, i.e., they accelerate the oxidation of methyl oleate in the model reaction of its thermal oxidation due to a high degree of its unsaturation [42]. The AOA value of the mice SHK brain lipids in the control groups also had negative values in the range from −855 h×ml/g to −590 h×ml/g in different series of experiments. During 2 days after the administration of potassium phenosan at both studied doses, the value of AOA of the brain lipids increased with an increase in doses by 215 h×ml/g and 375 h×ml/g, respectively. At the same time, an increase in AOA of the brain lipids was found in the groups of mice irradiated at a dose of 15 cGy and in those exposed to the combined action of potassium phenosan and irradiation; however, the scope of changes in this parameter significantly depended on its initial level in control. The most remarkable increase in AOA of lipids ($\Delta AOA = 480$ h×ml/g) was found in groups 3 and 4, where potassium phenosan was given at a dose of 10^{-15} mol/kg, i.e., in the groups, for which the most pronounced changes in the kinetic parameters and the efficiency of glycolytic enzymes in the structures of the mice brain were observed.

To clarify the possibility of preserving changes in the parameters of the glycolytic enzymes for a long time after the cessation of the exposure to ionizing radiation at low doses, a series of experiments were carried out aimed at studying the response of the biochemical properties of aldolase and LDH of the F_1 CBA/C57Bl mice brain structures characterized by a higher antioxidant status of tissues [5] to the action of the chronic γ-radiation at a total dose of 1.2 cGy. Figure 9.3 shows the dynamics of changes in V_{max} of aldolase and LDH in the cytoplasm and microsomes of the brain at different times after cessation of the radiation action.

As seen from the data presented, the dynamics of V_{max} of aldolase and LDH has a pronounced oscillatory character. So, there is an absence of changes in V_{max} of aldolase and a significant increase in V_{max} of LDH in the brain cytoplasm immediately after the termination of the irradiation/ while a reverse direction of changes in V_{max} of these enzymes is revealed in microsomes (Figure 9.3, point 0). There is a decrease in V_{max} of aldolase and LDH in the cytoplasm during 1 and 2 days after cessation of irradiation, a remarkable increase in V_{max} of both enzymes is observed in microsomes in this period. Further dynamics of the changes in V_{max} in the cytoplasm and microsomes of the brain is of the opposite nature. However, during 27 days after the termination of γ-irradiation, normalization of the maximum rate was found only for LDH in the cytoplasm (Figure 9.3). In all other cases, the value of V_{max} is significantly higher than that estimated immediately after the completion of irradiation. It is suggested an adaptive nature of changes of the enzyme parameters in the cell metabolism.

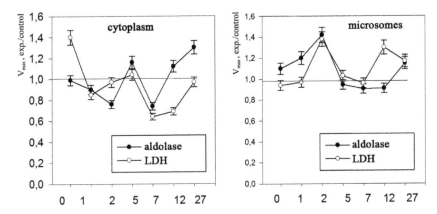

FIGURE 9.3 The relative changes in the values of the maximum rate (V_{max}) of aldolase and LDH in the cytoplasm and microsomes of the mice brain at the different times after cessation of the chronic γ-irradiation at the total dose of 1.2 cGy (0.6 cGy/day): the animals were decapitated during 1, 2, 5, 7, 12 and 27 days after cessation of exposure. The values of V_{max} of the enzymes in the cytoplasm and microsomes of the mice brain decapitated immediately after the termination of irradiation served as a control (point 0).

Consequently, the action of γ-radiation at low doses (15 cGy and 1.2 cGy) causes significant changes in the kinetic parameters of the glycolytic enzymes in the subcellular structures of the mice to the brain with the different antioxidant status of tissues, persisting for a long time after cessation of exposure. As in the case of the administration of potassium phenosan at the different doses and a combined action of AO and γ-radiation at the dose of 15 cGy in the early period after the exposure, the significant changes in the kinetic properties of aldolase and LDG are found in the subcellular structures of the mice brain; however, a response of aldolase and LDH to the exposure depends on differences in the sensitivity of enzymes in the studied subcellular structures of the mice brain.

These differences are more pronounced at the level of synaptosomes. For example, while in synaptosomes V_{max} of aldolase and LDH significantly increased both after irradiation at a dose of 15 cGy and under the administration of potassium phenosan at both doses followed by irradiation at a low dose, the effectiveness (V_{max}/K_m) of LDH significantly decreased. The function of synaptosomes requires a lot of energy, which is satisfied through glycolysis and oxidative phosphorylation with the ATP formation. Glycolysis is necessary to maintain transmission of the brain synaptosomes [19, 31]. As already mentioned, the terminal ends (synaptosomes) of the rat brain have 5 isoenzymes of LDH with a predominance of the LDH1 form, and the amount of LDH5 is significantly higher in neurons and astrocytes [30]. An aerobic form of LDH1 is predominant in the synaptosomal fraction of the primate forebrain cells [29]. Aldolase in synaptosomes is also represented by multiple isoforms. At the same time, aldolase, and LDH isozymes differ in their kinetic properties, which can influence their sensitivity to the various actions. Indeed, significant differences in the sensitivity of isoenzymes of LDH and aldolase in the cytoplasm of the F_1 CBA/C57Bl mice brain were previously detected after the chronic γ-radiation in the dose range of 0.6 cGy to 5.4 cGy [32].

Since the glycolysis enzymes are inversety linked to the lipid membranes of the subcellular structures, including the brain, this makes it possible to suggest that the changes in their activity could be associated with the state of membranes. An important role in the cellular metabolism is played by the physicochemical system of regulation of lipid peroxidation (LPO), whose functioning is shown at the membrane, cellular, and organ levels [43, 44]. In addition to AOA lipids and functional parameters, the composition of lipids, their ability to oxidize, and the structural state of membranes are considered among the parameters of this regulatory system. A high sensitivity of its parameters to the action of the chemical and physical factors at low doses was previously experimentally proved both in laboratory experiments with animals irradiated at low doses with different dose rates and in the tissues of wild rodents caught in the radioactively contaminated areas [1, 2, 17, 44–46].

Therefore, the next stage of the research was aimed at studying the composition of the mice SHK brain lipids in the experiments on the effect of potassium phenosan at the different doses, irradiation at a dose of 15 cGy and a combined action of these factors. The table presents the following generalized parameters of the lipid composition in the mice brain: the share of PL in the composition of the total lipids (%PL); the ratio of the main fractions of PL in the animal tissues phosphatidylcholine/phosphatidylethanolamine (PC/PE) reflecting the structural state of the membrane system of the organ, and the ratio of the sums of the more easily oxidizable and the more poorly oxidizable fractions of PL ($\sum$EOPL/$\sum$POPL), which characterizes the ability of lipids to oxidation [44].

The analysis of the data presented in Table 9.1 indicates certain differences in the parameters of the LPO regulation system in the brain of the control mice in the different experiments. Although the ratio of the main PL fractions is practically the same, the share of PL in the composition of the total lipids is slightly higher, and the ability of lipids to oxidation is significantly higher in the control group of mice in the experiments with potassium phenosan at a dose of 10^{-15} mol/kg. During 2 days after administration of potassium phenosan at both doses a significant increase in the ratio PC/PE and a remarkable reduction of the proportion of sums of the more easily and the more poorly oxidizable PL fractions was observed, which was the most pronounced after administration of AO at the dose of 10^{-5} mol/kg. In the early period after γ-irradiation of mice at a dose of 15 cGy, we observed an increase in the ratio PC/PE and a decrease of the ratio $\sum$EOPL/$\sum$POPL, as well as in the content of PL in the lipids of the brain of irradiated mice. The scope of changes varies depending on the time of the experiment performance. The prophylactic administration of potassium phenosan 30 min before the irradiation at a low dose had a normalizing effect on the composition of the mice brain PL, slightly increasing%PL in the total lipid composition when using AO at a dose of 10^{-15} mol/kg. The administration of potassium phenosan at a dose of 10^{-5} mol/kg only normalized the ratio of the main PL fractions in the mice brain as compared to the PC/PE ratio in the control group.

TABLE 9.1 Generalized Indices of the Lipid Composition of Brain in the Control and Experimental Groups of Mice SHK

Conditions of the experiment	Share of phospholipids in the total lipids (% PL)	PC/PE	$\sum$EOPL/$\sum$POPL
Control (group 1) for the potassium phenosan experiments at a dose of 10^{-5} mol/kg	55.8 ± 3.8	1.096 ± 0.040	2.12 ± 0.12
Administration of potassium phenosan at a dose of 10^{-5} mol/kg; decapitation during 2 days after administration	61.8 ± 1.2	4.27 ± 0.65	0.321 ± 0.011

TABLE 9.1   *(Continued)*

Conditions of the experiment	Share of phospholipids in the total lipids (% PL)	PC/PE	∑EOPL/∑POPL
γ-Irradiation at a dose of 15 cGy; decapitation during 1 day after irradiation	32.3 ± 1.7	1.338 ± 0.069	1.106 ± 0.080
Administration of potassium phenosan at a dose of 10^{-5} mol/kg 30 min before γ-irradiation at a dose of 15 cGy; decapitation during 1 day after irradiation	27.6 ± 0.9	1.087 ± 0.024	1.523 ± 0.081
Control (group 1) for the potassium phenosan experiments at a dose of 10^{-15} mol/kg	43.7 ± 1.6	1.053 ± 0.042	1.447 ± 0.055
Administration of potassium phenosan at the dose of 10^{-15} mol/kg; decapitation during 2 days after administration	27.3 ± 0.1	1.888 ± 0.076	1.125 ± 0.042
γ-Irradiation at a dose of 15 cGy; decapitation during 1 day after irradiation	37.8 ± 1.7	1.963 ± 0.033	1.303 ± 0.058
Administration of potassium phenosan at a dose of 10^{-15} mol/kg 30 min before γ-irradiation at a dose of 15 cGy; decapitation during 1 day after irradiation	54.3 ± 2.65	1.118 ± 0.014	1.407 ± 0.041

Such significant changes in the parameters of the system of regulating LPO in the brain of mice from experimental groups relative to their initial values in the groups of the control animals had an obvious impact on the structural state of the membranes of the brain subcellular structures. This was confirmed by the results of the studies on the changes in the microviscosity of the lipid bilayer of membranes in microsomes and synaptosomes of the brain in the mice of the same groups after the studied actions.

Figure 9.4 presents data on the changes in the rotational correlation times of probes 1 and 2 in different areas of the lipid bilayer in microsomes and synaptosomes of the mice brain after administration of potassium phenosan at different doses, exposure to γ-radiation at a dose of 15 cGy and their combined action in the early periods after the action.

During 2 days after administration, potassium phenosan (see Figure 9.4, columns 2) caused a significant increase in the microviscosity in the surface layer of the membrane lipids (probe 1) of microsomes and synaptosomes, which was more pronounced in microsomes after the AO administration at a dose of 10^{-15} mol/kg and in synaptosomes after the administration of phenosan potassium at a dose of 10^{-5} mol/kg. At the same time, the microviscosity of the near-protein layer of lipids (probe 2) in the membranes of microsomes remained almost at the level of the control values, and increased 1.5 times in synaptosomes after the administration of potassium phenosan only at a dose of 10^{-5} mol/kg.

In response to γ-irradiation of mice at a dose of 15 cGy changes in the microviscosity are mainly associated with the state of the membranes in the subcellular structures of the brain in the groups of the control animals (see Figure 9.4, columns 3). For example, in the experiment where AO was given at a dose of 10^{-5} mol/kg, the changes in the microviscosity of the membrane lipid bilayer were found only in synaptosomes: the increase in the rigidity of the synaptosome surface layer (probe 1) and a significant decrease of the microviscosity in the area of the near-protein lipids (probe 2). In the experiment with potassium phenosan at ULD, more significant changes in the structural state of the membranes were, on the contrary, found in microsomes after γ-irradiation of mice: an increase in the rotational correlation time of probe 1 is observed with a decrease in τ_c of probe 2. At the same time, a significant increase in the microviscosity of only the surface lipid bilayer of membranes (probe 1) was revealed in synaptosomes.

Under a combined action of the studied factors (Figure 9.4, columns 4), administration of potassium phenosan at a dose of 10^{-5} mol/kg before the irradiation had a less significant effect on the structural state of membranes of microsomes and synaptosomes as compared to the administration of AO at ULD. So, a more significant increase in the values of τ_c of both probes in microsomes and probe 1 in synaptosomes was found only after the prophylactic administration of potassium phenosan at a dose of 10^{-15} mol/kg.

Comparison of the effect of the investigated factors on the generalized indices of the composition of the mice brain lipids (Table 9.1) and the structural state of microsomes and synaptosomes in the same groups of animals (see Figure 9.4) in the early periods after the action allowed us to detect the following regularity: an increase in the rotational correlation time of the probe localized in the surface layers of the lipid phase of the brain subcellular structures is accompanied by an increase in the ratio of the main fractions of PL in the mice brain.

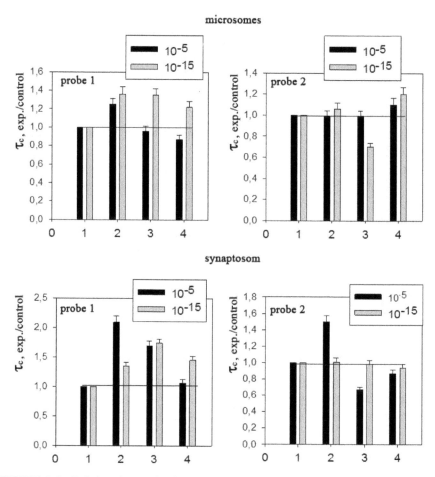

FIGURE 9.4 Relative changes in the rotational correlation times of probes 1 and 2 in a suspension of brain microsomes and synaptosomes from mice SHK during 2 days after the administration of potassium phenosan at a dose of 10^{-5} mol/kg (•) or ULD (o) (columns 2); during 1 day after γ-irradiation at a dose of 15 cGy (columns 3) and during 1 day after the termination of the irradiation under a combined action of potassium phenosan and γ-radiation (columns 4). Columns 1: the reference values.

The data on the changes in the microviscosity (fluidity) of the membrane lipid bilayer is presented above and related to the early periods after the exposure to AO and γ-irradiation at the dose of 15 cGy. The possibility of the preserving structural changes in membranes after the exposure to ionizing radiation at the low doses was a subject of our further studies. In this connection, the microviscosity of the surface lipid bilayer of the microsome membranes was estimated using probe 1 at the different times after the termination of the chronic γ-irradiation of mice F_1 CBA/C57Bl at a dose of 1.2 cGy. The results are shown in Figure 9.5.

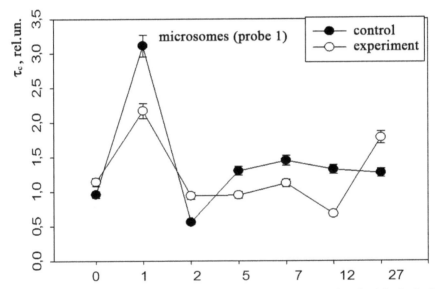

FIGURE 9.5 Relative changes in the rotational correlation times of probe 1 in the brain microsomes at the different times after the termination of γ-irradiation of mice F_1 CBA/C57Bl at a dose of 1.2 cGy. Animals were decapitated during 1, 2, 5, 7, 12, and 27 days after the cessation of exposure. Intact mice from the same group that was decapitated simultaneously with the irradiated animals served as a control. Point 0 corresponds to the time of the decapitation of mice immediately after the termination of the irradiation.

First of all, it should be noted that the changes in the microviscosity of the membrane surface lipid bilayer of the microsomes from the control (intact) animals in the investigated time period have an oscillatory character. This corresponds to the literature data on the significant effect of the experimental time on the parameters of the physicochemical system of the LPO regulation in the brain of the laboratory animals [46, 47]. Dynamics of changes in the structural state of the brain microsomes after the cessation of γ-irradiation also had an oscillatory character. Besides, we detected a significant decrease in the microviscosity of membranes in the experimental groups of mice during 1, 2, 5, 7, and 12 days after the cessation of exposure as compared to the corresponding values in the control groups. However, during 27 days after the termination of the chronic γ-radiation at the dose of 1.2 cGy, the rigidity of the microsomal membranes in the group of irradiated mice is significantly higher than in the intact animals. These obtained data about the structural changes in the membranes are important for the understanding of changes in the activity of both integral and peripheral enzymes that are associated with lipid membranes of the subcellular structures.

The ability of enzymes of glycolysis to form local complexes with cell membranes determines their ability to respond quickly to the energy needs of the cell. In this case, enzymes of glycolysis, in addition to their main function in the glucose metabolism through the glycolytic pathway, can form complexes with other proteins and enzymes of the cell. For example, aldolase interacts with F-actin, phospholipase D_2, α-tubulin, i.e., with proteins that are not associated with glycolysis. The interaction between aldolase and vacuolar H^+ ATPase is shown to play an important role in the regulation of proton pump in the cell [48], and M-LDH is physically associated with sarcolemmal K_{ATP} channels and mediates cytoprotection in heart embryonic H9C2 cells [49].

As shown on the model systems, PL can modulate the activity of the glycolytic enzymes [50, 51]. At the same time, the analysis of the literature data leads to the conclusion about an important role in the nature of PL in these processes. For example, LDH from muscles was not bound by liposomes prepared from PC in vitro, but interacted with liposomes of a mixed composition: phosphatidylserine-PC [52]. The correlation analysis of the relationships between the kinetic parameters of the glycolytic enzymes and the PL composition of the brain synaptosomes from mice after γ–irradiation at a dose of 15 cGy, i.e., in in vivo conditions, also revealed a number of interesting regularities [53]. So, the inverse correlations with the PE content ($p = 0.025$, $n = 9$) and sphingomyelin ($p = 0.018$, $n = 6$) were found for V_{max} of aldolase in the brain synaptosomes of mice. At the same time, a significant direct correlation of V_{max} of LDH was revealed only with the content of PE ($p = 0.05$, $n = 9$).

9.4 CONCLUSION

Thus, a set of the presented data corresponds to the ideas about the important role of the physicochemical system that regulates oxidative processes to ensure the functioning of enzymes associated with the lipid membranes that include the glycolytic enzymes of the mice brain under study. Also detected is a high sensitivity of aldolase and LDH and their different ability to normalize the kinetic parameters, a structural state of membranes of microsomes and synaptosomes, a composition of the brain lipids and an availability of AO to the action of potassium phenosan at the different doses, the low-intensity γ-irradiation at the low doses and the combined action of AO and γ-irradiation.

The scope and direction of the changes in the studied parameters significantly depend both on the dose of the preparation and the subcellular structures of the brain, as well as on the initial values of the studied parameters in the groups of control mice. While the effectiveness of aldolase is significantly reduced in the cytoplasm during 2 days after the administration of potassium phenosan at ULD, as well as in the cytoplasm and microsomes in the early period after the administration of the substance at a dose of 10^{-5} mol/kg and after γ-irradiation of mice at a dose of 15 cGy, the increase in the efficiency of this enzyme is detected in synaptosomes after administration of AO at ULD and after γ-irradiation of mice at a low dose and in the cytoplasm during 2 days after administration of potassium phenosan at a dose of 10^{-5} mol/kg. Multidirectional nature of changes both in the LDH effectiveness in the subcellular structures of the brain and in the microviscosity of the membranes in microsomes and synaptosomes, as well as in the PL content in the total lipid composition of the brain is detected in early periods after irradiation of mice at low doses and under a combined effect with potassium phenosan depending on the dose of the preparation and irradiation.

Preserving the changes in the activity of enzymes and the microviscosity of membranes in the brain structures after termination of the chronic γ-irradiation of mice at a dose of 1.2 cGy for a long time, as well as the need for a certain quantitative ratio of the PL fractions to maintain the activity of glycolytic enzymes of subcellular structures of the mice brain require further studies of the formation mechanism underlying the effects of potassium phenosan administered at the low and ULD, which can cause the significant disorders of the brain functions, especially after a combined action of factors of the different nature.

KEYWORDS

- aldolase
- lactate dehydrogenase
- microsomes
- microviscosity
- phospholipids
- synaptosomes

REFERENCES

1. Burlakova, E. B., (1994). The effect of ultra-low doses. *Herald of the Russian Academy of Sciences, 64*(5), 205–210.
2. Polyakova, N. V., & Shishkina, L. N., (1995). The effect of low dose rate on the lipid peroxidation processes in mice tissues. *Radiat. Biology. Radioecology, 35*(2), 181–188 (in Russian).
3. Burlakova, E. B., Konradov, A. A., & Maltseva, E. L., (2004). Effect of extremely weak chemical and physical stimuli on biological systems. *Biophysics, 49*(3), 522–534.
4. Burlakova, E. B., (2007). Bioantioxidants. *Russian Journal of General Chemistry, 77*(11), 1983–1993 (in Russian).
5. Burlakova, E. B., Alesenko, A. V., Molochkina, E. M., Palmina, N. P., & Khrapova, N. G., (1975). *Bioantioxidants in the Radiation Damage and Tumor Growth* (p. 214). Moscow: Nauka, (in Russian).
6. Goloshchapov, A. N., & Burlakova, E. B., (1975). Spin label study of microviscosity and structural transition in cell membrane lipid and protein. 1. Heat-induced structural changes in cell nuclei of mouse organs in normal conditions. *Biophysics, 20*(5), 816–821 (in Russian).
7. Belov, V. V., Maltseva, E. L., & Palmina, N. P., (2010). α-Tocopherol as modifier of the lipid structure of plasma membranes in vitro on a wide range of concentrations studied by spin-probes. In: Zaikov, G. E., & Kozlowski, R. M., (eds.), *Chemical Reactions in Gas, Liquid and Solid Phases: Synthesis, Properties and Application* (pp. 29–43). New York: Nova Science Publishers.
8. Binyukov, V. I., Alekseeva, O. M., Mil, E. M., Fattakhov, S. G., Goloshchapov, A. N., Burlakova, E. B., & Konovalov, A. I., (2011). Study of influence of phenosan, ICFAN-10 and melaphen on erythrocytes in vivo by the method of atomic-intensity microscopy. *Reports of Russian Academy of Sciences, 441*(1), 114–117 (in Russian).
9. Palmina, N. P., Chasovskaya, T. E., Belov, V. V., & Maltseva, E. L., (2012). Dose dependences of lipid microviscosity of biological membranes induced by synthetic antioxidant potassium phenosan salt. *Reports of Chemistry and Biophysics, 443*, 100–104.
10. Chasovskaya, T. E., Maltseva, E. L., & Palmina, N. P., (2013). Effect of potassium phenosan on structure of plasma membranes of mice liver cells in vitro. *Biophysics, 58*(1), 78–85.

11. Allen, R. G., & Tresini, M., (2000). Oxidative stress and gene regulation. *Free Radic. Biol. & Med. 28*, 463–499.

12. *20 Years after the Chernobyl Accident: Past, Present and Future* (2006). [Eds. Elena B. Burlakova and Valeria I. Naidich]. New York: Nova Science Publishers 358 p.

13. Kudyasheva, A. G., Shishkina, L. N., Shevchenko, O. G., Bashlykova, L. A., & Zagorskaya, N. G., (2007). Biological consequences of increased natural radiation background for *Microtus oeconomus* Pall. populations. *J. Environm. Radioactivity, 97*, 30–41.

14. *The Lessons of Chernobyl: 25 Years later* (2012). [Eds. Elena B. Burlakova and Valeria I. Naidich]. New York: Nova Science Publishers 331 p.

15. Burlakova, E. B., Mikhailov, V. F., & Mazurik, V. K., (2001). The redox homeostasis system in radiation-induced genome instability. *Radiat. Biology. Radioecology, 41*(5), 489–499 (in Russian).

16. Feinendegen, L. E., (2005). Evidence for beneficial low-level radiation effects and radiation hormesis. *The British Journal of Radiology, 78*, 3–7.

17. Tang, F. R., & Loke, W. K., (2014). Molecular mechanisms of low dose ionizing radiation-induced hormesis, adaptive responses, radioresistance, bystander effects, and genomic instability. *Intern. J. Radiation Biology, 9*(1), 1–15.

18. Calabrese, E. J., (2009). The road to linearity: Why linearity at low doses became the basis for carcinogen risk assessment. *Arch. Toxicol., 83*, 203–225.

19. Jang, S. R., Nelson, J. C., Bend, E. C., Rodriguez-Laureano, L., Tueros, F. G., Cartagenova, L., Underwood, K., Jorgensen, E. M., & Colon-Ramos, D. A., (2016). Glycolytic enzymes localize to synapses under energy stress to support synaptic function. *Neuron, 90*(2), 278–291.

20. Lall, R., Ganapathy, S., Yang, M., Xiao, S., Xu, T., Su, H., et al., (2014). Low-dose radiation exposure induces a HIF-1-mediated adaptive and protective metabolic response. *Cell Death and Differentiation, 21*, 836–844.

21. Bernal, A. J., Dolinoy, D. C., Huang, D., Skaar, D. A., Weinhouse, C., & Jirtle, R. L., (2016). Adaptive radiation-induced epigenetic alterations mitigated by antioxidants. *FASEB J., 27*, 665–671.

22. Shishkina, L. N., Polyakova, N. V., Mazaletskaya, L. I., Bespal'ko, O. F., & Kushnireva, Y. V., (1999). The radioprotective properties of Phenoxane under low dose and low-intensity γ-irradiation. *Radiat. Biology. Radioecology, 39*(2/3), 322–328 (in Russian).

23. Penhoet, E., & Rutter, W. J., (1971). Catalytic and immunochemical properties of homomeric and heteromeric combinations of aldolase subunits. *J. Biol. Chem., 246*, 318–323.

24. Markert, C. L., (1963). Lactate dehydrogenase isozymes: Dissociation and recombination of subunits. *Science, 140*, 1329–1330.

25. Markert, C. L., Shakelee, J. B., & Whitt, G. S., (1975). Evolution of a gene: Multiple genes for LDH isozymes provide a model of the evolution of gene structure, function and regulation. *Science, 189*, 102–114.

26. Knull, H. L., & Fillmore, S. J., (1985). Glycolytic enzyme levels in synaptosomes. *Comp. Biochem. Physiol., 81B*(2), 349–351.

27. Knull, H. L., (1990). Compartmentation of glycolytic enzymes in the brain and association with the cytoskeletal protein actin and tubulin. *Structural and Organizational Aspects of Metabolic Regulation*, 215–228.

28. Campanella, M. E., Chu, H., & Low, P. E., (2005). Assembly and regulation of glycolytic enzyme complex on the human erythrocyte membrane. *Proc. Natl. Acad. Sci. USA., 102*, 2402–2407.

29. Duka, T., Anderson, S. M., Holkins, L. B., Collins, Z., Raghanti, M. A., Ety, J. J., Hoff, P. R., Wildman, D. E., Goodman, M., Grossman, L. I., & Sherwood, C. C., (2014). Synaptosomal lactate dehydrogenase isoenzyme composition is shifted toward aerobic forms in primate brain evolution. *Drain Behav. Evol., 83*, 216–230.

30. O'Brain, J., Kla, K. M., Hopkins, I. B., Malecki, E. A., & McKenna, M. C., (2007). Kinetic parameters and lactate dehydrogenase isozyme activities support possible lactate utilization by neurons. *Neurochem. Res., 32*(4/5), 597–607.

31. Barros, L. F., & Deitmer, J. W., (2010). Glucose and lactate supply to the synapse. *Brain Research Reviews, 63*(1/2), 149–159.

32. Treshchenkova, Y. A., & Burlakova, E. B., (1997). Changes in the kinetic properties of aldolase and lactate dehydrogenase of the brain cytoplasm of mice exposed chronically to low dose γ-irradiation *Radiat. Biology Radioecology, 37*(1), 3–12 (in Russian).

33. Treshchenkova, Y. A., Goloshchapov, A. N., & Burlakova, E. B., (2003). Effect of low doses of phenosan on lactate dehydrogenase and microviscosity of microsomal membranes of brain cells. *Radiat. Biology Radioecology, 43*(3), 320–323 (in Russian).

34. Treshchenkova, Y. A., Goloshchapov, A. N., Shishkina, L. N., & Burlakova, E. B., (2015). Action of antioxidant phenosan and radiation at low doses on glycolytic enzymes in subcellular structures of the brain cells. *Bioantioxidants: Proceedings of IX International Conference (Moscow)* (pp. 135–140). Moscow: RUDN, (in Russian).

35. Pinto, P. V., Kaplan, A., & Van Dreal, P. A., (1969). Aldolase: Colorimetric method for its determination. *Clinical Chemistry, 15*(5), 339–349.

36. Bergmeyer, H. U., Bernt, E., & Hess, B., (1965). In: Bergmeyer, H. U., (ed.), *Methods of Enzymatic Analysis* (p. 737). London, New York.

37. Cornish-Bowden, A., (1979). *Principles of Enzyme Kinetics* (p. 280). Moscow: Mir, (Russian version).

38. Lowry, O., Rosenbrouch, N., Barr, A., & Randall, R., (1951). Protein measurement with the Folin Phenol reagent. *J. Biol. Chem., 193*(1), 265–275.

39. Kates, M., (1975). *The Technologue of Lipidology* (p. 322). (Russian version), Moscow: Mir, (in Russian).

40. Shishkina, L. N., Kushnireva, Y. V., & Smotryaeva, M. A., (2001). The combined effect of surfactant and acute irradiation at low dose on lipid peroxidation processes in tissues and DNA content in blood plasma of mice. *Oxidation Communications, 24*(2), 276–286.

41. Brin, E. F., & Travin, S. O., (1991). Modeling the mechanism of chemical reactions. *Chem. Phys. Reports, 10*(6), 830–837.

42. Shishkina, L. N., & Khrustova, N. V., (2006). The kinetic characteristics of lipids of animal tissues in autooxidation reactions. *Biophysics, 51*(2), 340–346 (in Russian).

43. Burlakova, Y. B., Palmina, N. P., & Maltseva, E. Y. L., (1991). A physicochemical system regulating lipid peroxidation in biomembranes during tumor growth. In: Carmen, V. P., (ed.), *Membrane Lipid Oxidation* (Vol. 3, pp. 209–236). Boston: CRC Press.

44. Shishkina, L. N., Kushnireva, Y. V., & Smotryaeva, M. A., (2004). A new approach to assessment of biological consequences of exposure to low-dose radiation. *Radiat. Biology Radioecology, 44*(3), 289–295 (in Russian).

45. Shishkina, L. N., Kudyasheva, A. G., Zagorskaya, N. G., & Taskaev, A. I., (2006). The regulation of the oxidative processes in the tissues of wild rodents caught in the

Chernobyl NPP accident zone. *Radiat. Biology Radioecology, 46*(2), 216–232 (in Russian).

46. Klimovich, M. A., Sergeichev, K. F., Karfidov, D. M., Lukina, N. A., & Shishkina, L. N., (2010). Biological effectiveness of X-ray irradiation at low doses under changing dose rate. *Technologies of Living System, 7*(8), 17–28 (in Russian).

47. Klimovich, M. A., Shishkina, L. N., Paramonov, D. V., & Trofimov, V. I., (2010). Interrelations between the physicochemical properties and composition of natural lipids and the liposomes formed from them. *Oxidation Communications, 33*(4), 965–973.

48. Jovanovic, S., Qingyou, D., & Sukhodub, A., (2009). M-LDH physically associated with sarcolemmal K_{ATP} channels mediates cytoprotection in heart embryonic H9C2 cells. *Int. J. Biochem. Cell Biol., 41*(11), 2295–2301.

49. Lu, M., Anmar, D., Ives, H., & Gluck, S. L., (2007). Physical interaction between aldolase and H+-ATPase is essential for the assembly and activity of the proton pump. *J. Biol. Chem., 282*(34), 24495–24503.

50. Gutowicz, J., & Terlecki, G., (2003). The association of glycolytic enzymes with cellular and model membrane. *Cell. Mol. Biol. Lett., 8*, 667–680.

51. Terlecki, G., Czapinska, E., & Gutowicz, J., (2002). The association of glycolytic enzymes with cellular and model membrane. *Cell. Mol. Biol. Lett., 7*, 895–903.

52. Terlecki, G., Czapinska, E., Rogozik, K., Lisowski, M., & Gutowicz, J., (2006). Investigation of the interaction of pig muscle lactate dehydrogenase with acidic phospholipids at low pH. *Biochim. Biophys. Acta, 1758*(2), 133–144.

53. Treschenkova, Y, A., Molochkina, E. M., Shishkina, L. N., & Burlakova, E. B., (2015). Alteration of glycolytic enzymes in mice brain synaptosomes after whole body gamma-irradiation at low dose of various intensities. *Low and Superlow Fields and Radiations in Biology and Medicine* (p. 201). Science proceeding of the Congress, Saint–Petersburg.

CHAPTER 10

Antioxidants in Aquatic Ecosystems: Role in Adaptation of the Organisms to the Changing of Global and Local Factors

IRINA I. RUDNEVA, IRINA I. CHESNOKOVA, TATYANA B. KOVYRSHINA, TATYANA V. GAVRUSEVA, and VALENTIN G. SHAIDA

The A.O. Kovalevsky Institute of Marine Biological Research of RAS, Nahimov Av., 2, 299011, Sevastopol, Russian Federation, E-mail: svg-41@mail.ru

ABSTRACT

Different kinds of low molecular weight antioxidants are present in aquatic organisms. They play an important role in their adaptation to the changing environment, including biotic, abiotic, and anthropogenic factors. They participate both in the antioxidant defense against ROS and play a role in many physiological functions of the aquatic organisms, namely reproduction, growth, development. Additionally, the content and composition of antioxidants of marine species are very important for people consumption. Fish, crustacean, mollusk tissues are rich in the antioxidants; therefore, they are the main sources of these essential compounds for people diet. The current study aimed to compare the low molecular weight antioxidants diversity in aquatic organisms and to determine the effects of environmental changes on antioxidant status of the different species, their response on these changes and the role of antioxidants in the adaptation related to harmful environmental conditions. This study will provide to establish a link between the global and local harmful factors of the marine environment and the role of antioxidants in marine animals response.

10.1 INTRODUCTION

The atmosphere of Earth was originally reduced, but cyanobacteria capable of oxygenic photosynthesis had evolved. As a result, molecular oxygen appeared in significant amounts in the atmosphere. Growth of its concentration and accumulation in the atmosphere and shallow aquatic habitats led to the evolution of aerobic eukaryotic organisms [1, 2]. Metabolism of the living organisms is associated with oxygen consumption involving it into the major biochemical and physiological pathways. However, O_2 has two unpaired electrons, and the univalent reduction of the molecular oxygen produces reactive oxygen species (ROS) such as superoxide radical (O_2^-), singlet oxygen ($^1O_2$), hydrogen peroxide (H_2O_2), hydroxyl radical (HO•) and finally water H_2O. Nitric oxide (NO•) may also form, and it plays an important role in the metabolic processes. ROS are generated by physical, chemical, and metabolic processes that convert O_2 into ROS.

Additionally, in marine environments, the absorption of solar radiation and especially of its UV-spectra by dissolved organic matter leads to the formation of phytochemical compounds, including ROS. Hydrothermal vents and the abundance of hydrogen sulfide and oxygen lead to the oxidation of the H_2S and generate ROS also as oxygen and sulfur-centered radicals [1]. They may harmful for biological systems, because they result in oxidative stress, the generation and accumulation of ROS in tissues, which damage cell membranes, lipids, proteins, and DNA. Lipids and particular polyunsaturated fatty acids (PUFA) are the main substrates of the oxidation and lipid peroxidation (LPO). LPO is a very important consequence of oxidative stress.

ROS production can be induced by external factors. Environmental pollution may well be one of the strong causes of the ROS generation in the organisms. Chemicals represent in the environment (transition metals Cu and Fe, which catalyze the production of hydroxyl radicals through Fenton reaction, biphenyls, quinones, nitroaromatics, and drugs can also induce production of superoxide by redox cycling and via biotransformation reactions [3–5].

At the other hand, ROS can also act in signal transduction and participate in the formation of signal molecules (prostaglandins, leukotrienes, etc.). They play an important role in the expression of several transcription factors, heat shock-inducing factor, nuclear factor, the cell-gene p53, nitrogen-activated protein kinase, etc. Oxidative stress also plays a role in apoptosis in two pathways, through the death-receptor and the mitochondrial way [1, 2, 4, 6–8].

The antioxidant system plays a key role in the inactivation of ROS, and thereby controls oxidative stress as well as redox signaling. The antioxidant system presents in all living organisms. They have an antioxidant defense to protect against the toxic effects of ROS. Antioxidant defense of aquatic organisms depends on many abiotic (temperature, season, salinity, pH, oxygen concentration) and biotic (feeding conditions, diet quality, and quantity), physiological status, age, and other both exogenous and endogenous factors [9, 10]. The antioxidant system, composed of nutrients, enzymes, and small molecular weight molecules, differs between tissues, is related to the lifestyle and depends on the evolutionary history of a species [11]. Low molecular weight antioxidants are non-enzymatic compounds such as vitamins A, E, K, and C, carotenoids, SH-containing amino acids and peptides (glutathione (GSH)) and small-molecule antioxidants such as uric acid, urea, mycosporine-like amino acids (MMAs), etc. [1, 3]. Several of them such as vitamin E, carotenoids are strictly obtained as a nutritional antioxidant in vertebrates and invertebrates, while the others (for instance, vitamins C, A) are synthesized in few fish species.

These components are present in aquatic organisms, and they play an important role in their adaptation to the changing environment. At the other side, the content and composition of antioxidants of organisms, especially marine species, are very important for people consumption, because seafood has been recognized as a high-quality, healthy, and safe food type. Fish, crustacean, mollusk tissues are rich in the antioxidants, and some of them are present in seafood in high concentrations. Therefore, they are the main sources of these essential compounds for people diet and one of the most important food commodities consumed worldwide.

However, some external factors, especially global climate change and local anthropogenic pollution, may cause stress to aquatic organisms, damage their health, growth, and reproduction. The researchers suggest that oxidative stress in various aquatic organisms can be correlated with several types of stressful conditions, such as: chemical pollution, hypoxia/anoxia, microbial, and viral infections and parasitic invasion, dietary influence, feed deprivation, acute temperature changes, algal blooms, UV-irradiation, etc. [10]. At this case, accumulation of the ROS products in the seafood can also be a source of harmful contaminants with potential to impact on human health.

Climate changes are accompanied by the temperature increase caused the emission of greenhouse gases (CO_2, CH_4, SO_2, nitrogen oxides, etc.). Seawater temperature of 2–3°C above long-term average summer temperature results stress in aquatic organisms, led oxidative damage. High temperature elicits a series of physiological responses, such as increasing of standard body metabolism. The temperature has been known to alter metabolic rate, enzyme activities, oxygen consumption, and oxidative stress, attributed with the changes of prooxidant/antioxidant ratio of the organism [12]. The synergistic effects of global climate changes and local anthropogenic pollution of aquatic ecosystems lead to oxidative stress in the organism and change the balance between the prooxidant and antioxidant processes, which is principal in physiological homeostasis.

Marine animals are frequently exposed to unfavorable impact, that induces ROS, including adverse situations (low oxygen, lack of food or water, temperature changes, salinity, and pH fluctuations, etc.). A hypometabolic state is a response to unfavorable living conditions, and it is an important survival strategy for many organisms, exposed to environmental stressors. At this case, the resistance and tolerance of marine animals to hypometabolic periods depends on the effectiveness of the antioxidant system, which provides protection against ROS [13].

The aim of the present study is to compare the diversity of low molecular weight antioxidants in aquatic animals and to show the response of the antioxidants in different species on the local and global environmental changes. The role of antioxidants in the adaptation of aquatic organisms related to harmful environmental conditions is discussed.

10.2 WATER SOLUBLE ANTIOXIDANTS

There are many hydrophilic non-enzymatic compounds, which are represented by ROS scavengers. Among them, ascorbic acid, thiol-containing amino acids, GSH, amines, and peptides, metallothioneins (MTs) are represented in the tissues and cells of the organisms and play an important role in the defense against ROS.

10.2.1 ASCORBIC ACID

Ascorbic acid (Figure 10.1) functions as a reductant source for many ROS, and it scavenges both H_2O_2 and O_2^-, HO•, lipid hydroperoxides without enzyme catalysis. Additionally, it participates in recycling α–tocopherol to its reduced form. Plants and some animals can synthesize vitamin C *de novo*, and other animals can obtain it through their food also [1]. The researchers reported, that plasma concentration of ascorbic acid changes caused physical activities and correlated directly with cortisol level [14].

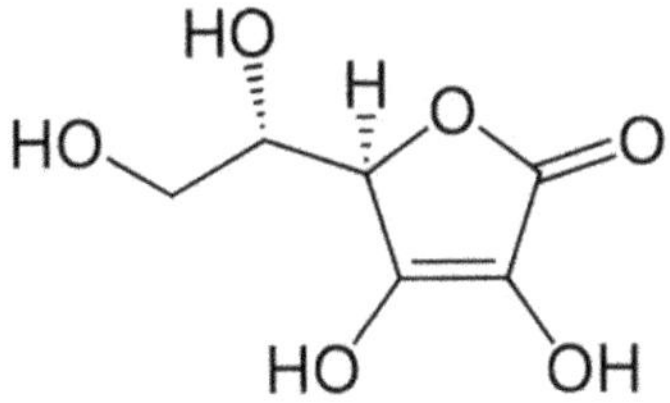

FIGURE 10.1   Ascorbic acid.

Ascorbic acid has been detected in many aquatic animals in different concentrations. The concentration of vitamin C found for the group of elasmobranchs fish species was lower than that for teleosts [11]. Vitamin C is an essential component for fish reproduction. High concentrations of the ascorbic acid have been found in steroid producing tissues, such as gonads and adrenal gland. The authors suggested that ascorbic acid may participate in the steroid synthesis and may act as a cofactor or as a regulator in the synthesis of estrogens in the follicle cells. During gametogenesis, ascorbic acid concentration increases in ovaries of the fish [15]. At the other side, the level of vitamin C in the ovaries of the female, eggs, and hatching larvae depended on the diet of the broodstock. The researchers showed, that in order to ensure high vitamin C level for successful ontogenetic developments of sea bass and sea bream embryos and yolk sac larvae, the diet of the broodstock needed to contain 2 g kg^{-1} feed every second day [15].

In the brain of freshwater Indian catfish *Heteropneustes fossils* during temperature exposure from 25°C (control) to 37°C at various periods from 60 min to 240 min ascorbic acid content was changed. It increased at 32°C and 37°C significantly from 85.99 ± 7.71 µg g^{-1} wet tissue in control to 180.08±8.21 µg g^{-1} wet tissue. The growth of the ascorbic acid level was directly correlated to the increase of LPO. The authors postulated that vitamin C might act as a protective agent against free radicals. In conditions of vitamin C deficiency symptoms of oxidative stress, including cell damage, might be expected to appear as the oxidant defense system are challenged to compensate for the lack of ascorbic acid. As temperature exposure resulted in a significant increase in LPO, the increase of ascorbic acid concentration was probably due to the reducing of ascorbic acid in the brain during temperature dependent free radical generation [12]. It is very important for the understanding of the protective function of ascorbic acid in primitive vertebrates, and to apply this parameter for the evaluation of oxidative stress in them at the case of climate changes.

The fluctuations of the level of ascorbic acid were shown to be changed related to environmental conditions and pollution. The authors displayed the increase of ascorbic acid level in the tissues of perch (*Perca fluviatile*) inhabiting waters contaminated bleached kraft mill effluents as compared with the animals from the non-polluted site. The increase of ascorbate concentration in fish tissues from the highly polluted site was the response to oxidative stress caused unfavorable living conditions [3].

10.2.2   GLUTATHIONE (GSH)

GSH (Figure 10.2) is an endogenous antioxidant tripeptide (GLU-CYS-GLY), which forms a thiol radical that interacts with a second oxidized GSH and forms a disulfide bond (GSSG). It represents the first line of defense against ROS, participating in many cellular metabolic pathways by directly neutralizing pro-oxidants or participating in enzymatic reactions, where it acts as a substrate. The ratio of GSH/GSSG is used as a biomarker of oxidative stress in the organism [4, 5].

GSH oxidizes H_2O_2 and organic hydroperoxides in oxidized form GSSG spontaneously or enzymatic by GSH peroxidase, then GSSG reduces again to GSH according to the equation:

$$GR\text{-}GSSG + NADPH^+ + H^+ \rightarrow 2GSH + NADP^+$$

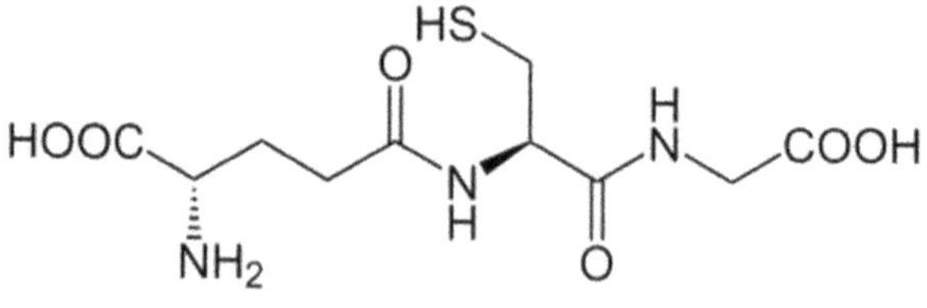

FIGURE 10.2   Glutathione (GSH).

In our previous studies, we demonstrated high variability of the GSH level in various aquatic organisms, which reflected the adaptive strategy of fish species and invertebrates to oxidative stress and their ability to cope with the environment [16, 17]. Its level depended on tissues specificity and ecological peculiarities of fish. For instance, in fish blood, GSH content was relatively higher in pelagic active fish such as horse mackerel and in sluggish benthic forms such as scorpion fish, while in other species belonging to suprabenthic and suprabenthic/pelagic groups the values were intermediate [17]. High GSH level in red blood cells in marine elasmobranch has been found, and it could compensate for the low level of their enzymatic status [16, 18]. The interspecies differences of the total GSH content in fish tissues as well as interspecies variations have been shown in red blood cells of teleosts [17]. GSH concentration in fish red blood cells ranged between 22.8 and 60.0 μg%. In muscle, liver, and gonads, the interspecies variations of the total GSH content were less than in blood. GSH in erythrocytes protects the hemoglobin against spontaneous oxidation to methemoglobin and xenobiotics damage. Erythrocytes depleted of GSH become very sensitive to environmental stress because fish hemoglobin has a higher tendency to oxidation as compared with other vertebrates [19]. At the other side, the researchers noted that in four commonly cultured freshwater fish species (tilapia *Oreochromis niloticus*, carp *Cyprinus carpio,* trout *Onchorhynchus mykiss*, and catfish *Clarias garipienus*) GSH concentration in the liver tissues were similar in most cases for all fish species, except a few differences [20].

We also found that the GSH content changed during early development of the aquatic animals. During *Artemia,* early development from dormant cysts to *nauplia* the GSH level decreased progressively from 60.25 μg g^{-1} wet weight to 13.70 μg g^{-1} wet weight [21].

Aquatic organisms are commonly used in biological monitoring of the water. Biomarkers in aquatic species are regarded as important for detecting stressors, such as the presence of pollutants and changes in environmental factors (pH, temperature, dissolved oxygen, etc.). GSH content may be one of the appropriate biomarker of oxidative stress in the animals caused negative factors, especially the hepatic ratio of oxidized to reduced glutathione (GSSG/GSH) [5].

The researchers showed the increase of GSH level in three crustacean species *Asellus aquaticus, Gammarus pulex* and *Potamon ibericum* collected from the polluted site as compared with the non-pollutes area, which was demonstrated the response of animals on oxidative stress [22].

Petroleum and its derivatives are the main pollutants of the aquatic systems. Several studies have demonstrated the effects of exposure to them on the antioxidant defense of fish. The authors have shown the increase in hepatic content of reduced GSH together with decreased glutathione reductase (GR) activity in freshwater fish *Prochilodus lineatus* after 24 and 96 h of exposure to a water-soluble fraction of gasoline (WSFG). In the gills, the GSH content increased after 24 h of WSFG exposure. As the gills is the first organ to be exposed to pollutants, the antioxidant defenses, including GSH and GSH-dependent enzymes were triggered immediately upon exposure to WSFG and were able to prevent the tissues against oxidative stress [23]. However, GSH level and GSH-Px activity were significantly decreased ($p < 0.05$) in gill, muscle, kidney, and heart tissues of *Capoeta trutta* from polluted site compared to those collected from the reference area. Because the apparent decrease in GSH detoxification system was observed in the gill, which directly contacted with xenobiotics, the GSH system of the gills is a good indicator for environmental stress [24]. The researchers also reported that Cu caused stress in fish gills and acclimation with the induction of GSH, which were important in the protection against metal damage [25]. Therefore, GSH plays an important role in the adaptation of the aquatic organisms to the environmental stress both as low molecular weight component and in the enzymatic GSH-containing system.

The data of GSH levels responses in fish species from 10 laboratory studies and 17 field studies was summarized in the review of Van der Oost [26]. A significant increase in GSH levels was observed in 60% of the laboratory studies and 35% of field studies. A significant increase in GSSG levels was observed in two of the laboratory studies (67%) and two of the field studies (50%). Strong increase (>500% of the control) was detected only in the liver of English sole caught in a highly polluted area. The authors have concluded that hepatic GSH and GSSG levels in fish cannot yet be considered as valid stress biomarkers for ERA purposes, but could be used as the potential biomarkers for oxidative stress [26]. Additionally, the changes in GSH metabolism and fluctuations of the ratio GSSG/GSSH in fish tissues were summarized in a review by Stoliar and Lushchak [5]. The authors postulated, that the elevation of GSH level and RI GSH reflects the adaptation of fish to unfavorable living conditions, while GSH depletion is usually associated to enhancing of peroxidation processes in the cell membrane, which result in toxic and pathologic consequences.

10.2.3   *AMINO ACIDS, AMINES, AND PEPTIDES*

The low molecular weight of nitrogenous constituents comprises between 0.5 and 1% of the total mass of the fish muscle. Over 95% of the non-protein muscle constituents of fish and shellfish consist of amino acids, imidazole dipeptides, guanidine compounds, trimethylamine oxide (TMAO), urea, betaines, nucleotides, and compounds related to nucleotides. In general, they participate in antioxidant defense mechanisms as synergists or primary antioxidants [27].

Low molecular weight scavengers namely amino acids, especially sulfur-containing amino acids (Figure 10.3), several amines, nucleotides, and peptides play a role in the conjugation with the metals and other xenobiotics and neutralize them. They believed to be important metal chelators present in fish and aquatic invertebrates. Amino acids are also suggested to have antioxidant properties as reaction products with carbonyls from oxidizing lipids [27].

Free amino acids in fish muscle participate in the regulation of osmotic pressure. Their concentrations were higher in marine species than in their freshwater counterparts, and the highest was observed in crustacean muscle [27]. In our studies of serum SH-groups of the amino acids in several Black Sea fish species, we found that their level ranged in serum in different fish and depended on their taxonomic position. We showed that the content of the total SH-groups was significantly higher in the serum of Black Sea elasmo-branch *Squalus acanthias* ($p < 0.01$) as compared with the teleosts, and the level of non-protein SH- groups were greater than the protein ones [16, 17].

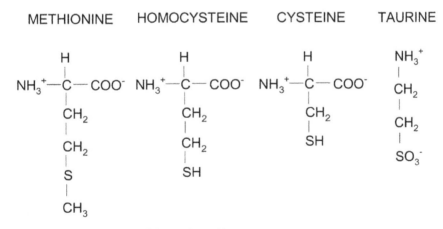

FIGURE 10.3 Sulfur-containing amino acids.

In the Black Sea teleosts, high values of the serum SH-groups were indicated in *Scorpaena porcus*, *Spicara smaris*, and *Merlangus merlangues euxinus,* while in other species they were less. In all tested teleost fish concentration of protein, SH-groups were higher than the content of non-protein, an elasmobranch, and the opposite trend was marked: in *S. acanthias* concentration of non-protein SH-groups was greater than the protein ones. The ratio of protein and non-protein SH-groups has also differed among examined fish species [16, 17].

Among non-enzymatic peptides, which contain high levels of SH-groups, MTs play an important role in the protective mechanisms of aquatic organisms. They are low molecular weight proteins which contain approximately 60 amino acids, among them 20 are cysteine, and none of them are aromatically occurring mainly in the cytosol and in the nucleus and lysosomes. In some cases, MTs contain high cysteine content that estimated at 30%. The thiol groups of cysteine residues enable MTs to bind particular cations of heavy metals [28]. Because they can conjugate with the transition metals, their participation in antioxidant defense can be explained by the high content of thiols and the particular metal binding/release dynamics intrinsic to these proteins. The participation of MTs in antioxidant defense can also be indirectly through the mechanism of the regulation of the ratio between metal ions within the cell in deposited form and unbound, potentially toxic form. Metal-keeping function and possible participation in the antioxidant defense, expressed by concentrations of complexes of MT with metal ions (MT-Me) and total level of MTs (MT-SH), can be alternative/complementary characteristics of MTs in aquatic animals inhabiting polluted environment [5].

MTs have been associated with zinc homeostasis and metal detoxification by binding to several (7–12) heavy metals. Tissues directly involved in metal uptake, storage, and excretion have a high capacity to synthesize MTs. In fish, these proteins have been identified in high concentrations in gills, liver, and some other tissues. MTs bind to an excess of essential or toxic metals and protect the organism against toxicity by restricting the availability of these cations at detrimental sites. Some toxic metals (Ag, Cd, Cu, Hg, and Zn) have a high binding affinity for cysteine [28, 29].

Recently the researchers have found the low molecular weight amino acids MMAs, which play the protective role against UV-radiation and its damage effects. Some MMAs show antioxidant activity. While these compounds are synthesized in photoautotrophs, they acquired by aquatic animals through their food. In corals, the mycosporine-glycine level decreases significantly upon exposure to prolonged high-temperature conditions, UV, and hyperoxia [1].

The antioxidant properties of carnosine and anserine are also observed, because they involve both the amino acid composition and the peptide linkage between β-Ala and His, whereas neither imidazole group methylation nor free carboxyl-group amidation seems to interfere greatly with the antioxidant activity. TMAO is known for its synergistic effect with tocopherol in the lipids. The polyamines, putrescine, spermidine, and spermine inhibit lipid oxidation by free radical inactivation and inhibition of iron-catalyzed reactions [30]. Antioxidant activity of the polyamines increases with an increasing number of amine groups as following: spermine > spermidine > putrescine [31].

Urea and uric acid (Figure 10.4) can quench both $^1O_2$ and HO. Uric acid inhibits oxidative reaction by quenching and/or scavenging singlet oxygen, iron chelation, and free radical scavenging [32]. It is found in high concentrations in marine invertebrates and in blood serum of elasmobranch species, and it can be a potent antioxidant [1]. In elasmobranch tissues the antioxidant properties of urea derivatives (hydroxyurea, dimethylurea, thiourea), together with GSH and vitamins C and E may contribute to their protection against ROS compared to tissues from other vertebrates [33]. These antioxidants appear to act synergistically to prevent oxidative damage in tissues of elasmobranchs.

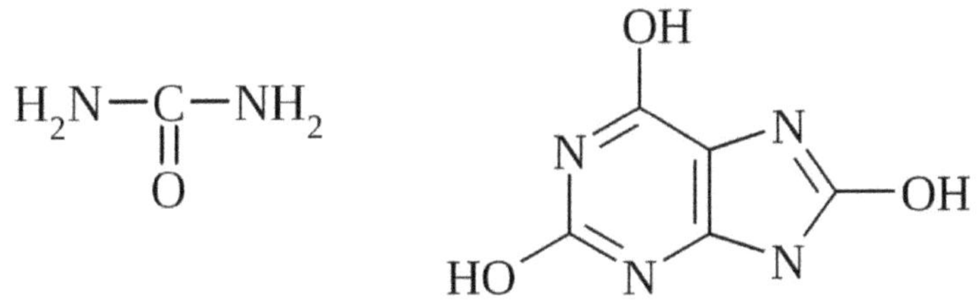

FIGURE 10.4 Urea and uric acid.

In aquatic animals, there are some other pathways for the protection against ROS. For example, Crustacea use a cooper-dependent oxygen transport system, which was developed during evolution and was linked with the induction of cooper-dependent oxygen transport system [1]. In sea urchin eggs, that is, the good model of the ROS studies in early development stages of the aquatic invertebrates, the transferrin-like, iron-chelating proteins were identified, that could potentially be very useful in preventing hydrogen peroxides radicals [1]. Different kinds of low molecular weight antioxidants are also found in aquatic organisms, but in some cases their nature is unknown. Generally, information of the antioxidant role of poly-amines and uric acid in skeletal muscle of aquatic organisms is limited and require further studies [30].

10.3 FAT-SOLUBLE ANTIOXIDANTS

Lipophilic compounds which recognized as antioxidants play an important role in the protection of cell membranes against ROS. It's a large group of low molecular weight compounds, which are synthesized both by plants, algae, and animals, or obtain with the diet.

10.3.1 *CAROTENOIDS*

Carotenoids (tetraterpenoids) (Figure 10.5) are lipid-soluble compounds that protect against $^1O_2$ because they have highly conjugated double bonds. They quench ROS and can prevent LPO in marine animals [34, 35]. Carotenoids are widely distributed naturally occurring pigments, usually red, orange or yellow. They are synthesized by plants, algae, several bacteria and fungi. Animals obtain the necessary carotenoids either directly from the diet or modify the dietary carotenoid precursors through metabolic pathways (oxidation, reduction, cleavage of double bonds or epoxy bonds, and transla-tion of double bonds) to fit their requirements [36].

There are over 600 known carotenoids, which are grouped into two classes, namely xanthophylls (which contain oxygen), and carotenes (which are purely hydrocarbons, and contain no oxygen) [36–38]. They play a role in growth and photosynthesis, and they are the main dietary source of vitamin A in animals. Carotenoids provide protection against several stressors, including UV radiation, ROS and free radicals, have important roles in vision, and act as precursors of transcription regulators and in the immune system. They can participate in the mechanisms, associated with reduced risk of oxidative stress [39].

Because carotenoids have a specific chemical structure, they play an important role in their oxygen scavenge properties [40]. The hydrocarbon carotenoids have oxygenated derivatives, or xanthophylls, where the hydroxy, keto, epoxy, or aldehyde oxygen-containing groups are responsible for their polarity, solubility, and overall chemical properties [36]. Carotenoids (espe-cially lutein and zeaxanthin) have the ability to absorb light in the visible range of the spectrum and provide protection against blue light and UV radiation [41]. Different kinds of carotenoids demonstrate different ability to protect against ROS. For instance, astaxanthin has free radical antioxidant activity several folds higher than that of β-carotene and α-tocopherol [42]. Astaxanthin has been reported to prevent the progression of some human deceases such as diabetic nephropathy mainly through the ROS scavenging effect in mitochondria of mesangial cells and to inhibit oxidative stress and inflammation and enhance immune system response [36, 43].

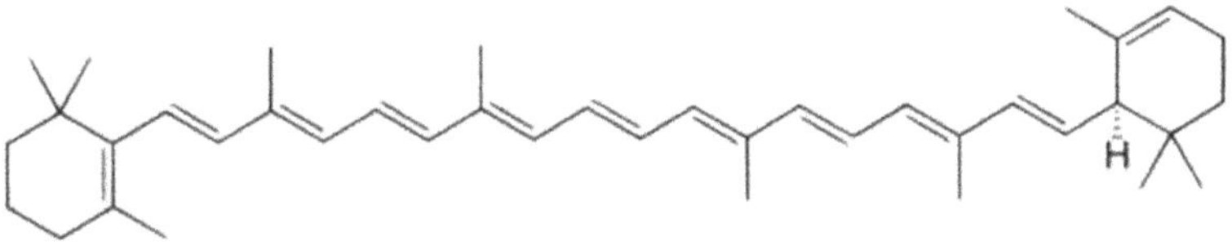

FIGURE 10.5   General chemical structure of the carotenoid molecule.

On the other hand, β-carotene may serve either as an antioxidant or as a prooxidant, depending on its intrinsic properties and diet concentration [44]. The researchers have shown that high doses of β-carotene result oxidative stress in the animal tissues [45]. The integrated antioxidant system, comprising endogenous antioxidant enzymes and dietary antioxidants, provide to cope with immune-mediated oxidative stress. The researchers tested the interaction between dietary supplementation with carotenoids and immune challenge on immune defenses and the activity of the antioxidant enzymes SOD and CAT, in the amphipod crustacean *Gammarus pulex* [46]. They found that dietary supplementation increased the concentrations of circulating carotenoids and hemocytes in the hemolymph, while the immune response induced the consumption of circulating carotenoids and a drop of hemocyte density. They suggested the presence of specific interactions of dietary carotenoids with endogenous antioxidant enzymes and potential importance of carotenoids in the evolution of immunity and/or of antioxidant mechanisms in crustaceans. They proposed that carotenoids may take over the action of the antioxidant enzymes, reducing their relative activity and therefore, their maintenance costs. However, carotenoids are known as free radical scavengers, and they can stimulate the activity of these antioxidant enzymes, which would enhance detoxification effectiveness upon immune activity. Therefore, carotenoids may increase the reduction of free radicals, alleviating the costs associated with oxidative stress [46].

Carotenoid pigments are known to offer protection from UV radiation to zooplankton crustaceans in transparent lakes. The extent of UV induced damage is largely modulated by the presence of competent photoprotection by, e.g., carotenoids, MAAs (see the Section 10.2) and other pigments, and photorepair systems [36]. The researchers noted that the carotenoid content in marine pelagic copepods is much lower than in freshwater species, while the content of other natural sunscreen compounds such as the transparent MAAs is similar [47]. It has been negatively correlated with the zooplankton mortality: the mortality rates due to predation may reach 10% per day in the ocean [47], and fish predation on marine pelagic copepods may be responsible for more than 69% of the mortality rate of the fish. Therefore, a preference for the non-pigmented MAAs over carotenoid pigments was explained by providing protection against UV radiation without increasing visual predation risk. It may be an adaptation of wild marine copepods to avoid predation [36].

In marine organisms, carotenoids are found in the gut, gonads, gills, liver, muscle, and other tissues. Filter feeding mollusks, namely bivalves, accumulate carotenoids obtained both directly from their dietary microalgae, crustaceans such as copepods and shrimp, can rapidly produce astaxanthin after absorption of precursors from dietary phytoplankton (mainly β-carotene and some zeaxanthin after digestion). Fish also obtain carotenoids through their diet directly at the case of herbivores species or through the modifications during later metabolic pathways [36].

Aquatic animals conspicuously accumulate carotenoids in their gonads in high concentrations, which is assumed to be essential for their successful reproduction and further survival and development of their eggs and early larvae [17, 36, 48]. In our studies, we observed the variability of carotenoid concentration in different Black Sea fish species [17]. The greatest content of carotenoids was shown in *M. barbatus ponticus* tissues (13.46 μg g^{-1} lipid) while in other tested teleosts it varied from 2.09 μg g^{-1} lipid in *S porcus* to 6.93 μg g^{-1} lipid in *S. smaris*.

Crustaceans cannot synthesize carotenoids *de novo,* and a dietary source of these pigments is required. During sexual maturation, most crustaceans accumulate carotenoids in the hepatopancreas, ovarian, and during vitello-genesis, these are transported in the hemolymph as carotenoglycolipoproteins to accumulate in the eggs as part of the lipovitellin protein [49, 50]. The highest concentrations of vitamin A and β-carotene were also found in the hepatopancreas of crustacean when compared to the muscle and ovarian [48].

Among the carotenoids, the red pigment astaxanthin is the main pigment in the body and eggs of herbivorous crustacean zooplankton, shrimp, and salmonoid fish, providing pigmentation and antioxidant protection [37]. Aquatic organisms must either acquire it directly through their diet or modify the dietary carotenoids [36]. Decrease of astaxanthin concentration at the basis of the food web may decrease antioxidant protection of predatory fish to oxidative stress. The researchers observed the important role of astaxanthin in the processes of reproduction, maturation, growth, and development and survival of larvae.

Carotenoid-rich diet is very important for fish and crustacean in aquaculture, because the optimal concentration provides the high survival of the early life stages, growth, and development. *Artemia nauplia* is the most effective start diet for the aquaculture animals, and the carotenoids content in the *nauplia* is an important commercial characteristic of *Artemia* strains. However, carotenoids content in *Artemia* varies during the life cycle, depends on the geographical strain and the feeding of microalgae composition. Additionally, during its development, the carotenoids composition can be changed, caused chemical reactions and modifications [36].

In our study, we also demonstrated the changes in carotenoids concentration during the early development of *Artemia*. As we described previously, carotenoid concentration increased in hydrated cysts as compared with the dormant cysts (40.85 µg g^{-1} lipid and 36.65 µg g^{-1} lipid correspondingly), and decreased in hatching *nauplii* (31.96 µg g^{-1} lipid) [21]. We also showed the progressively decreasing of carotenoid concentration during the embryogenesis of the Black Sea sculpin *Proterorhinus marmoratus*, and we found the negative correlation between the carotenoids content in the fish eggs in different stages of their development and antioxidant enzyme activities [21, 51]. We suggested, that in early stages of embryonic development of marine animals, the low molecular weight antioxidants, including carotenoids play an important role against oxygen damage. During embryogenesis, the concentration of low molecular weight antioxidants decreases, while an enhancement of antioxidant enzyme activities was detected, possibly as a result of gene expression and in hatching larvae induction of antioxidant enzyme is the main mechanism of protection against oxidative stress. During the next period of their life cycle larvae feed extensively on microalgae rich in carotenoids, vitamins, and other antioxidants and mobilize them from microalgae and accumulate in the body, especially in eggs and gonads.

10.3.2   VITAMIN A

Carotenoids (especially α-carotene, β-carotene, and β-cryptoxanthin) are also important precursors of vitamin A (Figure 10.6), which is essential for normal growth and development of the aquatic animals, immune system function, and vision [36]. Several investigators suggest that vitamin A is probably inessential, but carotenoids, which is its precursors, may be essential in crustacean [48]. Together with carotenoids, vitamin A is a critical factor of crustacean and fish reproduction, and they cannot be synthesized in these taxonomic groups, and they obtain these essential components through food webs.

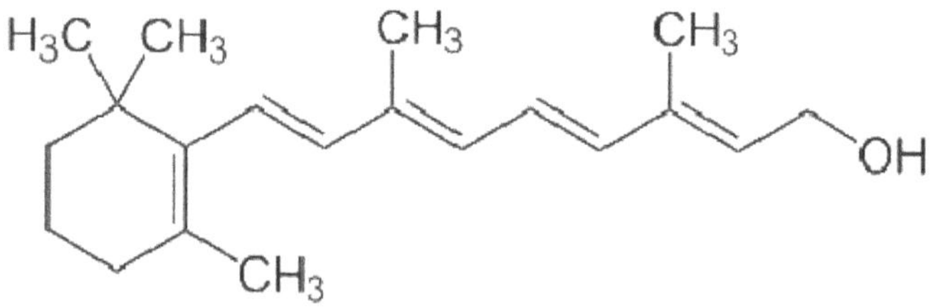

FIGURE 10.6   Vitamin A.

According to our previous data, vitamin A content varied insignificantly in Black Sea teleost fish tissues [17]. At the other side, unfavorable environmental conditions may change vitamin A level in the aquatic organisms. Hypoxia is a common negative event in aquatic environments. Decrease of oxygen concentration (O_2) in water has resulted in hypoxic stress and significant changes in antioxidant defense systems in fish and invertebrates. However, no differences were observed in the concentration of vitamin A in the muscle, liver, and kidney tissues of juvenile rainbow trout under different oxygen levels (3.5 mg O_2/l, 4.5 mg O_2/l, and 7 mg O_2/l) as compared with the levels of vitamin C and vitamin E [52]. Metal pollution may also change the metabolism of vitamin A in the fish. The researchers showed that yellow perch (*Perca flavescens*) from Cd-contaminated lakes had significantly higher concentrations of liver dehydroretinol and dehydroretinol esters than did animals from reference site [53]. The authors suggested that the enzymes and the binding proteins involved in vitamin A metabolism are inhibited by the presence of Cd, because the yellow perch living in contaminated lakes display high liver vitamin A_2, providing protection against oxidative stress induced by metal exposure. The increase in the tissue of wild fish yellow perch vitamin A levels could serve better counteract the oxidative stress caused by Cd exposure. Their findings illustrated that vitamin A_2 homeostasis could be altered as a consequence of chronic exposure to low Cd concentrations. They proposed also that the percentage of fish liver free dehydroretinol can be considered as a biomarker of for in situ Cd exposure and be used for environmental risk assessment [53].

10.3.3 TOCOPHEROL (VITAMIN E)

Tocopherols, especially α-tocopherol (vitamin E, Figure 10.7), are lipid-soluble antioxidants which are located within the bilayers of cell membranes and protect them against ROS [7]. The term vitamin E includes two families of compounds, tocopherols, and tocotrienols, which have a similar structure and differ only in their saturation; α-tocopherol is the most active and abundant component in cell membranes of the animals [11]. The biological role of α-tocopherol is widely be recognized as a lipophilic antioxidant, protecting cellular macromolecules (DNA, proteins, lipids), membranes of cells, and organelles from oxidation by scavenging organic free radicals. Its activity may be especially important during normal health and life functions, such as growth, development, reproduction in fish and crustacean or under the conditions of the oxidative challenge such as infection, stress, and pollution. In most fish, the liver is the main storage organ of vitamin E.

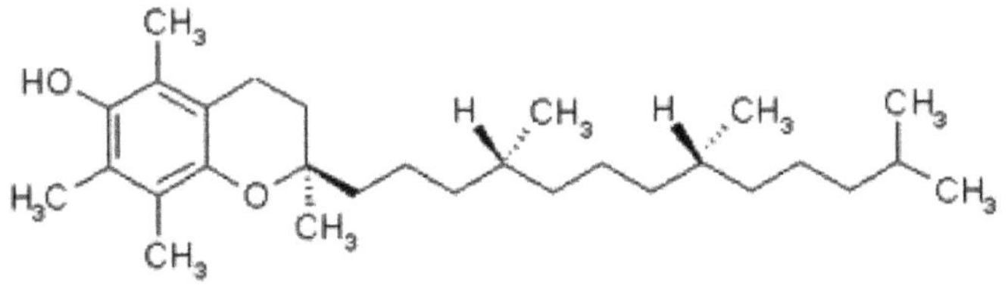

FIGURE 10.7 (RRR)-α-tocopherol – the original and the most biological active structure among the tocopherols family.

It is a multifunctional antioxidant, which is the result of its ability to quench both $^1O_2$ and peroxides. Additionally, vitamin E protects vitamin A and essential fatty acids from oxidation in the body cells and prevents the damage of the tissues [48]. In fish tissues, vitamin E plays an important role in antioxidant defense in vivo, and its content is highly related to the stability of lipids and fats. Together with vitamin C, vitamin E can act as antioxidants in a synergistic manner. Vitamin C acts as a terminal element in the protection against tissue damage caused by free radicals, but when both vitamins are present, the major function of vitamin C is restoration of vitamin E. However, vitamin E is strictly obtained as a nutritional antioxidant both in vertebrates and invertebrates, while vitamin C is synthesized in few fish species [11].

In our studies of the tissues of several Black Sea fish species vitamin E concentration ranged between 0.37 µg g^{-1} wet weight in *Scorpaena porcus* to 7.01 µg g^{-1} wet weight in *Mullus barbatus ponticus* [17]. The authors also noted that the concentration of vitamin E was lower in the muscle of the group of elasmobranchs compared with that of teleosts. In their study, the concentration of vitamin E in elasmobranch and teleost fishes were within the range of 0.011–0.038 µg g^{-1} [11]. At the other side, in several fish species, the concentration of the vitamin E is higher: in the liver of *Scophthalmus maximus*, it was estimated as 160 µg g^{-1}, in *Hippoglossus hippoglossus* 71 µg g^{-1} and in *Sparus aurata* 54 µg g^{-1} [54]. Walking catfish (*Clarius brachysoma*) having omnivorous diet composition showed the highest total α-tocopherol (29.65 mg kg^{-1}) level in the muscle. However, tilapia (*Tilapia niloticus*) showed very low tocopherol concentration (0.08 mg kg^{-1}). Freshwater shark (*Wallago Attu*) is carnivorous feeder, and it showed 4.02 mg kg^{-1} α-tocopherol in muscle. Similarly, rohu (*Labeo rohita*) having herbivorous feeding habit contained 6.61 mg kg^{-1} α-tocopherol. The authors concluded, that tocopherol content in the muscle of the examined freshwater fishes was not significantly influenced by their feeding habits and taxonomic position, and it depended on the composition of food [55]. The results of the Tocher [54] agree with the previous data, and the authors postulated that the vitamin E content in the liver, muscle, and blood of different fish species depended by their diet and lipid intake. Therefore, because vitamin E functions as a lipid-soluble antioxidant protecting biological membranes and lipoproteins against oxidation, it has been demonstrated to be an essential dietary nutrient for all fish studied.

At the other side, the level of vitamin E in different fish species is also related to the environment inhabited by them. The authors noted that in two species of Antarctic fish, the concentration of vitamin E was up to six times more than the similar temperate species [56]. A less common isomer of vitamin E, α-tocomonoenol, was found in Antarctic fish. Therefore, the concentration of various forms of vitamin E differs between species and depends on the specificity of their abundance [56].

Vitamin E plays an important role in the success of aquatic organism reproduction, maturation, growth, and development. Larvae diet of the shrimp *Macrobrachium rosenbergii* supplemented of vitamin E, plays a positive role in the development and survival of post-larvae stages of the shrimp [57]. The dietary supplementation of vitamin E significantly decreased the oxidative stress in crayfish *Astacus leptodactylus* in hepatopancreas, ovarian, and muscle and increased the number of eggs in ovarian [58].

Tocopherol is used for prevention of the autooxidation in several commercial fish species during their processing. It was shown that 1% of α-tocopherol is therefore regarded as the preferred concentration for controlling oxidation of kilka oil based on efficacy [59].

Additionally, vitamin E plays a role in the protection of aquatic animals against the unfavorable environmental impact. The researchers have shown, that vitamin E derivative protected against lead-induced oxidative stress in Nile tilapia *Oreochromis niloticus*. They found that the intake of vitamin E in Pb-exposed fish inhibited the metal accumulation in tissue and enhanced the fish growth. Because vitamin E demonstrates the ability to protect the organism against oxidative damage induced by heavy metals, it could have potential use as a preventive or therapeutic agent in fish exposed to toxic elements [60]. The protective role effect of lycopene and vitamin E on oxidative stress in *Oreochromis niloticus* exposed to diazinon (DZN, organophosphate insecticides, which are widely used in agriculture) was shown in the study of Ibrahim [61]. DZN significantly led to a decline in total antioxidant capacity. However, vitamin E supplementation plays a positive role in detoxification of DZN toxicity because it decreased the toxic effect of diazinon, LPO level, and DNA fragmentation. The dietary supplementation of vitamin E and propolis of juvenile rainbow trout with antioxidants alleviated the flow stress-induced oxidative damages, induced the experimental hypoxic conditions [62].

10.3.4   VITAMIN K

Vitamin K (Figure 10.8, also known as phylloquinone) can protect the organism against ROS [63]. It participates in the prerequisites for blood coagulation and for controlling binding of calcium in bones and other tissues.

In our previous studies, the highest level of vitamin K was indicated in Black Sea elasmobranch *Squalus acanthias* muscle, which was significantly greater than in examined teleost species. The least content of vitamin K was shown in scorpion fish *S. porcus* tissues [17]. Generally, the information on vitamin K and its levels in aquatic animals is very limited.

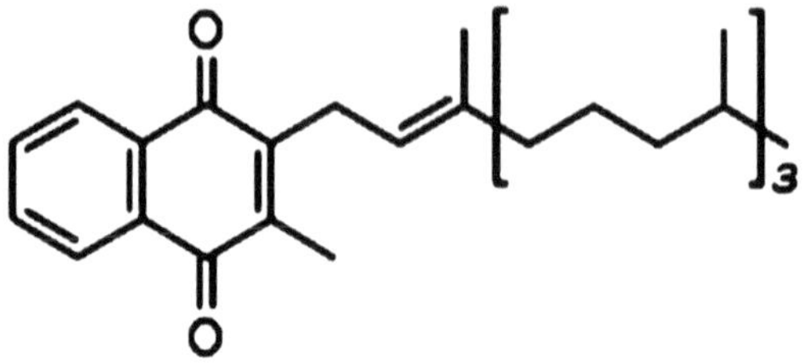

FIGURE 10.8   Vitamin K.

10.4   OTHER ANTIOXIDANTS

Coenzyme Q10 or ubiquinone is a group of homologous quinones (Figure 10.9), which is the lipophilic antioxidant that can be synthesized "*de novo*" in animal cells. Ubiquinone is widely distributed in animals, plants, and microorganisms, and almost every cell of a living organism contains CoQ. An enzymatic mechanism mitochondrial and microsomal electron-transport systems that can regenerate the antioxidant from its oxidized form resulting from its inhibitory effect on LPO was shown. Ubiquinol-QH may act as a chain-breaking antioxidant by hydrogen donation to reduce peroxyl radicals (ROO•): and the recycling of the phenoxyl radical (TO•) of vitamin E [64]. Ubiquinone plays a role in energy metabolism, immunological competence, and antioxidant protection.

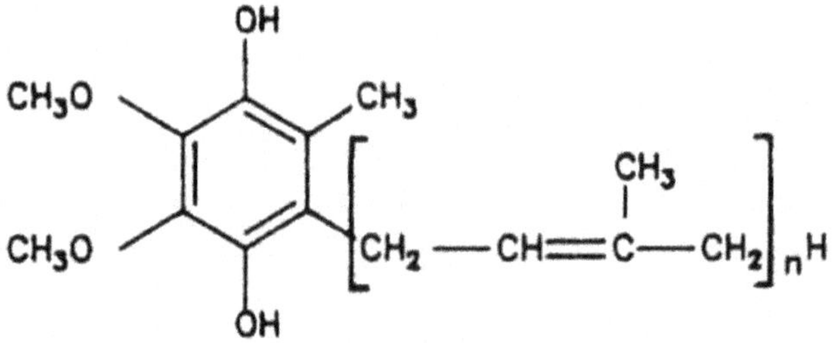

FIGURE 10.9   Ubiquinone.

The researchers demonstrated the antioxidant effect of various phospholipids (PL) involved in cell membranes, but the information is contradictory. Physical and structural differences among and within each PL group may in part contribute to the differences observed in their antioxidant properties. They are believed to function as synergists and metal chelators and may also bring about decomposition of hydroperoxides. Secondary antioxidant effects of PL arise from the synergistic activity in mixtures with natural tocopherols and synthetic antioxidants [27].

10.5 CONCLUSIONS

Based on the data described above, we can conclude that low molecular weight scavengers play an important role in the antioxidant defense of the aquatic animals. They also participate in the other significant functions of the organisms, namely reproduction, survival, growth, and development. Some of them, such as vitamin A and carotenoids, which are a critical factor in determining reproductive fitness and life functions, cannot be synthesized by fish and crustaceans like vitamin E. Their concentrations in the body tissues of fish and invertebrates depend on food composition and the level of these nutrients in it. However, high content of these compounds in the diet promotes the oxidative stress in the organisms and in this case, they play the role of pro-oxidants. Therefore, although vitamins A, E, and β-carotene are one of the most important factors for the successful rearing of aquatic organisms and these nutrients are interactions in terms of function and metabolism, this matter has yet to be investigated for the crustacean. The combination of vitamins C and E in fish contributes to improving growth, metabolism, and resistance to stress; vitamin C plays a role in the process of absorption and regenerates of vitamin E and metabolism of lipids.

There are some antioxidants, which aquatic animals consume from the algae and zooplankton and accumulate in their body. The function of these antioxidants is to control prooxidants, scavenge free radicals and inactivate ROS. Low molecular weight peptides and amino acids and especially MTs play an important role in the process of the conjugation of transition metals, which induce oxidative stress in the organisms. However, there some compounds, isolated from aquatic organisms, which demonstrate antioxidant activity, but this mechanism is unknown yet [65]. Generally, all antioxidants and their complex provide the ability of the organism to enhance its resistance against unfavorable environmental factors, including anthropogenic pollution and climate change, and keep their life in the changing living conditions.

KEYWORDS

- anthropogenic impact
- aquatic organisms
- climate change
- low molecular weight antioxidants
- oxidative stress
- pollution

REFERENCES

1. Lesser, M. P., (2006). Oxidative stress in marine environment: Biochemical and physiological ecology. *Ann. Rev. Physiol., 68*, 253–278.
2. Lane, N., (2011). The evolution of oxidative stress. In: Pntopoulos, E., & Schipper, H. M., (eds.), *Principles of Free Radical Biomedicine* (pp. 1–17). Nova Science Publ. Inc.
3. Winston, G. W., & Di Giulio, R. T., (1991). Prooxidant and antioxidant mechanisms in aquatic organisms. *Aquatic Toxicol., 19*, 137–161.
4. Livingstone, D. R., (2001). Contaminant-stimulated reactive oxygen species production and oxidative damage in aquatic organisms. *Mar. Pollut. Bull., 42*, 656–665.
5. Stoliar, O. B., & Lushchak, V. I., (2012). Environmental pollution and oxidative stress in fish. In: Lushchak, V. I., (ed.), *Oxidative Stress – Environmental Induction and Dietary Antioxidants*. URL: http://www.intechopen.com/books/oxidative-stress-environmentalinduction-and-dietary-antioxidants/environmental-pollution-and-oxidative-stress-in-fish (Accessed on 29 May 2019).
6. Winston, G. W., (1991). Oxidants and antioxidants in aquatic organisms. *Comp. Biochem. Physiol., 100*(1/2), 173–176.
7. Burlakova, E. B., (2005). Bioantioxidants: Yesterday, today, tomorrow. *Chemical and Biological Kinetics: New Approach* (Vol. 2, pp. 10–45). Moscow: Chemistry Publ. (in Russian).
8. Vladimirov, J. A., & Proskurina, E. V., (2007). *Lectures of Medical Biophysics* (p. 432). Moscow University Publ.: Academkniga. (in Russian).
9. Martinez-Alvarez, R. M., Morales, A. E., & Sanz, A., (2005). Antioxidant defenses in fish: Biotic and abiotic factors. *Fish Biology and Fisheries, 15*, 75–88.
10. Ferreira, P. L. F., (2014). *Growth Performance, Antioxidant and Innate Immune Responses in European Seabass Fed Probiotic Supplemented Diet at Three Rearing Temperatures* (p. 72). Thesis, University Porto.
11. Vélez-Alavez, M., Méndez, R. L. C., De Anda, M. J. A., Mejía, H., Galván, M. F., & Zenteno, S. N., (2014). Vitamins C and E concentrations in muscle of elasmobranch and teleost fishes. *Comp. Biochem. Physiol., Part A, 170*, 26–30.
12. Dubey, A. K., (2013). Temperature stress induces neuronal damage via a mechanism leading to free radical production and altering endogenous antioxidant defenses in freshwater Indian catfish *Heteropneustes fossilis*. *The Journal of Free Radicals and Antioxidants, 139*, 204–209.

13. De Almeida, E. A., & Di Mascio, P., (2011). Hypometabolism and antioxidative defense systems in marine invertebrates. In: Anna, N., & Michał, C., (eds.), *Hypometabolism: Strategies of Survival in Vertebrates and Invertebrates* (pp. 39–55). Kerala, Research Signpost.

14. Clarkson, P., & Thompson, H. S., (2000). Antioxidants: What role do they play in physical activity and health? *American J. of Clinical Nutrition, 72*(2), 637S–646S.

15. Terova, G., Saroglia, M., Papp, Z. G., & Cecchini, S., (1998). Ascorbate dynamics in embryos and larvae of sea bass and sea bream, originating from broodstocks fed supplements of ascorbic acid. *Aquaculture International, 6*, 357–367.

16. Rudneva, I. I., (1997). Blood antioxidant system of Black Sea elasmobranch and teleosts. *Comp. Biochem. Physiol. C., 118*, 255–260.

17. Rudneva, I. I., (2012). Antioxidant defense in marine fish and its relationship to their ecological status. In: Seam, P. D., (ed.), *Fish Ecology* (pp. 31–59). Nova Science Publ. Inc.

18. Rocha-e-Silva, T. A. A., Rossa, M. M., Rantin, F. T., Matsumura, T. T., Tundisi, J. G., & Degterev, I. A., (2004). Comparison of liver mixed-function oxygenase and antioxidant enzymes in vertebrates. *Comp. Biochem. Physiol., 137C*, 155–165.

19. Li, H. C., Zhou, Q., Wu, Y., Fu, J., Wang, T., & Jiang, G., (2009). Effects of waterborne nano-iron on medaka (*Oryzias latipes*): Antioxidant enzymatic activity, lipid peroxidation and histopathology. *Ecotoxicology and Environmental Safety, 72*, 3684–3692.

20. Atli, G., Esin, G., Canli, E. G., Eroglu, A., & Mustafa, C. M., (2016). Characterization of antioxidant system parameters in four freshwater fish species. *Ecotoxicology and Environmental Safety, 126*, 30–37.

21. Rudneva, I. I., (1999). Antioxidant system of black sea animals in early development. *Comp. Biochem. Physiol. Part C., 122*, 265–271.

22. Akbulut, M., Selvi, K., Kaya, H., Duysak, M., Akcay, F., & Celik, E. S., (2014). Use of oxidative stress biomarkers in three Crustacean species for the assessment of water pollution in Kocabas Stream (Canakkale, Turkey). *Mar. Sci. Tech. Bull., 3*(2), 27–32.

23. Simonato, J. D., Fernandes, M. N., & Martinez, C. B. R., (2011). Gasoline effects on biotransformation and antioxidant defenses of the freshwater fish *Prochilodus lineatus*. *Ecotoxicology, 20*, 1400–1410.

24. Yildirim, N. C., Benzer, F., & Danabas, D., (2011). Evaluation of environmental pollution at Munzur river of Tuncel applying oxidative stress biomarkers in *Capoeta trutta* (Henkel, 1843). *The Journal of Animal & Plant Sciences, 21*(1), 66–71.

25. Fırat, Ö., & Kargın, F., (2010). Response of Cyprinus carpio to copper exposure: Alterations in reduced glutathione, catalase and proteins electrophoretic patterns. *Fish Physiology and Biochemistry, 36*(4), 1021–1028.

26. Oost Van Der, R., Beyer, J., & Vermeulen, N. P. E., (2003). Fish bioaccumulation and biomarkers in environmental risk assessment: A review. *Environmental Toxicology and Pharmacology, 13*, 57–149.

27. Bragadóttir, M., (2001). *Endogenous Antioxidants in Fish* (p. 63). A literature review submitted in partial fulfillment of the requirements for the degree of master of science in food science Department of Food Science University of Iceland.

28. Sarkar, A., Ray, D., Shrivastava, A. N., & Sarker, S., (2006). Molecular biomarkers: Their significance and application in marine pollution monitoring. *Ecotoxicology, 15*, 333–340.

29. Razo Del, L. M., Quintanilla-Vega, B., Brambila-Colombres, E., Calderon-Aranda, E. S., Manno, M., & Albores, A., (2001). Stress proteins induced by arsenic. *Toxicol. Applied Pharmacol.*, *177*(2), 132–148.

30. Decker, E. A., & Xu, Z., (1998). Minimizing rancidity in muscle foods. *Food Technol.*, *52*(10), 54–59.

31. Løvaas, E., (1991). Antioxidative effects of polyamines. *JAOCS, 68*, 353–358.

32. Halliwell, B., Aeschbach, R., Loelinger, H., & Aruoma, O. E., (1995). The characterization of antioxidants. *Food Chem. Toxic., 33*, 601–617.

33. López-Cruz, R. I., Zenteno-Savín, T., & Galván-Magaña, F., (2010). Superoxide production, oxidative damage and enzymatic antioxidant defenses in shark skeletal muscle. *Comp. Biochem. Physiol. A., 156*, 50–56.

34. Winston, G., (1990). Physicochemical basis for free radical formation in cells: Production and defenses. *Stress Responses in Plants: Adaptation and Acclimation Mechanisms* (pp. 57–86). Louisiana State University, Baton Rouge.

35. Hussein, G., Sankawa, U., Goto, H., Matsumoto, K., & Watanabe, H., (2006). Astaxanthin, a carotenoid with potential human health and nutrition. *J. Nat Prod., 69*, 443–449.

36. De Carvalho, C. C., & Caramujo, M. J., (2017). Carotenoids in aquatic ecosystems and aquaculture: A colorful business with implications for human health. *Frontiers in Marine Science, 4*, 93.

37. Maoka, T., Yokoi, S., & Matsuno, T., (1989). Comparative biochemical studies of carotenoids in nine species of cephalopods. *Comp. Biochem. Physiol., 92*, 247–250.

38. Caramujo, M. J., De Carvalho, C. C., Silva, S. J., & Carman, K. R., (2012). Dietary carotenoids regulate astaxanthin content of copepods and modulate their susceptibility to UV light and copper toxicity. *Mar. Drugs, 10*, 998–1018.

39. Lali, S. P., & Lewis-McCrea, L. M., (2007). Role of nutrients in skeletal metabolism and pathology in fish – an overview. *Aquaculture, 267*, 3–19.

40. Britton, G., (1995). Structure and properties of carotenoids in relation to function. *FASEB J., 9*, 1551–1558.

41. Krinsky, N. I., Landrum, J. T., & Bone, R. A., (2003). Biologic mechanisms of the protective role of lutein and zeaxanthin in the eye. *Annu. Rev. Nutr., 23*, 171–201.

42. Kurashige, M., Okimasu, E., Inoue, M., & Utsumi, K., (1990). Inhibition of oxidative injury of biological membrane by astaxanthin. *Physiol. Chem. Phys. Med. NMR, 22*, 27–38.

43. Miyachi, M., Matsuno, T., Asano, K., & Mataga, I., (2015). Anti-inflammatory effects of astaxanthin in the human gingival keratinocyte lineNDUSD-1. *J. Clin. Biochem. Nutr., 56*, 171–178.

44. Palozza, P., Serini, S., Di Nicuolo, F., Piccioni, E., & Calviello, G., (2003). Prooxidant effects of b-carotene in cultured cells. *Mol. Aspects Med., 24*, 353–362.

45. Russell, R. M., (2004). The enigma of b-carotene in carcinogenesis: What can be learned from animal studies. *J. Nutr., 134*, 262S–268S.

46. Babin, A., Saciat, C., Teixeira, M., Troussard, J. P., Motreuil, S., Moreau, J., & Moret, Y., (2015). Limiting immunopathology: Interaction between carotenoids and enzymatic antioxidant defenses. *Developmental and Comparative Immunology, 49*, 278–281.

47. Hylander, S., & Hansson, L. A., (2013). Vertical distribution and pigmentation of Antarctic zooplankton determined by a blend of UV radiation, predation and food availability. *Aquat. Ecol., 47*, 467–480.

48. Barim Ö z, Ö., Ebru, B. E., & Kamil, O. N., (2011). The effect of different levels of vitamin E in the diet on vitamins A, E and β-carotene concentrations in the tissues of mature freshwater crayfish *Astacus Leptodactylus* (Eschscholtz, 1823). *Turkish Journal of Science & Technology*, *6*(1), 25–33.

49. Zhang, P., & Omaye, S. T., (2001). β-Carotene: Interactions with α-tocopherol and ascorbic acid in microsomal lipid peroxidation. *Journal of Nutritional Biochemistry*, *12*, 38–45.

50. Linan-Calello, M. A., Paniagua-Michel, J., & Zenteno-Savin, T., (2003). Carotenoids and retinal levels in captive and wild shrimp, *Litopenaeus vannamei*. *Aquaculture Nutrition*, *9*, 383–389.

51. Rudneva, I. I., (2013). *Biomarkers for Stress in Fish Embryos and Larvae* (p. 206). CRC Press. USA.

52. Keleştemur, G. T., (2012). The antioxidant vitamin (A, C, E) and the lipid peroxidation levels in Some tissues of juvenile rainbow trout (*Oncorhynchus mykiss*, W. 1792) at different oxygen levels. *Iranian Journal of Fisheries Sciences*, *11*(2), 315–324.

53. Defo, M. A., Pierron, F., Spear, P. A., Bernatchez, L., Campbell, P. G. C., & Couture, P., (2012). Evidence for metabolic imbalance of vitamin A2 in wild fish chronically exposed to metals. *Ecotoxicology and Environmental Safety*, *85*, 88–95.

54. Tocher, D. R., Mourente, G., Van Der Eecken, A., Evjemo, J. O., Diaz, E., Bell, J. G., Geurden, I., Lavens, P., & Olsen, Y., (2002). Effects of dietary vitamin E on antioxidant defense mechanisms of juvenile turbot (*Scophthalmus maximus* L.), halibut (*Hippoglossus hippoglossus* L.) and sea bream (*Sparus aurata* L.). *Aquacult. Nutr.*, *8*, 195–207.

55. Devadason, C., & Jayasinghe, C., (2017). Effects of feeding habits and nutrition status of freshwater fishes on muscle with lipid fatty acid composition and tocopherol contents. *International Journal of Fisheries and Aquatic Studies*, *5*(2), 492–497.

56. Wilhelm-Filho, D., (2007). Reactive oxygen species, antioxidants and fish mitochondria. *Front. Biosci.*, *12*, 1229–1237.

57. Dandapat, J., Chainy, G. B. N., & Rao, K. J., (2003). Improved post-larval production in giant prawn *Macrobrachium rosenbergii*, through modulation of antioxidant defense system by dietary vitamin E. *Indian J. of Biotechnology*, *2*, 195–202.

58. Barim, O., (2009). The effects of dietary vitamin E on the oxidative stress and antioxidant enzyme activities in their tissues and ovarian egg numbers of freshwater crayfish *Astacus leptodactylus* (Eschscholtz, 1823). *J. Animal and Veterinary Advances*, *8*(6), 1190–1197.

59. Bagheri, R., & Ali, S. M., (2013). Comparison between the effects of tocopherol and BHT on the lipid oxidation of Kilka Fish. *World Applied Sciences Journal*, *28*(9), 1188–1192.

60. El-Shebly, A. A., (2009). Protection of Nile Tilapia (*Oreochromis Niloticus*) from lead pollution and enhancement of its growth by α-tocopherol vitamin E. *Research Journal of Fisheries and Hydrobiology*, *4*(1), 17–21.

61. Ibrahim, A. T. A., (2015). Protective role of lycopene and vitamin E against diazinon-induced biochemical changes in *Oreochromis niloticus*. *African Journal of Environmental Science and Technology*, *9*(6), 557–565.

62. Kelestemur, G. T., Seven, P. T., & Seval, Y. S., (2012). Effects of dietary propolis and vitamin E on growth performance and antioxidant status in blood of juvenile Rainbow

trout, *Oncorhynchus mykiss* (Teleostei: Salmoniformes) under different flow rates. *Zoologica, 29*(2), 99–108.

63. Hardy, R. W., (2001). Nutritional deficiency in commercial aquaculture: Likelihood, onset and identification. In: Lim, C., & Webster, C. D., (eds.), *Nutrition and Fish Health* (pp. 131–148). N.Y., Oxford.

64. Kagan, V. E., Nohl, H., & Quinn, P. J., (1996). Coenzyme Q: Its role in scavenging and generation of radicals in membranes. In: Cadenas, E., & Packer, L., (eds.), *Handbook of Antioxidants* (pp. 157–201). Marcel Dekker, New York.

65. Chai, T. T., Law, Y. C., Wong, F. C., & Kim, S. K., (2017). Enzyme-assisted discovery of antioxidant peptides from edible marine invertebrates: A review. *Mar Drugs, 15*(2), 42.

CHAPTER 11

Antioxidant Properties of Safflower Culture (*Carthamus tinctorius* L.), Introduced Into the Central Region of the Russian Federation

SULUCHAN K. TEMIRBEKOVA,[1] YULIYA V. AFANASYEVA,[2]
DMITRY A. POSTNIKOV,[3] SVETLANA M. MOTYLEVA,[2] and
NATALIA E. IONOVA[4]

[1]*All-Russian Research Institute of Phytopathology, Odintsovo Regom, Bolshie Vyazemy, 5, Institute St., Moscow Region 143050, Russia, E-mail: sul20@yandex.ru*

[2]*All-Russia Horticultural Institute of Vegetable of Breeding, Agrotechnology, and Nursery, 4, Zagoryevskaya St., Moscow 115598, Russia*

[3]*Russian State Agrarian University, Moscow Timiryazev Agricultural Academy, Timiryazevskaya St., 49, Moscow 127550, Russia*

[4]*Kazan Federal University, 18, Kremlyovskaya St., Kazan 420008, Russia*

ABSTRACT

The safflower as the cultural plant was introduced into the Central region of the Russian Federation. The results of many years of research have a fundamental and applied significance. Vegetation period from germination to maturity in the years with different meteorological conditions is 94–114 days. Duration of flowering is about a month. The stem is erect, branching, naked, and height is about 83–90 cm. Leaves are with small spines. One plant can range from 5–7 to 20–50 and more baskets. Seed's coat is hard, and it is difficult to crack, reach 40–50% from the mass of seeds. The weight of 1000 seeds – 48–51 g. Productivity in contrast zone is 0.7–1.7 t/ha. Safflower seeds contain 32–38% of fat; the yield of oil was 240 L kg/ha. Absolute fat content in the treated seeds reaches more than 60%, and it is fit for food.

Leaves of safflower *Carthamus tinctorius* L. by new the cultivar 'Krasa Stupinskaya,' showed higher total antioxidant activity in all years of study. It is noted that the synthesis and accumulation of antioxidant activity occur in the active photosynthetic plant organs with a depth of 5 cm seeding and seeding index of 10–14 kg/ha. The total antioxidant activity of alcohol extract from safflower with white petals has higher values compared to samples with red and yellow petals, in aqueous extracts, the highest antioxidant activity has a yellow color petal. The obtained results extend the understanding of the physiologically active antioxidant substances in the various organs of safflower.

11.1 INTRODUCTION

N. I. Vavilov attached special importance to the issue of new cultures, fuller use of the world's wild flora both within our country and abroad. Following his ideas, scientists learn and introduce in the production cultures, previously unknown in our agricultural science and practice. The problem of the new crops introduction is becoming increasingly important due to the fact that the provision of Russia by vegetable oils and biologically active substances is carried out mainly by imports. To such new cultures applies safflower [1].

Adaptive plant breeding aimed to harvest size and quality increasing by their better adaptation to the environment, including the ability to stand the effects of abiotic and biotic stressors. Therefore, global mobilization and adaptation of plant resources is the basis of progress in the more northern areas biologically possible and economically viable cultivation of new crops. Thanks to this were greatly expanded ranges of sunflower, corn, winter wheat, peas, rape, some types of fruit, berry, and other cultures effective cultivation [2].

The aim of our study was the introduction of safflower in the Central region of Russia, the study of its biological characteristics, creation of adaptive varieties for using in agricultural production and food processing and the development of adaptive technology of cultivation recommendations.

Safflower (*Carthamus tinctorius* L.) belongs to the family Asteraceae. Homeland is Egypt, India. As a result of many years work (2005–2012) in All-Russia Selection-Technological Institute of Horticulture and Nursery was created a new safflower cultivar 'Krasa Stupinskaya' [3]. 'Krasa

Stupinskaya' is included in the State Register of Selection Achievements in 2012 [3]. It is recommended for all regions of the Russian Federation [4].

At the moment, we have reached a clear understanding that human health and life expectancy largely determines the nature of food. With the violation of human health is the disturbance of proper nutrition and the extremely low level of human energy consumption [5]. Directly the deficiency of antioxidants leads to a sharp decrease of the body's resistance to adverse environmental factors through dysfunction of the antioxidant defense system in humans and plants. In connection with this plant, food is the primary and most affordable source of antioxidants for humans.

Antioxidants constitute a large group of chemical compounds of different nature, able to neutralize free radicals and active oxygen species generated in the cells of living organisms under the action of biogenic and abiogenic stressors. Herbal antioxidants are phenolic compounds such as betacyanin, ascorbic acid, and other compounds that perform many physiological functions in the body. Evaluation and selection of crops with the highly effective antioxidant system, studying the structure, content, and physicochemical properties of water-soluble antioxidants, their mechanisms of action are relevant and necessary for developing functional food products that enhance human health and reduce the risk of various diseases.

The aim of our research was the study of the antioxidant activity of the new cultivar of safflower 'Krasa Stupinskaya,' introduced in the Moscow Region, and the conduct of various its cultivation methods. Comparison of the results of cultivation of this cultivar in the Moscow and in other regions Russia, as well as in Central Tajikistan was also carried out.

11.2 MATERIALS AND METHODOLOGY

The studies have been conducted in the Center of Gene Pool and Bioresources of plants of All-Russia Horticultural Institute for Breeding, Agrotechnology, and Nursery, in 2012–2015 in the All-Russian Research Institute of grain crops (Zernograd, Rostov Region), then we were continued work in the Educational-Experimental of Moscow Agricultural of Farm "Mummovskoe" of Timiryazev Agricultural Academy, in Saratov region, and in Central Tajikistan, Gissar region, experimental production farm of the Tajik Research Institute of Food and Agriculture. The object of research was safflower cultivar 'Krasa Stupinskaya' and collection samples from Tajikistan.

Phenological observations and biometric assessment were conducted during the growing season in accordance with the Methodology of State Testing of Agricultural Cultures [6].

Harvest definition was carried out with using of sample plots in 3 replicates, accounting plot area – 10 m². Determination of oil content in the seeds was conducted in accordance with GOST (State Standard of the Russian Federation) 10857 "Oilseeds" [7].

Determination of oil content, the fatty acid composition of the oil were made in accordance with GOST 30623–98 "Vegetable oils and margarine. The detection method of falsification" [8].

The research objects were the leaves and petals of safflower 'Krasa Stupinskaya' collected in the phases of branching, budding, flowering, and maturation (Figure 11.1).

FIGURE 11.1   On the left is a whole plant safflower 'Krasa Stupinskaya,' on the right is a flower. Yellow petals are clearly visible.

Was conducted the study of antioxidant activity depending on farming methods: seeding index was 10, 12, and 14 kg/ha, depth of sowing – 3, 5, and 7 cm. The leaves were selected from the middle of the shoot, with the control plots in different vegetation phase of the plants, the petals of the central flowers in the flowering stage. Total antioxidant activity of aqueous and alcoholic extracts was determined by spectrophotometer Helios by the method of DPPH (2,2-diphenyl-1-picrylhydrazyl) [9, 10]. This physicochemical method based on the interaction of substances – antioxidants with 2,2-diphenyl-1-picrylhydrazyl. Samples of leaves were crushed and prepared the extracts with distilled water and methanol by the continuous stirring on the shaker for 12 h. The resulting solutions were filtered. As a background solution was used 0.0025% solution of DPPH. Total antioxidant activity was calculated as a relative value and was determined by the ratio of extinction of the experimental sample and DPPH solution during 10 min after the reaction beginning.

11.3   RESULTS AND DISCUSSION

11.3.1   AGRONOMIC CHARACTERISTICS OF SAFFLOWER 'KRASA STUPINSKAYA' IN CONTRASTING SOIL AND CLIMATIC CONDITIONS

Comparative study of the growth zone influence on the vegetation period and the main economically valuable signs of safflower grown in four regions was made: Central Federal District (Moscow region, Mikhnevo), Volga Federal District (Saratov region) and the Southern Federal District (Rostov region) and Central Tajikistan [11].

'Krasa Stupinskaya' is an annual herbaceous plant with a well-developed tap root system that penetrates into the soil to 10–20 cm, in the southern regions up to 1.5–2 m (and Central Tajikistan) [12].

The stem is glabrous, erect, branchy, and height is about 83–90 cm. Leaves are sessile, lanceolate, oval or elliptical lancet, on the edges with small teeth, ending with small spines. The inflorescences are many-baskets, 1.5–3.5 cm in diameter. The number of baskets on the plant is from 5–7 to 20–50. The flowers are tubular, with five separate corolla yellow or orange color. Fruit is achene, brilliant, reminiscent of sunflower achenes. Its hard shell, it is difficult to split, is 40–50% of the seed weight. The seeds not crumble can germinate at a temperature of 1–2°C, but there are better and friendly germinate when the soil warms up to 5–6°C and more at a depth of 10 cm.

The sowing is carried out annually in each region in the following terms depending on climatic features: Mikhnevo 7.05–11.05, Saratov region 7.05, Rostov region 26.04, Central Tajikistan 20.12–25.12 and 10.03–15.03 (spring landing). Seedlings always have been friendly and appeared in 3–8 days. The period from the beginning of budding until flowering was within 18–23 days. Flowering lasted about 29–35 days. Harvest ingathering was carried out in next dates: Mikhnevo – 23.08, Rostov region – 12.08, Saratov region – 16.08, Central Tajikistan 7.04–10.04 (at the winter sowing) and 28 June – 2 July (during the spring sowing).

Vegetation period from germination to maturation was 96 days in Moscow region (versus 110–115 days in 2010–2012, 2014–2015), 93–95 days in the Rostov region and 89–103 days in the Saratov region and in Central Tajikistan – 110 days. In all regions, the duration of the vegetative period of the safflower culture was almost the same (Table 11.1).

TABLE 11.1   Environmental Studies of Harvest Indicators of Safflower 'Krasa Stupinskaya' for 2010–2015

Years of study	Weigh of 1000 seeds, g	Productivity, t/ha	Vegetative period, days
Moscow region			
2010	50.0	0.9	112
2011	51.1	0.8	115
2012	48.0	0.7	115
2013 (atypical wet year)	30.3	0.4	96
2014	45.2	0.8	113
2015	44.7	0.8	105
Averages	44.8	0.7	109
Rostov region			
2012	42.3	1.3	93
2013	53.4	0.6	95
2014	42.6	1.1	95
2015	46.1	0.9	94
Averages	46.1	0.9	94
Saratov region			
2013	30.9	0.9	94
2014	48.1	2.0	103
2015	43.8	0.9	89
Averages	40.9	1.2	95

TABLE 11.1 *(Continued)*

Years of study	Weigh of 1000 seeds, g	Productivity, t/ha	Vegetative period, days
Central Tajikistan			
2014	34.2	1.7	109
2015	34.5	1.8	111
Averages	34.3	1.7	110

The calculation of basic safflower harvest indicators gave the following results: the number of plants per 1 m² (p/m²) was in Mikhnevo 26 p/m², in the Rostov region 30 p/m² (planted for seeds), and in the Saratov region 62 p/m² (planted for feeding purposes). Plant height ranged from 63–80 cm in all regions.

The mass of 1000 seeds and yields for 2010–2015 studies in different soil and climatic conditions are presented in Table 11.1. The high ecological plasticity of the 'Krasa Stupinskaya' cultivar is noted in various growing regions, which is supported by the stable high mass of 1000 seeds and by stable high yield sufficient for each region.

So the average productivity of safflower in 2010–2015 was 0.7 t/ha in the Moscow region and 0.9 t/ha in the Rostov region, 1.2 t/ha in Saratov region, and in the conditions of the Central Tajikistan 1.7 t/ha with an average weight of 1000 seeds 34.3 g.

On the basis of these high indices obtained during testing of the variety in different regions 'Krasa Stupinskaya' is recommended as sideral, phytosanitary, fodder, and ornamental and promising oilseed culture. Best of all as a green manure safflower manifests itself on soddy-podzolic soils.

Green manure crops (so-called green manure) are an important source of organic matter replenishment in soils. For example, white mustard sown in one hectare of land is the same that makes it 20 tons of manure for one season; on 1-hectare lupine can save up to 160 kg of nitrogen, which is equivalent to 30–35 tons of manure [13]. Green manure is really conducive to the rapid enrichment of soil organic matter, reduces its acidity and reduces the content of mobile aluminum, and increasing microbiological activity. In addition, using green fertilizers tissue and fruit plants are not contaminated by chemicals (fertilizers and pesticides).

Traditional green manure crops are white mustard, blue lupine, winter rye, etc. However, there is a special culture safflower, which has a number of unique properties for the Moscow region conditions.

Green manure crops have a diversified impact on the nutrient status of soddy-podzolic soil. Plowing under the white mustard post-cut root residues provides involvement optimization of the accumulated nitrogen into the circulation in the topsoil after the decomposition of root residues. Phosphorus accumulation in the root mass of the white mustard is 0.7%, but due to the small mass of the roots, their effect on the content of phosphorus in the soil is insignificant. The decomposition rate of the white mustard post-cut root residues decelerates in response to the separation of monocarboxylic acid in plant roots, and, probably, there is no complete decomposition of white mustard root mass at the time of soil samplings [14, 15].

Potassium content in the soil after the burial of the white mustard post-cut root residues increases as compared with the control by 1%. As a result of the burial of the total white mustard plant mass, soil nitrogen is back into the soil up to 40 kg/ha. As for phosphorus, its value is to 20 kg/ha.

In the context of complete green manuring of blue lupine, soil nitrogen is back into the soil up to 140 kg, allowing keeping a positive balance of this element after harvesting the major crop. The positive dynamics of the content of available phosphorus is in the topsoil should be noted [16].

In view of plowing the safflower post-cut root residues, nitrogen re-entry in the soil can be up to 8.5 kg, and during burial of the whole green mass from the root mass to the aboveground part, 120 kg/ha of biological nitrogen returns in the soil in total, that is relatively similar to the blue lupine on average. In terms of P_2O_5, plowing of safflower green mass corresponds to an average return to the soil up to 40 kg/ha. Plowing of safflower at a flowering stage has a positive effect on the exchange of potassium content in the root layer of the soil.

Yields of the dry aboveground part of safflower in different years were as follows: from 9.1 to 10.0 t/ha, and the root part was from 1.3 to 1.6 t/ha.

In general, while analyzing the values obtained by the effect and aftereffect of green manuring on contents of basic food elements in the soil, it is important to emphasize that this technique should be considered in complex with other agrochemical techniques for making compost, fertilizers, and meliorates.

The white mustard usage as a green manure is of interest to those crop rotations, where the crop is used as between crop and sown in early spring before planting potatoes or autumn after harvesting of the major crop. Under conditions of the southern part of Moscow region, effective tool for enhancing the activity of the soil microflora and increasing the isolated circulation of substances in agrocenosis is a burial, which in turn should be used along with the traditional cultures and with introduced plant-culture – safflower [15].

We have considered various application schemes of examined green manure crops and revealed that in the case of white mustard full green manuring is optimal in the context of the agronomy, and in case of growing lupine and safflower green material can be used for feeding purposes. In the absence of livestock specialization, farms should carry out the full burial of aboveground mass of green manure crops, which provide indicators improving effective soil fertility.

In modern agricultural systems, green manuring should not base on the only one culture, because the functional integrity of the intensive agroeco-systems is defined by the set of cultivated species, which may belong to different families and thus have diverse effects on agrocenosis generally. Expanding of the green manure crop range creates the right conditions for further domestic agricultural greening, for the integrated and sustainable development of the agricultural sphere in whole [16].

11.3.2   SAFFLOWER AS AN OILSEED

At the moment, seeds oil content increasing selection have become a major asset of our agricultural production; such as the property becomes breeding for oil quality change.

It has been shown that each variety and even the shape of the population are composed of a larger or smaller number of biotypes differing by a number of features, including the concentration of the fatty acid oil [17].

The basis in the selection to the quality of oil for technical and food use is the knowledge of genotypic variability of the fatty acids composition in the range of cultivated species and wild relatives.

N. I. Vavilov attached great importance to the study of differentiation within the species for chemical signs of quality grades [18] repeatedly emphasizing need to identify genetic differences that can be seen in the study in the same conditions of different varieties in different geographical locations.

It was observed that linoleic acid in sunflower oil contains near 67%, and in safflower oil was 80% under the experimental conditions of Kuban station of All-Russian Institute of Plant Industry.

Qualitative differences are determined by genetic characteristics of oil varieties and forms. Oil quality features can be enhanced at content increasing of main fatty acids (oleic and linoleic).

Large variability in the content of linoleic acid in the various years of cultivation, probably due to a stretched period of flowering and late maturing of certain cultures. Fluctuations in temperature cause changes in the partial pressure of oxygen in the cells, which affects the processes of oxidative hydrogenation.

It has been determined the influence of environment on the unsaturated fatty acids accumulation rate – of oleic and linoleic. In all sunflower varieties, intensive accumulation of linoleic acid observed in the more northern growing area compared with the southern zone (71.7–72.0 and 53.7–59.0%, respectively). Thus, high linoleic acid 46 content is combined with a low concentration of oleic (16.9–17.9 and 29.0–36.0%, respectively).

The oils of different crops include fatty acids, mainly with C_{16} to C_{22} chains, saturated or unsaturated with double bonds (one, two, three). Within various crop species identified individual grades and within individual varieties biotypes are differing by increased or reduced content of typical fatty acids. These biotypes features caused by various factors (e.g., mutation) are also inherited. Individual variability (for plants) is the basis of selection to increasing of concentrations and a decrease in some – other fatty acids, functionally related to each other.

A comparative analysis of the seeds oil content of safflower 'Krasa Stupinskaya' for three years, grown in Moscow and Rostov regions showed that the content of the seeds mass fraction of oil in Rostov region (2013) was 19.0%, which is 4.4% higher than in 2012; in 2014, 23.7%, which is 9.2% higher than the seeds oil content in 2012. The oil content in the seeds grown in the Moscow region in 2012, showed – 22.9%, which is 8.4% higher than in the Rostov region. In atypical weather conditions in 2013 seeds, oil content of safflower grown in the Rostov region was 12.7% higher than that in safflower grown in the Moscow region – 6.4%. In 2014, safflower seed oil content in the Moscow region amounted to 30.2%, which is 6.5% higher than in safflower from the Rostov region (23.7%) (Table 11.2).

TABLE 11.2   The Oil Content in the Seeds of Safflower Cultivar 'Krasa Stupinskaya' When Tested in Different Regions, 2012–2014

Rostov region			Moscow region		
2012	2013	2014	2012	2013	2014
14.5	19.0	23.7	22.9	6.4	30.2

We marked that the accumulation of oil content depends not only on the quantity of precipitation, but also by the temperature factor. Moderate rainfall and temperatures above 18°C (phase of flowering and ripening) have a positive effect on the formation of oil content.

It should be noted a direct correlation depending on the proportion of oil mass accumulation in safflower seed culture on the amount of precipitation during the growing season and temperature regime.

It should be noted a direct correlation between the oil content in the seeds of safflower and the amount of precipitation during the growing season and temperature [19].

Noted the influence of agro-biological factors on the oil content of seeds safflower cultivation in contrasting years. The analysis of the seed oil content at five reproductions 2010–2015 years safflower cultivar 'Krasa Stupinskaya' grown in the Moscow region is presented in Table 11.3.

TABLE 11.3 The Influence of Agro-Biological Factors on Seed's Oil of 'Krasa Stupinskaya' in Contrasting Conditions, 2010–2014

Region, years of study	Oil content (fat mass fraction), %	Precipitation, mm		Temperature, t °C	
		Average during several years	Average during the growing season	Average during several years	Average during the growing season
Moscow, 2010	31.2	264	154.4	15.1	18.8
Moscow, 2011	29.0	264	285.5	15.1	17.8
Moscow, 2012	22.3	264	245.8	15.1	17.8
Moscow, 2013	6.4	264	335.8	15.1	18.4
Moscow, 2014	30.2	264	184.1	15.1	16.4
Central Tajikistan, 2015	34.3	510	306.8	16.8	20.5

In 2010, the acutely arid, characterized by high air temperature 18.8°C (long-term average 15.1°C) and low precipitation 154.4 mm during the growing season, the accumulation of a mass fraction of oil in the seeds was 31.2% and in a more humid 2011 was 285.5 mm rainfall during the growing season, temperature 17.8°C in 2012 (optimal warmth, less humid) was 245.8 mm during the growing season, temperature 17.8°C was 29.0 and 22.3%, respectively. In 2013, when the rainfall during the growing season fell to 335.8 mm (at a rate of 264 mm) and a temperature of 18.4°C, the mass fraction of fat was only 6.4% in 2014 precipitation during the growing season was 184.1 mm, average temperature – 16.4°C, oil content was 30.2% in grade 'Krasa Stupinskaya.' In 2015, the oil content of seeds 'Krasa Stupinskaya' was 34.3% in Central Tajikistan and in the Central region of the Russian Federation – 30.9%.

We conducted a comparative analysis of the oil content determination in the safflower seeds 'Krasa Stupinskaya' for three years, obtained from the Rostov region. Mass fraction of oil in the seeds was 19.0% in 2013, at which is 4.5% higher than in 2012, while the oil content in seeds was 23.7% in 2014 g.

In the Moscow region, safflower seed oil yield in 2012 was 8.4% higher than in the Rostov region. In 2013 safflower seed oil content in safflower grown in the Moscow region was 12.7% higher than in the Rostov region. In 2014, safflower seed oil grown in the Moscow region accounted for 30.2%, 2.0% lower than in safflower from the Rostov region (28.0%). These results show the dependence of the mass fraction of fat accumulation in the safflower seeds under weather conditions, Tables 11.3 and 11.4.

It has been determined the influence of agro-biological factors on seed's oil in contrasting years. Analysis of seed's oil content in five safflower reproductions 2010–2015 grown in the Moscow region (see Table 11.3) shown that in conditions of 2010 year, characterized by increased air temperature up to 18.8°C long-term (average 15.1°C) and low rainfall up to 154.4 mm during the growing season, the accumulation of oil in seeds was 31.2%. At the same time in 2011 (285.5 mm of rainfall during the growing season, temperature 17.8°C) oil content was 29.0%, and in 2012 (optimal heat 17.8°C, less humid 245.8 mm) – 22.3%. In 2013 characterized with high precipitation (near 335.8 mm at a temperature 18.4°C) oil mass fraction was only 6.4%. In 2014 precipitation during the growing season was 184.1 mm, the average temperature was 16.4°C, 'Krasa Stupinskaya' oil content was 30.2%. In Central Tajikistan (2015) oil content was 34.3%. In samples 'Moldir' in 2014, the accumulation of oil in seeds was 24.0%, by the 'Moldir 2008' was 22.2% and 'VIR 2933' was 21.7%, respectively. It is important for the creation of breeding varieties with different fatty acids ratio in the oil (Table 11.4).

TABLE 11.4 Mass Fraction of Fatty Acids, % to Total Content of Fatty Acids, 2013, 2014

Fatty acids	'Mahalli 260' (Tajikistan), 2013	'Krasa Stupinskaya,' 2013	'Krasa Stupinskaya,' 2014	Norms in accordance with State standard of RF 30623–98 [8]
$C_{14:0}$ (myristic)	0.1	0.1	0.1	< 1.0
$C_{16:0}$ (palmitic)	7.6	7.7	9.94	2.0–10.0
$C_{16:1}$ (palmitoleic)	0.2	0.1	0.55	< 0.5
$C_{18:0}$ (stearic)	2.6	2.0	2.48	1.0–10.0

TABLE 11.4 *(Continued)*

Fatty acids	'Mahalli 260' (Tajikistan), 2013	'Krasa Stupinskaya,' 2013	'Krasa Stupinskaya,' 2014	Norms in accordance with State standard of RF 30623–98 [8]
$C_{18:1}$(oleic)	13.2	13.6	16. 9	7.0–42.0
$C_{18:2}$ (linoleic)	75.6	75.7	65.9	55.0–81.0
$C_{18:3}$ (linolenic)	0.2	0.1	–	< 1.0
$C_{20:0}$ (arachidic)	0.3	0.4	–	< 0.5
$C_{20:1}$ (gondoinovaya)	0.2	0.3	–	< 0.5

According to the content of linoleic acid that is not synthesized in the human body, this 'Krasa Stupinskaya' is not inferior to the southern 'Mahalli 260.' The content of oleic acid is 16.89%, responsible for provides the freshness of the oil over a long period. At this indicator, the cultivar exceeds other cultivars.

The higher content of saturated fatty acids, particularly palmitic characterized cultivar 'Krasa Stupinskaya.' 'Krasa Stupinskaya' has oil yield near 240 kg/ha (at a plant density of 250–300 thousand/ha and seed yield – 0.8 t/ha). In Central Tajikistan, oil output amounted near 940 kg per hectare (at a plant density of 160 thousand plants per hectare and crop seeds 1.7 t/ha).

The productivity increases, and safflower product quality depends on farming practices of cultivation. It is necessary to adhere to morpho-biological features of crops and varieties, keeping the complex soil-climatic conditions of the region, a specific agricultural production, and hydrothermal regime during the growing season. Technical equipment is of great importance, financial condition, and agronomic management frames.

Therefore, the potential yield and economic effect of the new culture introduction will largely depend on the use of cultivation technology adapted to local conditions, considering all these factors. All agricultural practices that are recommended for the cultivation of crops should be carried out at one time, because the omission or wrong application of one of the elements will affect the yield and quality of seeds.

11.3.3 ANTIOXIDANT PROPERTIES OF SAFFLOWER CULTURE (CARTHAMUS TINCTORIUS L.)

In this chapter, we present a study of antioxidant activity in leaves of cultivar 'Krasa Stupinskaya.' In 2013 was the determination of antioxidant activity in the leaves of safflower collected in the phase of flowering and ripening. In the flowering stage, the antioxidant activity of alcohol-soluble substances was equal to 71.1%, water-soluble – 59.5%. In the phase of maturation observed the higher antioxidant activity of the alcohol-soluble substances is 88.7% and reduction of water-soluble substances – 52.5%. It was established the increase alcohol-soluble substances in the leaves of safflower in the process of vegetation due to active photosynthesis, as also noted in the research M.S. Gins and V.K. Gins [5]. It should be noted that it is also associated with the greater ionic activity of methanol, which removes a greater number of substances – antioxidants. It should be noted that it is also associated with the greater ionic activity of methanol, which removes a greater number of substances – antioxidants (Table 11.5).

TABLE 11.5 Determination of Antioxidant Activity (%) in Leaves of Safflower 'Krasa Stupinskaya' in 2013

Extractant	Recurrence	Flowering phase	Maturation phase
Methanol	I	64.6	88.9
	II	71.4	92.6
	III	77.4	84.7
	Average	71.1	88.7
Water	I	61.5	56.2
	I	59.1	47.9
	III	58.0	53.4
	Average	59.5	52.5
LSD_{05}		0.1	0.2

LSD – smallest significant difference.

In 2014, the study was conducted regarding the antioxidant activity rising and accumulation in the leaves of plants according to the phases of development of culture in more details.

From Table 11.6, it is seen that during the growing season – the phases of branching, budding, and flowering, the greatest antioxidant activity of alcohol-soluble compounds made up from 89.9 to 91.6%. The content of water-soluble compounds has increased from the phase of budding (61.0%) to the maturation phase (67.9%).

TABLE 11.6 Antioxidant Activity (%) in the Leaves of Safflower Plants in Different Phases of Vegetation in 2014

Extractant	Young growth (10–15 cm)	Beginning of branching	Branching	The beginning of budding	Budding	Flowering	Maturation	LSD$_{0.5}$
Methanol	93.6	92.0	89.9	90.5	91.6	90.5	85.1	1.1
Water	65.8	84.2	76.6	31.1	61.0	78.6	67.9	6.6

LSD – smallest significant difference.

The study of antioxidant activity in the petals of safflower and the identification according to the total antioxidant activity from the color of the petals was held. Alcohol-soluble compounds in the yellow petals contain 15% more than in red, and water-soluble compounds are approximately the same in both variants (Table 11.7). In General, the antioxidant activity in the petals (yellow and red) in water extracts 2–3 times higher than in alcohol solutions.

TABLE 11.7 Antioxidant Activity (%) of Safflower Petals in 2014

Extractant	Yellow petals	Red petals
Methanol	33.2	18.6
Water	67.9	65.3

In 2015 studied the dependence of total antioxidant activity from agronomic practices: seeding rate and seeding depth. Observed high antioxidant activity in ethanol extracts in the phase of branching (90.7%) when seeding depth was 5 cm (Figure 11.2).

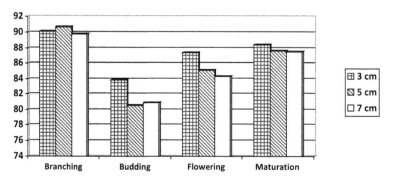

FIGURE 11.2 Changes in total antioxidant activity (%) in ethanol extract from safflower leaves depending on the development phase and depth of seeding in 2015.

In water extracts antioxidant activity was increased from the phase of branching to the phase of budding – 52.06 and 64.40%, respectively, when the seeding depth was 5 cm (Figure 11.3). At a depth of seeding 3 and 7 cm observed a decrease of antioxidant activity in both extracts.

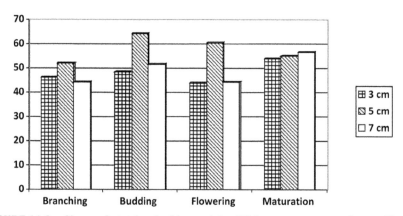

FIGURE 11.3 Changes in total antioxidant activity (%) in aqueous extracts from safflower leave depending on the vegetation phase and depth of seeding in 2015.

When the seeding index was equal to 14 kg/ha observed higher antioxidant activity in ethanol extracts (Figure 11.4).

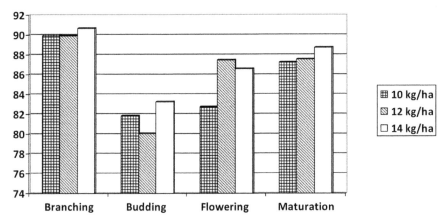

FIGURE 11.4 Changes in total antioxidant activity (%) in ethanol extract from safflower leaves depending on the vegetation phase and seeding rates 2015.

In water extracts the highest result was obtained at the seeding rate of 10 kg/ha in all phases of the growing season: branching, budding, flowering, and ripening – from 56.4 to 60.7% (Figure 11.5).

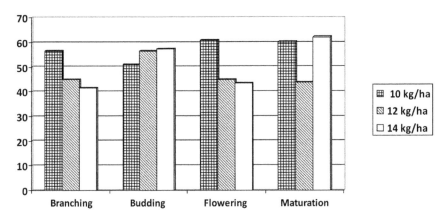

FIGURE 11.5 Changes in total antioxidant activity (%) in aqueous extracts from safflower leave depending on the vegetation phase and seeding rates in 2015.

Analysis of the antioxidant activity of cv 'Krasa Stupinskaya' petals showed the largest accumulation high in the antioxidant activity (Table 11.8).

TABLE 11.8　Antioxidant Activity (%) in Safflower Petals Depending on the Color in 2015

Extractant		Red color	Yellow color	White color
Methanol	I	30.9	51.8	73.3
	II	30.6	51.3	74.9
	III	31.1	52.2	72.5
Average		30.9	51.8	73.6
Water	I	84.3	88.5	71.4
	II	82.1	86.4	71.1
	III	86.4	90.3	71.7
Average		84.2	88.4	71.4

In an alcohol extract of petals of red color antioxidant activity in 2 times less than that of the petals of white color and was equal to 30.9%. The antioxidant activity of the water extract of the red lobe was higher by 12.9% (84.2%) than in the white petals (71.4%). The antioxidant activity of the yellow petals methanol extracts in 1.4 times less than in the white petals and 1.6 in times higher than in the red petals (51.8%). Antioxidant activity of the yellow petals in the aqueous extract is determined within 88.4% (that is 1.2 times higher than in the white petals and to 4.2% higher, than in the red petals). Total antioxidant activity of alcoholic extract from safflower white petals has a higher value compared to samples of the red and yellow petals (73.6%). In water extracts, the highest antioxidant activity has the yellow petals (88.4%).

11.4　CONCLUSIONS

1. The vegetation period from full germination to full maturity safflower culture in the Central Federal District in an excessively wet, atypical in 2013 was 96 days, 2010–2012, 2014 research 114 and in 2015 105 days, average for 5 years 111 days, in the Southern federal District 94 days, in the Volga federal District 95 days in the Central Tajikistan 110 days (the average for all years of study).

2. Safflower plant height in all regions ranged from 63.1 to 70.8 cm. Weight of 1000 seeds in different years was 30.3–53.4 g of crop seeds, safflower cultivar 'Krasa Stupinskaya' per 1 ha Moscow region for 2010–2015 amounted to 0.7 t/ha, in the Rostov region for the 2012–2015 biennium – 0.9 t/ha, in the Central Tajikistan – 1.7 t/ha

(based on the seed purposes) and in the Saratov region for 2013–2015 years – 1.2 t/ha (based on feed purposes).

3. In all regions, noted a sufficient accumulation of oil content in favorable on agro-meteorological conditions during filling and maturation of seeds. The greatest accumulation of fat mass fraction (in terms of dry matter) in the seeds of safflower dye cultivar 'Krasa Stupinskaya' (Moscow region in reproduction) noted in the harvest 2010, 2011 and 2012 from 22.3 to 31.2% (with the exception of 2013 where the harvest are 6.4%). In the Rostov region, the harvest was from 14.5 to 23.7%, in Central Tajikistan – 34.3%. According to the content of linoleic and oleic acids has not reached the level cultivar 'Krasa Stupinskaya,' which is important for practical purposes.

4. The leaves of safflower cultivar 'Krasa Stupinskaya' have high total antioxidant activity in all years of study.

5. The concentrations of antioxidant activity in aqueous and alcohol extracts of safflower cultivar 'Krasa Stupinskaya' was higher from the phase of branching to the end of the growing season, when sowing seeds at planting depth of 5 cm (in comparison with the planting depth in 3 cm and 7 cm).

6. It is shown that at the seeding index of 14 kg/ha increased the antioxidant activity in ethanol extracts, while in water extracts the highest result was obtained at the seeding index of 10 kg/ha in all phases of the growing season: branching, budding, flowering, and ripening from 56.4 to 60.7%.

7. Total antioxidant activity of safflower white petal's alcoholic extracts has a higher value compared to samples of the red and yellow petals (73.6%). In water extracts, the highest antioxidant activity have the yellow petals (88.4%).

KEYWORDS

- **2,2-diphenyl-1-picrylhydrazyl**
- **fatty acids**
- **green manure**
- **oil accumulation**
- **sod-podzolic soils**

REFERENCES

1. Gorbatenko, L. E., Ozerskaya, T. M., & Stishonkova, N. A., (2007). *N.I. Vavilov – The Creator of the Introduction of the Theory of World Collection of VIR* (pp. 15–17). The genetic resources of cultivated plants in the XXI century. Condition, problems and prospects. Vavilov II International Conference. St. Petersburg (in Russian).
2. Zhuchenko, A. A., (2011). Adaptive strategy for sustainable development of the Russian agriculture in the XXI century. *Theory and Practice* (p. 415). Moscow: Agrorus, I. (in Russian).
3. Temirbekova, S. K., Ionova, N. E., Kulikov, I. M., Kurilo, A. A., Metlina, G. V., Norov, M. S., & Postnikov, D. A., (2012). *Safflower Cultivar Krasa Stupinskaya.* State register of selection achievements approved for use application no. 8755720 with a priority date of 30.08.2012 (in Russian).
4. Temirbekova, S. K., Kulikov, I. M., Ionova, N. E., Metlina, G. V., Postnikov, D. A., & Afanasyeva, Y. V., (2014). Introduction and features of cultivation of safflower seeds in the conditions of the Central area of the Nonchernozem zone. *Herald of the Russian Academy of Agricultural Sciences, 1,* 41–43 (in Russian).
5. Gins, M. S., & Gins, V. K., (2011). *Physiological and Biochemical Bases of Introduction and Breeding of Vegetable Crops* (p. 128). Moscow: Russian University of Peoples' Friendship, (in Russian).
6. The method of state variety testing of agricultural crops. Oilseeds, essential oil, medicinal and technical crops, mulberry, silkworm, (1983). *The State Commission for Variety Testing of Agricultural Products at the Ministry of Agriculture of the USSR* (Vol. 3, p. 184). Moscow: Standartinform, (in Russian).
7. State Standard of Russian Federation (GOST) 10857–64, (1964). *Seeds of Olives: Methods for Determining the Oil Content* (p. 74). Moscow. Standartinform (in Russian).
8. State Standard of Russian Federation (GOST) 30623–98, (2010). *Vegetable Oils and Margarine Products: Fatty Acid Composition of Specific Vegetable Oils and Margarines (by Groups)* (p. 9). Appendix B. Introduction. 01/01/2000. – Moscow. Standartinform, (in Russian).
9. Khasanov, V. V., Ryzhova, G. L., & Maltseva, E. V., (2004). Antioxidants research methods. *Chemistry Plant Row Material, 3,* 63–75 (in Russian).
10. Gutteridge, V., Westekmarck, T., & Halliwell, B., (1986). Oxidation of oxygen in biological system. *Free Radical: Aging and Degenerative Disease* (p. 142). New York.
11. Temirbekova, S. K., Afanasyeva, Y. V., & Kurilo, A. A., (2016). Adaptive technology of cultivation of oilseed safflower cultivar Krasa Stupinskaya in bio-organic agriculture. *Recommendations* (p. 64). Moscow: Agrorus, (in Russian).
12. Norov, M. S., (2009). *Technology of Cultivation of Safflower in Conditions of Bogary of the Republic of Tajikistan (Recommendation)* (p. 38). Dushanbe: Irfon, (in Russian).
13. Pryanishnikov, D. N., (1965). *Selected Works* (Vol. 2, 379–395). Moscow. Sel'khozgiz (in Russian).
14. Postnikov, D. A., Neumann, G., Romkheld, F., & Chekeres, A. I., (2001). Accumulation of phosphorus with white mustard and rape when various forms of phosphates are introduced into the soil. *Izvestiya of Timiryazev Agricultural Academy. Scientific-Theoretical Journal of Russian Timiryazev State Agrarian University, 1,* 113–124 (in Russian).

15. Temirbekova, S. K., Kurilo, A. A., Afanasyeva, Y. V., & Konovalov, S. N., (2015). Using the new introduced species – safflower under the conditions of the Central region of non-chernozem zone of Russia. *Kormoproizvodstvo (Fodder Cropping in Rus.)*, *6*, 24–28 (in Russian).

16. Kurilo, A. A., & Temirbekova, S. K., (2010). Phytomeliorative influence of white mustard, blue lupine and safflower to effective fertility of sod-podzolic soils. *Fruit and Berry-Culture in Russia* (Vol. XXIII, pp. 275–282). Moscow: All-Russia Horticultural Institute for Breeding, Agrotechnology and Nursery, (in Russian).

17. Ermakov, A. I., & Popova, E. V., (1972). Sunflower breeding for improving the technological properties of food and oil. *Works of Applied Botany, Genetics and Breeding*, *48*(3), 171–172 (in Russian).

18. Vavilov, N. I., (1934). *Selection as a Science* (p. 16). Moscow, Leningrad: Sel'khozhiz (in Russian).

19. Temirbekova, S. K., Afanasyeva, Y. V., Konovalov, S. N., Postnikov, D. A., Metlina, G. V., Vasilchenko, S. A., & Norov, M. S., (2016). The results of introduction and "north promotions" of a new oil-bearing crop safflower in the Russian Federation. *Food Safety of Agriculture in Russia in the twenty-first century, Zhuchenkov Readings II* (Vol. 11, No. 59, pp. 55–68). Moscow: All-Russian Williams Fodder Research Institute (in Russian).

CHAPTER 12

Influence of Low Temperatures on the Catalase Activity of Psychrotolerant Bacteria of the Genus *Bacillus*

OLGA V. DOMANSKAYA,[1] NINA A. BOME,[2] NATALIA N. KOLOKOLOVA,[2] VLADIMIR O. DOMANSKII,[3] and NATALIA V. POLYAKOVA[1]

[1]*Tyumen State University, International Institute of Cryology and Cryosophy, Volodarsky St., 6, Tyumen 602003, Russia, E-mail: olga-nv@bk.ru*

[2]*Tyumen State University, Institute of Biology, Volodarsky St., 6, Tyumen 625003, Russia*

[3]*Tyumen Industrial University, Volodarsky St., 38, Tyumen 625000, Russia*

ABSTRACT

The current chapter describes the relationships between the temperature of cultivation of Bacteria from *the Bacillus* genus, extracted from Arctic permafrost, and the activity of oxidation stress ferments at the example of catalase. Bacteria were cultivated under 5, 22, and 45°C. The catalase activity of cultures under study varied from 0.73 to 12.35 units×mg^{-1} of protein and from 0.62 to 14.16 mg^{-1} under temperatures of 45°C and 22°C, respectively. The catalase activity of cultures, cultivated under 5°C, was several times higher than under 45 and 22°C, varying from 10.46 to 26.79 units×mg^{-1} of protein. The increase of catalase activity under low positive temperatures indicates high antioxidant protection under hypothermia stress.

12.1 INTRODUCTION

The first information about living microorganisms in the cryolitosphere appeared at the end of the XIX century when studying findings of mammoths in Siberia. Currently, viable microorganisms for sure are detected in all regions of permafrost spreading: until the depth of 400 m (Canadian Arctic), under mean annual soil temperatures down to –25°C (Antarctic) and salinity up to 170–300 g/L (Kolyma plain) [1, 2]. The presence of microbiota under extreme conditions promotes the manifestation of its unique characteristics, in particular, the activity of specific enzymes, such as protein tyrosine phosphatase, α-amylase, and β-galactosidase and aminopeptidase. Microorganisms develop protective mechanisms using some antioxidant enzymes, such as catalase, peroxidase, and superoxide dismutase (SOD), to protect the cell from oxidative stress. One of the ways to preserve active life in low-temperature conditions is high catalase activity. It is known that catalase is a psychrophilic enzyme characterized by high activity in the temperature range from 0 to 30°C and protects microorganisms from exogenous and endogenic oxidative stresses, neutralizing free oxygen radicals. Toxic substrate – superoxide ion (O^{-2}), formed in cells as a result of metabolic processes, is converted into hydrogen peroxide (H_2O_2) by the SOD enzyme [3]. In its turn, hydrogen peroxide is decomposed by catalase into molecular oxygen and water.

Most of the Earth's biosphere is characterized by low temperatures, about 70% are oceans, with an average temperature of 4°C. About 24% of the land of our planet is occupied by permafrost deposits [4].

Microorganisms are an integral part of ice or other frozen rocks. Exposure of microbiota under extreme conditions (for example, low temperature) promotes the manifestation of their unique characteristics, in particular, the activity of specific enzymes. Temperature is one of the most important parameters of the regulation of the activity of microorganisms, since the temperature affects the cell response and its adaptation to temperature fluctuations. The effect of low temperatures on Bacteria can be manifested in three directions: biochemical modifications, physiological changes, and control over the response to the effect of low temperature. In recent years, a lot of scientific articles have appeared on the possibility of the existence of microorganisms under extreme conditions and on the study of biochemical modifications and physiological changes taking place in the bacterial cell [5, 6].

One of the earliest reactions to cooling is oxidative stress, which results in the formation of active forms of oxygen. Active oxygen forms (oxygen ions, free radicals, and peroxides) affect all biological macromolecules: DNA, RNA, proteins, and lipids. As a result of the rupture of the DNA chain and modification of the nitrogenous bases and carbohydrate residues, replication is disrupted and completely blocked [7]. The main antioxidant enzymes in bacterial cells are catalase, peroxidase, SOD, DNA repair system, and various substrates involved in the neutralization of free radicals. Catalase and SOD protect microorganisms from exogenous and endogenous oxidative stresses by neutralizing free oxygen radicals. Oxidative stress also occurs during freezing and thawing, and such enzyme as catalase can participate in the cryotolerance of a bacterial cell [8].

The aim of this research was to study the relationship between the temperature of cultivation and the activity of enzymes of oxidative stress; for example, catalase in Bacteria of the *Bacillus* genus, isolated from the Arctic permafrost in Western Siberia. According to the scheme of permafrost zonation, Western Siberia is characterized by the intermittent and insular distribution of permafrost [9]. The features of the relief were formed in the middle, late Pleistocene, and Holocene under the influence of marine transgressions and the accumulative activity of rivers and lakes, as well as permafrost processes. Discovered deposits are supposedly to be Zyrianka horizon [10].

12.2   MATERIAL AND METHODOLOGY

The Bacteria of the genus *Bacillus* from samples of permafrost, taken from two wells with a depth of 30 m, drilled near the Dremuchee field, located near the city of Tarko-Sale (Yamal-Nenets Autonomous District) were used in the study.

Well No. 1 is located on a ridge (relatively low forms of relief) in the lake-alas complex, with moss and vegetation cover. The ground cover is dominated by green mosses, wild rosemary (*Ledum paluster* L.), willow shrubbery *(Salix* sp.*)*, dwarf birch *(Betula nana* L.*)*, cowberry *(Vaccinium vitis-idaea* L.*)*.

Well No. 2 is located in a sparse forest represented by Siberian pine (*Pínus sibírica*), Scots pine (*Pínus sylvéstris* L.), larch (*Lárix* sp.), the understory is represented by wild rosemary (*Ledum paluster* L.), dwarf birch (*Betula nana* L.), reindeer moss (*Cladonia* sp.).

The central part of the sample was selected under aseptic conditions from the obtained frozen core of undisturbed composition for microbiological studies according to the procedure described by Zvyagintsev D.G. and Gilichinsky D.A. [11].

The counting and isolation of microorganisms were carried out by means of serial dilutions on solid nutrient media of the following composition: meat-peptone agar for various chemo-organo-heterotrophic Bacteria; potato-glucose agar for Actinomycetes and Bacteria; Czapek's medium for fungi; the cultivation was performed at the temperatures of +5°, +16°, and + 36°C [12].

We investigated Bacteria of the genus *Bacillus* isolated from frozen deposits [13]. As a result of microbiological analysis of the samples, 9 pure cultures of *Bacillus spp.* were extracted. Species identification of these strains was carried out according to cultural-morphological and physiological-biochemical properties. To confirm the taxonomic status, bacterial strains were identified using an analysis of the sequences of fragments of the 16SrRNA gene obtained in the PCR (polymerase chain reaction) reaction using primers 8f (AGA GTT TGA TCC TGG CTC AG), 926r (CCG TCA ATT CCT TTR AGT TT) and 1492r (GGT TAC CTT GTT TAC GAC TT). Sequencing was carried out in the State Research Institute of Genetics and Selection of Industrial Microorganisms of the National Research Center Kurchatov Institute (Moscow, Russia). The resulting nucleotide sequences were compared with the sequences from the international NCBI databank using the BLAST software package. For the editing and alignment of the nucleotide sequences of 16SrRNA genes, the ClustalW2 program was used, and for the phylogenetic analysis, we applied MEGA software, version 5.2.

For comparative experiments we applied a strain of *Bacillus* sp. 3 M (VKPM B-10130) isolated from frozen rocks of Yakutia and presented by the colleagues from Tyumen Scientific Center of the Siberian Branch of the Russian Academy of Sciences [14].

Bacterial strains of *Bacillus* sp. were grown on a shaker (180 rpm) in 100 mL Erlenmeyer flasks with 50 mL of nutrient medium. Cultivation was conducted on a nutrient medium of the following composition (g/l): K_2HPO_4 1.0; $MgSO_4 \cdot 7H_2O$ 0.5; $(NH_4)_2 SO_4$ 2.0; peptone 2.0; yeast extract 5.0; glycerin 5.0. For inoculation, a daily culture grown on a medium of the same composition was used. The cultivation temperature of 22 and 45°C was maintained on a thermostated shaker BioSan ES-20 (Latvia). The cultivation was carried out at 5°C in a refrigerating chamber of Polair (Russia). Samples were taken every hour of cultivation at a temperature of 22°C

(45°C) and every 3 h at 5°C, then the optical density of the cell suspension was determined on a UNICO-2800 (US) spectrophotometer at 540 nm, the cell thickness was 10 mm. The kinetic parameters of the periodic growth of microorganisms were determined according to recommendations of S. J. Pirt [15].

The activity level of catalase was estimated by the method of [16], based on the ability of peroxide to form a colored complex with ammonium molybdate with an absorption maximum at 405 nm.

All measurements were performed in triplicate. Mathematical processing of the experimental data was carried out using the R software for data analysis and visualization, version 3.2.0 (Free Software Foundation, Inc. Boston, MA).

12.3 RESULTS AND DISCUSSION

Strains of Bacteria isolated from permafrost were identified and classified as *B. megaterium* (3 strains), *B. cereus* (4 strains.), *B. simplex* (1 strain) and *B. subtilis* (1 strain). According to the results of phylogenetic analysis of strains 1562 TS, 875 TS, 1257 TS, and 630 TS, the closest relative of the species is *B. cereus* with a similarity of 99%. It was found that strains 206 TS, 312 TS, and 629 TS are genetically close to *B. megaterium* with a similarity of 97%, 99%, and 98%, respectively. The closest strains of the 948-P1 TS and 948P strains are *B. simplex* and *B. subtilis*, respectively, with the similarity of 99% and 98%, respectively (Table 12.1).

TABLE 12.1 Results of Identification of Cultivated Bacteria Isolated from Frozen Rocks of Western Siberia (Dremucheye Deposit) (Year 2015)

Well No	Well depth, m	Strain	Taxonomic position	Similarity, 16S pRNA, %
1	2.5	1562 TS	*Bacillus cereus*	99
		875 TS	*Bacillus cereus*	99
	10.0	206 TS	*Bacillus megaterium*	97
	30.3	1257 TS	*Bacillus cereus*	99
2	4.2	312 TS	*Bacillus megaterium*	99
	10.0	629 TS	*Bacillus megaterium*	98
		630 TS	*Bacillus cereus*	99
	12.3	948P-1 TS	*Bacillus simplex*	99
		948P TS	*Bacillus subtilis*	98

The effect of the cultivation temperature on the growth parameters of *Bacillus* genus strains were evaluated in the simple periodic conditions with using of sowing material namely cells from lag-phase, bypassing such a way the phase of spores germination. It is known that the periodic culture of ia passes in its development four main phases: lag-phase (adaptation to the new environment), exponential phase (increasing growth rate), stationary (decreasing of growth rate) and senescence (degeneration of the population) phases. The synthesis of secondary metabolites by ia increases at the end of the exponential phase and reaches a maximum in the stationary phase [17].

The results of the study showed a variable relation to the temperature of strains under study. The multiplication of most of the strains under study started just after sowing, the duration of lag-phase varied from 30 min to 4 h in dependence on the cultivation temperature (Figure 12.1).

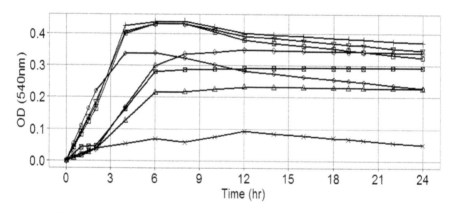

□ B. cereus 1257 △ B. cereus 630 × B. megaterium 312 ▽ B. simplex 948P 1
◇ B. cereus 1563 + B. cereus 875 ◇ B. megaterium 629 ▣ B. subtilis 948P

FIGURE 12.1   The dynamics of multiplication of *Bacillus* genus bacteria strains under the temperature of cultivation of 45°C. On the X-axis – time, hours; on the Y-axis – the density (OD 540 nm).

Under 45°C the duration of lag-phase of most of the cultures amounted 30 min, then an intensive growth of ial strains was observed during 3.5 h, and by 4 h the cultures entered into stationary development phase (see Figure 12.1).

The cultivation of cultures under 22°C led to the prolongation of lag-phase by 2–4 h. The exponential phase was observed after 4–6 h and continued on average 3.7 h. The phase of stationary growth of the cultures under study was reached after 10–12 h of cultivation (Figure 12.2). By this time considerable biomass accumulation was observed – almost 4 times higher than under 45°C, which follows from the measurements of the optical density.

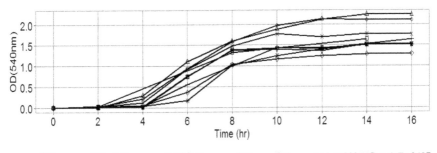

□ B .cereus 630   △ B. cereus 1563   × B. cereus 875   ▽ B. megaterium 629   ∗ B. subtilis 948P
◇ B. cereus 1257   + B. cereus 630   ◇ B. megaterium 312   ▣ B. simplex 948P 1

FIGURE 12.2   The dynamics of multiplication of *Bacillus* genus ial strains under the temperature of cultivation of 22°C. On the X-axis – time, hours; on the Y-axis – the optical density (OD 540 nm).

According to the results of the study under the cultivation temperature of 5°C, the slowing of the growth of cultures was observed in all phases. The lag-phase in most of the cultures continued from 15 to 68 h, which indicates a longer adaptation period. The exponential phase of growth continued from 27 to 117 h. The stationary phase was observed already after 72 h; the maximum density of cells was observed for *B. simplex* 948P1 and *B. meganterium* 312 (Figure 12.3).

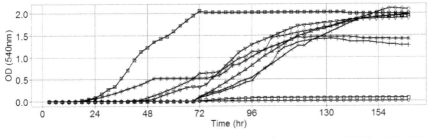

□ B. cereus 1257   △ B. cereus 630   × B. megaterium 206   ▽ B. megaterium 629   ∗ B. subtilis 948P
◇ B. cereus 1563   + B. cereus 875   ◇ B. megaterium 312   ▣ B. simplex 948P 1

FIGURE 12.3   Dynamics of multiplication of *Bacillus* genus bacteria strains under the temperature of cultivation of 5°C. On the X-axis – time, hours; on the Y-axis – the optical density (OD 540 nm).

Thus, the growth of ial strains of *Bacillus* genus depends on the temperature of cultivation. The detected regularities of growth of microorganisms allowed determining optimal temperatures of cultivation, which provide the highest productivity of the process. It was found that the maximum density of cells was reached under a temperature of 5 and 22°C, however, under 5°C longer time was necessary for reaching a maximum density of cells. The cultivation of strains under 45°C was not accompanied by considerable biomass increase. Earlier it was assumed that *B. cereus* is mesophilic, but recent studies also detected psychrotrophic forms of *B. cereus* strain, able to grow under a temperature range of 7–30°C [18, 19]. Our results showed the psychrotrophic nature of the strains: *B. cereus* 875, *B. cereus* 1257, and *B. cereus* 630.

The reaction of most of the organisms on the effect of extreme factors is displayed in the increasing activity of enzymes of the antioxidant defense system. The high catalase activity of microorganisms is one of their ways to keep active metabolism under low positive or negative temperatures. It is assumed that catalase psychrophilic enzyme characterized by high activity within the temperature range from 0°C to 30°C [20].

Maximum catalase activity under 45°C was observed in the cultures *B. cereus* 875 (12.35 units·mg^{-1}), *B. megaterium* 312 (12.24 units·mg^{-1}), and under temperature 22°C in *Bacillus sp.* B-10130 (14.16 units·mg^{-1}). Catalase activity of strains cultivated under 5°C was several times higher than under 45°C and 22°C varied from 10.46 to 26.79 units·mg^{-1} of protein, respectively; maximum activity was observed in *B. megaterium* 206 and *B. cereus* 1257 (Figure 12.4).

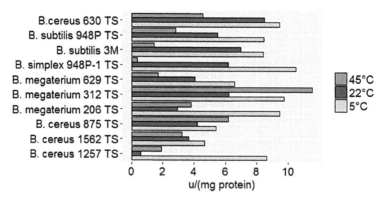

FIGURE 12.4   Catalase activity of *Bacillus* genus is under different temperature regimes of cultivation.

Our results on the increasing of catalase activity under low positive temperatures correspond to the general regularity for psychoactive microorganisms [21–23].

12.4   CONCLUSIONS

The results obtained are of interest, as they showed that the antioxidant catalase is the factor giving to ia of *Bacillus* genus the possibility to grow under low positive temperatures. It was found in the present study that temperatures of 5 and 22°C are optimal for the growth of ia cultures of *Bacillus* genus, which indicates the psychrotolerant character of isolates and the possibility of their existence and multiplication under broad temperature range. The active synthesis of specific enzymes is the universal mechanism of adaptation under conditions of low positive and negative temperatures. The enzymes with high activity under low positive temperatures allow compensating the disturbances of cell metabolism, which, in its turn, allows to ia cell continues its growth, although with a lower rate. The presence of antioxidant enzyme plays an important role in the survival of *Bacillus* genus in the natural ecosystems. Our results extend the knowledge about the limits of tolerance for growth and enzyme activity of microbiota from permafrost to the factor of temperature.

KEYWORDS

- **cryotolerance**
- **oxidative stress**
- **psychrophilic enzymes**

REFERENCES

1. Vorobyova, E. A., Gilichinskiy, D. A., & Soina, B. S., (1997). Life in the cryosphere: View of the problem. *Cryosphere of the Earth, 1*(2), 45–51 (in Russian).
2. Gilichinskiy, D. A., (2002). *Cryobiosphere of Late Age of Mammals: Permafrost as the Medium for Preserving Viable Microorganisms* (p. 58). Thesis for the degree of doctor of philosophy in horticulture presented and submitted into Tyumen Earth Cryosphere Institute (in Russian).

3. Fridovich, I., (1995). Superoxide anion radical and superoxide dismutase. *An. Rev. Biochem., 64,* 97–112.
4. Williams, P. J., & Smith, M. W., (1989). *The Frozen Earth* (p. 306). Cambridge University Press, Cambridge.
5. Baraúna, R. A., Freitas, D. Y., Pinheiro, J. C., Folador, A. R. C., & Silva, A. A., (2017). Proteomic perspective on the ial adaptation to cold: Integrating OMICs data of the psychrotrophic Bacterium *Exiguobacterium antarcticum* B7. *Proteomes, 5*(9), p. 11.
6. Nikrad, M. P., Kerkhof, L. J., & Haggblom, M. M., (2016). The subzero microbiome: Microbial activity in frozen and thawing soils. *FEMS Microbiol. Ecology, 92*(6), p. 16.
7. Sies, H., (1993). Damage to plasmid DNA by singlet oxygen and its protection. *Mutat. Res., 299,* 183–191.
8. Azizoglu, R. O., & Kathariou, S., (2010). Temperature-dependent requirement for catalase in aerobic growth of *Listeria monocytogenes* F2365. *Applied and Environmental Microbiology, 76*(21), 6998–7003.
9. Yershov, E. D., (1989). *Geocryology of USSR, West Siberia* (p. 454). Moscow: Nedra Pub., (in Russian).
10. Cherkasova, E. N., Kazbakova, K. T., Pakhomova, A. S., Gubarkov, A. A., Fadeyev, S. V., Smorygin, O. G., Oleynik, E. V., Samsonova, V. V., & Brushkov, A. V., (2008). Complex study of engineering-geological properties of soils in the Tarko-Sale region. *Proceedings of International Conference: Cryogenic Resources of Polar and Mountainous Regions* (p. 422–425). State and perspectives of engineering geocryology (in Russian).
11. Zvyagintsev, D. G., Gilichinskiy, D. A., Blagodatskiy, S. A., Vorobyova, E. A., Khlebnikova, G. M., Arkhangelov, A. A., & Kudryavtseva, N. N., (1985). Duration of the preservation of microorganisms in the permanently frozen sedimentary rocks and buried soils. *Microbiology, 54*(1), 155–161 (in Russian).
12. Netrusov, A. I., (2005). *Moscow Workshop on Microbiology* (p. 608). Academy, (in Russian).
13. Domanskaya, O. V., Domanskiy, V. O., & Kulakova, A. Y., (2015). Study of microbiological and biochemical activity of permafrost in Western Siberia. *Modern Problems of Science and Education,* 5. URL: https://www.science-education.ru/ru/article/view?id=21761 (date of the application: 18.08.2016) (in Russian).
14. Brushkov, A. V., Melnikov, V. P., Sukhovey, Y. G., Griva, G. I., Repin, V. E., Kalenova, L. F., et al., (2009). Relict microorganisms of cryolitozone as possible objects of gerontology. *Successes in Gerontology, 22*(2), 253–258 (in Russian).
15. Pirt, S. J., (1975). *Principles of Microbe and Cell Cultivation* (p. 274). Halsted Press, Division of John Wiley and Sons. New York.
16. Korolyuk, M. A., Ivanova, L. I., Mayorov, I. G., & Tokarev, V. E., (1988). Methods of determining of catalase activity. *Laboratory Practice, 1,* 16–18 (in Russian).
17. Timmusk, S., Nicander, B., Granhall, U., & Tillberg, E., (1999). Cytokinin production by Paenibacillus polymyxa. *Soil Biol. Boichem., 31,* 1847–1852.
18. Francis, K. P., Mayr, R., Stetten, F., Stewart, G., & Scherer, S., (1998). Discrimination of psychrotrophic and mesophilic strains of the *Bacillus cereus* group by PCR targeting of major cold shock protein genes. *Appl. Environ. Microbiol.,* 3525–3529.
19. Montanhini, M. T. M., Montanhini, R. N., Pinto, J. P. N., & Bersot, L. S., (2013). Effect of temperature on the lipolytic and proteolytic activity of *Bacillus cereus* isolated from dairy products. *Intern. Food Res. J., 20*(3), 1417–1420.

20. Gerday, C., Aittaleb, M., Arpigny, J. L., Baise, E., Chessa, J. P., Garsoux, G., Petrescu, I., & Feller, G., (1997). Psychrophilic enzyme: A thermodynamic challenge. *Biochim. Biophys. Acta, 1342*, 119–131.
21. Frank, H. A., Ishibashi, S. T., Reid, A., & Ito, J. S., (1963). Catalase activity of psychrophilic bacteria grown at 2°C and at 30°C. *Appl. Microbiol., 11*, 151–153.
22. Yumoto, I., Ichihashi, D., Iwata, H., Istokovics, A., Ichise, N., Matsuyama, H., Okuyama, H., & Kawasaki, K., (2000). Purification and characterization of a catalase from the facultatively psychrophilic bacterium *Vibrio rumoiensis* S-1 T exhibiting high catalase activity. *J. Bacteriol., 182*(7), 1903–1909.
23. Kimoto, H., Yoshimune, K., Matsuymaand, H., & Yumoto, I., (2012). Characterization of catalase from psychrotolerant *Psychrobacter piscatorii* T-3 exhibiting high catalase activity. *Int. J. Mol. Sci., 13*(2), 1733–1746.

CHAPTER 13

Biological Methods Increasing the Productivity of the Winter False Flax *Camelina silvestris* L.

SARRA A. BEKUZAROVA,[1] LARISSA I. WEISFELD,[2] and TUGAN A. DULAEV[1]

[1]*Gorsky State Agrarian University, Kirov St., 37, Vladikavkaz, Republic of North Ossetia – Alania 362040, Russia*

[2]*Emanuel Institute of Biochemical Physics of Russian Academy of Sciences, 4, Kosygin St., Moscow 119334, Russia, E-mail: liv11@yandex.ru*

ABSTRACT

The method of pre-sowing seed treatment of false flax (*Camelina silvestris* L. (Granz, 1762), family Brassicaceae) includes seed inoculation with nitrogen-fixing bacteria from the root system of incarnate clover. The most significant number of nodule bacteria in the root system of legumes develops in the phase of budding – flowering. The seeds of the false flax before sowing are inoculated with strains receiving from nodule bacteria of the roots of the clover of incarnate *Trifolium incarnatum* L. (family Fabaceae). Nodule bacteria are isolated from the roots, mixed with 10–15% sugar syrup, prepared in 0.1% aqueous solution of *p*-aminobenzoic acid (PABA) and add soil from the sown area. An increase in winter hardiness, seed productivity of the *Camelina silvestris* (L.) (Granz, 1762) plant and oil content in seeds compared with the control preparation of meadow clove based on rhizotorphin was revealed.

13.1 INTRODUCTION

To increase the yield and quality of seeds of cereals and winter oil–bearings crops, such as false flax, mustard, rape, recently use biological methods that exclude the application of agricultural chemistry. This task is carried out with the help of nitrogen-fixing legumes plants.

On the roots of legumes, crops are formed thickenings, called nodules. They contain soil bacteria of the genus *Rhizobium* (family Rhizobiaceae) and perform symbiotic nitrogen fixation. *Legumes* (family Fabaceae) have antioxidant activity and are capable of satisfying up to 60–90% of their nitrogen demand thanks to biological nitrogen fixation up to 60–90% of their demand for nitrogen content [1–4]. To increase the sowing qualities of seeds and getting decent harvests of the tested plants, their seeds are enveloped by crushed nodule bacteria selected from the rhizosphere of leguminous plants, for example, alfalfa, red clover or galega [3]. Liquid microbiological drug Baikal EM-1 containing a complex of microorganisms is also added to inoculates of bacteria [5, 6]. In some works also added the extract of stinging nettle dioeciously [7]. Nettle has antioxidant properties, rich in vitamins B_2, B_6, C, and K, contains a lot of chlorophyll, carotene, tannic compounds, as well as the cardiac glycoside urticin [8]. For the treatment of legumes seeds with solid shell was used as a solution of ragweed. *Ambrosia artemisiifolia* (family Asteraceae) is a widespread malignant weed, readily available for collection [9, 10]. This plant possesses a complex of chemical substances: essential oils, glycosides, macroelements, stimulating the germination of seeds.

To stimulate the growth of plants, the seeds were also treated with a solution of sodium selenate (Na_2SeO_4) [11–13], which has antioxidant qualities, and is used in plant growing, veterinary medicine and medicine as a mineral additive. In the latest inventions is used to stimulate the growth of plants and seeds of cereals and oilseeds.

In the invention, No. 2264069 [14], a method for soaking seeds with a preparation of effective microorganisms of Baikal EM-1 is offer. In this case, the seeds of spring wheat, rye, maize, and barley with reduced sowing qualities were processed. Before soaking, the seeds of spring wheat and barley were shaken in a shooter apparatus to improve germination. Over the next two hours, the seeds were kept in an aqueous solution of Baikal EM-1 preparation (dilutions 1:500, 1:1000, and 1:2000). To this mixture was added a 5% solution of nettle (*Urtica dioica* L.). Seed processing by this method significantly increased the number of sprouted seeds, energy germination of seeding, having initially low sowing qualities. The method is complicated by the fact that the experiments were carried out without contact with the soil in Petri dishes and by preliminary shaking of the material. In addition, the stinging nettle has a small distribution area (along roadsides) and a limited time for collection (June–July).

It has been published No. 2479974 [15], which include soaking of alfalfa seeds in a wormwood solution at a concentration of 0.2–0.3% for 6–8 h for softening the hard shell of seeds. From well–developed alfalfa plants of the 2nd–3rd year of life, nitrogen-fixing bacteria are taken from the upper part of the root system in the range of 0–20 cm during its maximum development (budding – flowering phase). Sampling from 5–8 plants is sufficient to obtain a biological preparation per 1 hectare of sowing. Nodules can be selected, without separating from the ground together with the roots. They can be stored in a warm place for a year at a temperature of 10–15°C. Before sowing, wet alfalfa seeds treated with ragweed are wrapped in crushed roots with bacteria and soil and sown in the amount of 300 g per 1 hectare. The method allows increasing seed germination (up to 96%) and germination energy (up to 92%), as well as to reduce the hardness of alfalfa shells seeds due to preliminary treatment with ambrosia solution. The disadvantage of the method is the threat of suppression of nodule bacteria under the influence of ambrosia ester oils, which possess bactericidal properties, and their ability to suppress growth processes.

In the invention No. 2524066 [16], seeds of cereal crops: winter wheat, foxtail millet (green foxtal) and panic were soaked in an aqueous solution of sodium selenate at a concentration of 0.1% for 0.5–1 h. In low doses, selenium stimulates the growth and development of leguminous plants. Under conditions of North Ossetia (Foothills of the North Caucasus), the wet seeds of winter wheat were treated with sodium selenate. These seeds were then inoculated with live bacteria *Rhizobium galegae*, isolated under the soil-climatic conditions of Belarus from *Galega orientalis* Lam. (goat's eastern), a cultivar of Nesterka. Then wheat seeds were mixed with soil taken from the goat's eastern rhizosphere at a depth of 15–20 cm. Wheat seeds were mixed with soil taken from the goat's rhizosphere at a depth of 15–20 cm, where most bacteria were concentrated. The seeds of winter wheat so treated were sown to the soil. The result was a significant increase in the germination of seeds, grain yield, and the amount of biological nitrogen in the soil (up to 25%). Similarly, the seeds of the cereal culture of the panic were treated. For the eastern goat, 'Bimbolat' cultivar, the soil was selected in the same way as in the experiment with the 'Nesterca' cultivar, then their mixed with the inoculum from the nodule bacteria. Panic seeds 20 kg/ha were treated in quantity 100 mL of inoculum per 200 g of soil. Similarly, the seeds of the foxtail millet ("chemise" in Russian) were treated. In all three variants, a significant increase in seed germinated (89%), grain yield (25%) and nitrogen content in the soil (69 kg/ha) were obtained.

The limitation of this method is that the selenium used for prolonged exposure and at high doses is toxic [11], which can lead to a decrease in activity of nodule bacteria. In addition, the large volume of soil introduced with the inoculant (in a ratio of 2:1), complicates the sowing when using a seeder.

Considering the described achievements and limitations, further work was aimed at expanding biological methods for such valuable oilseeds, which are antioxidants, like winter false flax, winter mustard, winter rape and other winter oil crops, in order to improve the quality of their seeds and the yields.

In this chapter, a method of using nitrogen-fixing cultures by treating seeds with a solution of sugar syrup and para-aminobenzoic acid (*p*-amino-benzoic acid, PABA) is proposed. PABA is a physiologically active natural compound, refers to group B vitamins, has antioxidant properties, and is necessary for the growth of many microorganisms, in particular bacteria [17, 18]. The antioxidant properties of PABA and its sodium salt have been extensively researched in a biochemical research in work [18].

Beginning in the 1960s of the XXth century, on the initiative and under the leadership of I. A. Rapoport various scientific and agricultural institutions of the USSR widely used the PABA on cultivated plants. Field and laboratory tests have shown that the PABA positively influences the signs: seed germination, early plant development, the survival of plants, yield's indicators and survival in the field conditions of various agricultural institutions. The Ministry of Agriculture of USSR allowed the use of PABA in the agricultural practice of n the Moscow Region [19]. In North Ossetia, the PABA is widely used for testing and growing field crops.

13.2 MATERIALS AND METHODOLOGY

The seeds of the winter cultivar 'Penzyak' before sowing are inoculated with a strain of nodule bacteria. The one-year-old plant – clover (red clover), *Trifolium incarnatum* L. was used. In the period of budding – of the beginning flowering (in late May) in the rhizosphere on roots are formed thickenings (nodules), where a significant number of active nitrogen-fixing bacteria accumulate. The rhizosphere of clover is extracted at a depth of 15–20 cm of soil, where the largest number of nodules accumulates, and then they were transferred into test tubes with a nutrient medium. A nutrient medium was used on de Man, Rogosa, and Sharpe (MRS-agar) recommended for the cultivation of *Lactobacillus*. For the preparation of inoculum, 3–4 clover plants with a total mass of nodules of 30–40 g per 1 hectare of sowing winter rye are sufficient. Since the nodules of legumes are developing in 2.5–3 months earlier (the period of budding –ripening of the plant), than the sowing occurs, the bacteria are stored in a laboratory environment in a nutrient medium before sowing the seeds of the false flax winter, those until the end of August – beginning of September. Before sowing, a solution for processing the seeds of the false flax winter prepared. In 1 L of hot water (70–80°C), 10 g of PABA powder need to dissolve. For receive solution 100–150 g of sugar must add to the solution of PABA. Since the sugar solution becomes sticky at such a concentration, the sowing is difficult because of the low flowability of the seeds. For better flowability of seeds, when sowing, to the prepared mixture, the soil of the plot where is planned to be sown is added. In the solution cooled to 25°C, an inoculant was placed – an incarnate clover strain. The culture of the bacteria was mixed with the seeds in the amount necessary for sowing on the planned area. The seed rate for of false flax winter is 6 kg/ha in one experiment and 8 kg/ha in the second experiment. 100–150 grams of topsoil was added from the site where they planned to sow. The whole mass was mixed, dried to the state when the mixture flows loosely on a plot, treated with inoculum and sowed on an area of 1 hectare. As a control, the seeds were coated with a biological fertilizer from factory prepared rhizotorphin, obtained on sterile peat, including nodules of clover meadow (*Trifolium pratense* L.). In the following spring, after the winter treatment the of false flax seeds, in spring were evaluated winter hardiness of plants, the yield of seeds and the oil content in them.

13.3   RESULTS AND DISCUSSION

The object of the experiment was the oil crop of genus *Camelina* – an herbaceous plant of the family Brassicaceae. False flax winter (*Camelina silvestris* (L.) Granz 1762) is a small-seeded oil crop of winter sowing, having a comparatively short vegetation period. It is one of the least demanding to conditions for the cultivation of oilseeds, characterized by cold resistance, drought resistance.

False flax winter cultivar Penzyak is included in the State Register of the Russian Federation in 2002 for all areas of cultivation of culture [20]. This cultivar is patented for food, fodder, and technical purposes. The patent owner is the "Penza Scientific Research Institute of Agriculture" [21–23]. A characteristic feature for flax false is undemanding to the conditions for the cultivation [22]. Over the period of 2012–2017, four varieties of false flax were created which were included in the State Register of the Russian Federation [20]. False flax oil contains glucosinolates, many polyunsaturated acids – up to 36% Omega-3 and up to 18% Omega-6, besides it contains magnesium, vitamins A, E, and D and carotenoids, also contain chlorophyll, phospholipids, and phytosterols. By stock of vitamin E, it overtakes flax seed, sunflower, cedar, and mustard oil. Most of the substances contained in oil of the seeds of lax false possess antioxidant properties [24, 25].

The seeds of the Penzyak winter variety before the sowing are inoculated with a strain from nodule bacteria of clover incarnate. Seeds are mixed with 10–15% sugar syrup, prepared in 0.1% aqueous solution of PABA, and add plot from land in the amount of 10%.

PABA is a precursor in the biosynthesis of important cofactors, is as bacterial vitamin H_1, vitamin B_{10} participates in the synthesis of purines and pyrimidines, which are part of the nucleic acids [17, 18].

PABA (chemical formula: $C_7H_7NO_2$) is an amino acid, a crystalline substance, is a poorly soluble white powder that dissolves in water well only at a temperature of 70–80°C (Figure 13.1).

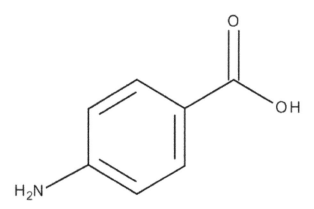

FIGURE 13.1   The structural formula of PABA.

I. A. Rapoport [18] designated the drug as a phenotypic activator of enzymes. PABA interacts with the genetic apparatus of the cell, has a wide impact on different taxonomic units, enhances enzymatic catalysis, participates in the repair of DNA nucleotides and amino acids, but is not a mutagen [19]. Due to its chemically active features, PABA is a growth factor for many bacterial species, including nodule bacteria [26, 27], used in this work to inoculate the seeds of winter false flax.

The antioxidant properties of PABA have wide practical applications. In North Ossetia – Alania, PABA is entered into the majority of agricultural inventions. PABA includes patents that ensure the increase yields of potato tubers [28, 29]; pre-planting treatment and for the cultivation of potatoes with the use of "snow technology" when processing shoots in the early stages [29]; for cultivation of triticale green mass [30]; presoving treatment of seeds [31]; for stimulate growth of clover, the seeds of which have a hard shell [32], for propagation of dogwood by cuttings [33] for defoliation in the cultivation of cotton plants [33], for presowing treatment of amaranth seeds [34], for treatment of fruit and berry crops [35], for the presowing treatment of chickpea seeds to increase the amount of nitrogen-fixing bacteria and to increase the yield of chickpea kernels [36]; for stimulation growth of winter false flax, which is not traditional for agriculture of North Ossetia and North Caucasus Foothills [37].

PABA is included in the invention for soil purification from oil contamination [38]. Water solution PABA in the concentration 0.03% was used to preserve and develop plants in the Winter Garden [39, 40].

In the present work, it was posed a task with the help of bacterial strains from the rhizosphere of the clover incarnate; increase the valuable qualities of the winter false flax. In the experiment, the seeds of the new crop were used. Before sowing, seed inoculation was carried out bacterial strains, with the inclusion of the sugar syrup dissolved in PABA, what can serve as a nutrient solution for the plant of clover incarnate and for the microflora of the soil where the seeds of winter false flax sown. The parameters of winter hardiness of plants, seed productivity, and oil content in seeds after different variants of seed treatment were compared (Table 13.1). In the control variant, a culture of nodule bacteria from *Trifolium pratense* L. reproduced in sterile peat for the best adhesion of the inoculum to seeds. It names a rhizotorphin.

To assess the effect of PABA without adding sugar syrup on plant development, red winter seed was treated with a 0.1% aqueous solution of PABA (Figure 13.2).

It is clearly seen that PABA activates the growth and development of young plants.

Table 13.1 shows that the lowest results were obtained with the use of rhizotorphin on the basis a meadow clover *Trifolium incarnatum* L. Inoculation of the seeds of the false flax using nodule bacteria at the addition of 5% and 15%th solutions of sugar dissolved in PABA gives an increase in all the indices. The best result was achieved about seed productivity of winter wheat (t/ha) and oil content in seeds (%).

TABLE 13.1   Winter Hardiness and Productivity of Seeds of Plants *Camelina silvestris* (L.) (Granz 1762) and the Content of Oil in Seeds After Inoculating of the Seeds with Bacteria from the Roots *Trifolium incarnatum* L. and after Processing with the Sugar Dissolved in PABA

	Winter hardness, %	Seminal productivity, t/ha	Containment of oil in seeds, %
Mixing the seeds of false flax with rhizotorphin (control)	78.3	1.48	37.8
Inoculation of the false flax seeds with a bacterial strain from roots of clover incarnate in mixture with a 5–9%th solution of sugar	82.6	1.56	38.4
Inoculation of the false flax seeds with a bacterial strain from roots of clover incarnate in mixture with a 10% solution of sugar	84.2	1.62	39. 6
Inoculation of the false flax seeds with a bacterial strain from roots of clover incarnate in mixture with a 15% solution of sugar and with the addition of 10% soil	85.1	1.68	40.5
Treatment of seeds of the redhead seeds with an aqueous solution of 0.1% solution of PABA supplemented with 10% soil	79.2	1.52	39.4
Inoculation of seeds of false flax with a mixture of bacterial strain from roots of clover incarnate and with 15% sugar solution	89.1	1.76	40.8
Inoculation with a strain of bacteria from roots of incarnate clover *Trifolium incarnatum* L., selected in budding phase – the beginning of flowering. Bacteria are mixed with 10–15%th sugar syrup, prepared in a 0.1% aqueous solution of PABA and with soil addition of the sowing plot in an amount of 10% of the volume of the solution	87.9	2.1	41.6

FIGURE 13.2   Development of young plants after seed treatment *Camelina silvestris* (L.) (Granz, 1762) with solution PABA (photo, made by S. A. Bekuzarova) (Left – leaves and roots of young plants after treatment of intact seeds in a solution of PABA, on the right – without PABA).

In the most complete treatment option: inoculation with nodule bacteria +10%th solution of sugar, dissolved in PABA solution, and soil addition gives the highest index of winter plant hardiness (87.9%), which significantly higher than in control (78.3%) and also than in other variants. Although this data is somewhat lower than at using 15% sugar. However, in such concentrated sugar syrup, the seeds develop poorly. It is also important to add the soil in a volume of 10% to the total volume of the solution, as this increases the flowability of the seeds, which is so necessary when sowing in large areas with seeder.

13.4 CONCLUSIONS

The proposed biological approach gives positive results when cultivating winter false flax, increases its productivity and quality of seeds, and increases winter hardiness of plants. The method can be used in the cultivation of various crops of winter sowing with small seeds. The biological agents described in this study have antioxidant activity.

An important conclusion is a possibility of obtaining a positive result by inoculating the seeds of false flax by bacterial strains from the roots of leguminous plants in combination with other ingredients of a natural nature without the chemicalization of the process.

The main trend of works in this area is to increase resistance to harmful anthropogenic factors in the plant environment, increase their endurance and yield without chemicalization when growing cultivated plants.

KEYWORDS

- Galega
- alfalfa
- incarnate clover
- nitrogen-fixing bacteria
- oil-bearing crops
- *p*-aminobenzoic acid (PABA)

REFERENCES

1. Vasilieva, G. G., (2004). Active forms of oxygen and antioxidant enzymes in the initial stages of pea interaction with nodule bacteria (*Rhizobium legumonozarum*). *PhD Thesis in Biology* (p. 23). Irkutsk. (In Russian).
2. Bekuzarova, S. A., Haniyeva, I. M., & Gishkayeva, L. S., (2014). Receiving of the new forms of red clover for growing in North Ossetia, Kabardino-Balkaria, and Chechen Republic. In: Opalko, A. I., Weisfeld, L. I., Bekuzarova, S. A., Bome, N. A., & Zaikov, G. E., (eds.), *Ecological Consequences of Increasing Crop Productivity, Plant Breeding and Biotic Diversity* (pp. 23–27). Toronto – New Jersey. Apple Academic Press.
3. Bushuyeva, V. I., (2014). The use of the regularities of manifestation of hereditary variability characteristics in legumes in the selection of *Galega orientalis*. In: Opalko, A. I., Weisfeld, L. I., Bekuzarova, S. A., Bome, N. A., & Zaikov, G. E., (eds.), *Ecological Consequences of Increasing Crop Productivity, Plant Breeding and Biotic Diversity* (pp. 277–287). Toronto – New Jersey. Apple Academic Press.
4. Kartyzhova, L. E., (2015). Selection of competitive and efficient strains of *Rhizobium Galegae*. Weisfeld, L. I., Opalko, A. I., Bome, N. A., & Bekuzarova, S. A., (eds.), *Biological Systems, Biodiversity and Stability of Plant Communities* (pp. 241–250). Apple Academic Press. Canada – USA.
5. Shablin, P. A., (2000). Effective microorganisms are the planet's hope. A collection of materials on the practical application of Baikal EM–1. In: Suhammera, S. A., (ed.), *Collection of Materials* (p. 68). Moscow. URL: http://st.arqo.ru/6/1828/465/Baykal_ EM._Biotehnologiya_21_veka.pdf (Accessed on 29 May 2019) (in Russian).
6. Blinov, V. A., (2003). *EM Technology for Agriculture* (p. 205). Saratov. Saratov State Agrarian University (in Russian).
7. Balagozyan, E. A., (2017). Pharmacognostic examination of rhizomes with roots of nettle (*Utrica dioca* L.). *PhD Thesis in Pharmaceutical Sciences* (p. 24). Samara (in Russian).
8. Soshnikova, O. V., (2006). Study of the chemical composition and biological activity of plants of the nettle genus (*Urtica*). *PhD Thesis in Pharmaceutical Sciences* (p. 23). Kursk (in Russian).
9. Luchinsky, S. I., & Makoveev, A. V., (2008). Biological features of ragweed. *Proceedings of the Kuban State Agrarian University* (Vol. 6, No. 15, pp. 92–95). Krasnodar (in Russian).
10. Glusheva, T. N., & Karpushina, E. N., (2009). *Allelopathy of Ragweed Amnioticum (Ambrosia artemisiifolia* L.) (Vol. 11, No. 66, 5–10). Belgorod. Belgorod State University (in Russian).
11. Golubkina, H. A., & Papazian, T. T., (2006). Selene in nutrition. *Plants, Animals, People* (p. 250). Moscow, Published City (In Russian).
12. Vikhreva, V. A., (2001). The influence of selenium on the growth and development and the adaptive potential of the goatskin of the eastern goat (*Galega orientalis*). *PhD Thesis in Biology* (p. 28). Moscow (in Russian).
13. Blinokhvatov, A. F., Vikhreva, V. A., & Kleimenova, T., (2012). *Selen in the Life of Plants* (p. 224). Penza. Publishing house BIBKOM (in Russian).
14. Blinov, V. A., & Burshina, S. N., (2005). Invention No. 2264069. *Method for Low-Grade Seedcorn Pretreatment.* Date of publication: 20.11.2005. Bulletin No. 32. Int. Cl. A 01 C 1/100 (in Russian).

15. Zherukov, B. X., Khanieva, I. M., Khaniev, M. K., Magomedov, K. G., Bekuzarova, S. A., & Boziev, A. L., (2013). Invention 2479974. *Method for Presowing Treatment of Alfalfa Seeds.* Date of publication: 27.04.2013. Bulletin #12. Int. Cl. A01C 1/100 (2006/01) A01N 65/00 (2009.01) (in Russian).

16. Bekuzarova, S. A., Bushuyeva, V. I., Tsomartova, F. T., Lushchenko, G. V., Kartyzhova, L. E., Avramenko, M. N., & Gazaeva, L. V., (2014). Invantion 2524066. *Method for Presowi Treatment of Seeds of Grain Varieties.* Date of publication: 27.07.2014 Bulletin #21. Int. Cl. *A01C1/100*(2006/01) URL: http://www.freepatent.ru/images/img_patents/2/2524/2524066/patent–2524066.pdf (Accessed on 29 May 2019) (in Russian).

17. Para-Aminobenzoic acid (*p*-Aminobenzoic acid), (2017). *Great Encyclopedia of Oil and Gas.* URL: http://www.ngpedia.ru/id84163p1.html (Accessed on 29 May 2019) (in Russian).

18. Rapoport, I. A., (1989). The action of the PABA in connection with the genetic structure. *Chemical Mutagens and Para-Aminobenzoic Acid in Increasing the Yield Productivity of Agricultural Crops* (pp. 3–37). Moscow. Nauka (Science, in Rus.) (in Russian).

19. Sirota, T. V., Lyamina, N. E., & Weisfeld, L. I., (2017). Antioxidant properties of para-aminobenzoic acid and its sodium salt. *Biophysics, 62*(5), 691–695. doi: 1134/S0006350917050219.

20. *State Register of Protected Selection Achievements,* (2017). (Vol. 1, p. 180) Cultivars of plants. *Winter Cameline.* Moscow (in Russian).

21. Kshnikatkina, A. N., & Prakhova, T. Y., (2017). Winter cameline – as an alternative to oilseeds. *The Farmer of the Chernozem Region. Regional Journal.* http://vfermer.ru/rubriki/konsultacii/918–ryzhik–ozimyy–kak–alternativa–maslichnym–kulturam.html (Accessed on 29 May 2019) (in Russian).

22. Prakhova, T. Y., (2013). *Camelina silvestris* (p. 209). Monograph Penza. (in Russian).

23. Prakhova, T. Y., (2013). Productivity of winter cameline depending on techniques of cultivation. *The Young Scientist, 6,* 783–784 (in Russian).

24. Zelenina, O. N., & Prakhova, T. Y., (2009). Fatty acid composition of winter cameline in seeds of winter cultivar Penzjak. *Oilseeds. Pustovojt Scientific and Technical Bulletin of the All- Russian Scientific Research Institute of Oilseeds Cultures, 2*(141), 119–122 (in Russian).

25. Krylov, A. P., (2016). Qualitative indicators of cameline oil. Innovative technologies in the agroindustrial complex: Technology and practice. 4th All–Russian Scientific and Practical Conference. *Collection of Articles* (pp. 60–63). Penza Scientific Research Institute, Penza (in Russian).

26. Volobueva, O. G., (1992). PABA and synthetic growth regulators as a factor of increase efficiency of legume-rhizobia symbiosis. *PhD Thesis in Biology* (p. 26). Moscow (in Russian).

27. Beletsky, Y. D., (1993). *Para-Aminobenzoic Acid is a New Biologically Active Compound* (p. 61). Rostov-on-Don. Publishing house of Rostov University (in Russian).

28. Bekuzarova, S. A., Farniev, A. T., Basieva, E. B., Gasiev, V. I., & Kalitseva, D. T., (2011). Invention 2416186. *Method to Stimulate Growth and Development of Clover.* Date of publication: 20.04.2011. Bulletin #11. Int. Cl. A01C 1/06 (2006.01) (in Russian).

29. Ikaev, B. V., Marzoev, A. I., Bekuzarova, S. A., Basaev, I. B., Bolieva, Z. A., & Kizinov, F. I., (2010). Invention 2386558. *Method for Pre-Planting Treatment of Potato Tubers.* Date of Publication: 20.04.2010. Int. Cl. A01C 1/00 (2006.01). (In Russian)

30. Bekuzarova, S. A., Bolieva, Z. A., Basiev, S. A., & Basieva, A. S., (2018). Snow technology of potato cultivation. In: Weisfeld, L. I., Opalko, A. I., & Bekuzarova, S. A., (eds.), *Horticulture for Sustainable Development and Environment* (pp. 35–45). N.Y., US. Apple Academic Press.

31. Bekuzarova, S. A., Antonov, O. V., & Fedorov, A. K., (2003). Invention 2212777. *Method for Increasing Protein Content in Green Mass of Winter Triticale.* Date of publication: 27.09.2003. Bulletin #27. Int. Cl. (in Russian).

32. Bekuzarova, S. A., Abieva, T. S., & Tadeeva, A. A., (2006). Invention 2270548. *Method for Presoving Treatment of Seeds.* Date of publication: 20.04.2006. Bulletin No. 6. Int. Cl. A01C 1/06 (2006.01). (In Russian)

33. Tsabolov, P. K., Bekuzarova, S. A., Tigieva, I. F., & Tadtaeva, E. A., (2007). Invention 2294619. *Method for Propagation of Dogwood by Cuttings.* Date of publication: 10.03.2007. Bulletin # 7. Int. Cl. A01G 17/00, A01C 1/00, A01N 33/06, A01N 37/10, A01N. 37/22, A01N 37/44 (2006.01) (in Russian).

34. Basaev, B. B., Bekuzarova, S. A., Farniev, A. T., Gazdanov, A. U., Juldashev, M. A., & Kushch, D. A., (2006). Invention 2267925. *Method for Defoliation of Cotton Plants.* Date of publication: 20.01.2006. Bulletin No. 2. Int. Cl. A01N 55/00, A01N 57/24 (2006.01) (in Russian).

35. Bekuzarova, S. A., Abaev, A. A., Gasiev, V. I., & Abieva, T. S., (2016). Invention 2575043. *Method for Presowing Treatment of Amaranth Seeds.* Date of publication: 10.02.2016. Bulletin # 4. Int. Cl. A01C 1/00 (2006.01) (in Russian).

36. Bekuzarova, S. A., Opalko, O. A., Pakhomenko, A. V., & Gagloeva, L. C., (2017). Invention 2626731. *Method for Treatment of Fruit and Berry Crops.* Date of publication: 31.07.2017. Bulletin 22. Int. Cl. A01G 17/00 (2006.01), A01G 7/06 (2006.01) (in Russian).

37. Bekuzarova, S. A., Tedeeva, V. V., Abaev, A. A., Tedeeva, A. A., & Khokhoeva, N. T., (2017). Invention 2622665. *Method for Inoculation of Chickpea Seeds.* Date of publication: 19.06.2017. Bulletin #17. Int. Cl. A01C 1/00 (2006.010) (in Russian).

38. Bekuzarova, C. A., Buyankin, B. I., & Dulaev, T. A., (2016). Stimulants for the growth and development of winter flax false. *News of Gorski State Agrarian University, 53*(3), 16–20 (in Russian).

39. Bekuzarova, S. A., Alexandrov, E. N., Weisfeld, L. I., & Pliev, I. G., (2015). Invention 555595. *Method for Reclamation of Oil Contaminated Soil.* Date of publication 0.07.2015. Bulletin #9. Int. Cl. B09C 1/10 (2006.01), A62D 3/02 (2007.01), A01B 79/02 (2006.10) (in Russian).

40. Opalko, A. I., Weisfeld, L. I., Bekuzarova, S. A., Burakov, A. E., Opalko, O. A., & Tatarinov, F. A., (2018). Tolerance improvement of indoor plants temperate. In: Weisfeld, L. I., Opalko, A. I., & Bekuzarova, S. A., (eds.), *Horticulture for Sustainable Development and Environment* (pp. 119–149). N.Y. US, Canada, Apple Academic Press.

Glossary

Absorption spectroscopy or reflectance spectroscopy in the ultraviolet-visible spectral region uses light in the visible and adjacent ranges. The absorption or reflectance in the visible range directly affects the perceived color of the chemicals involved.

Adaptogens are a pharmacological group of preparations of natural or artificial origin, which are capable of increasing the nonspecific resistance of the organism to a wide range of harmful effects of physical, chemical, and biological nature. The term was first introduced into the literature by N. V. Lazarev (1952) to medicine for the correction radiation stress.

Alanit is the zeolite clay of North Ossetia deposits contain 51–53% silicon, aluminum 16–17%, iron 56%, 30–33% calcium, potassium 0.07%, phosphorus 0.38%, manganese 0.04% sulfur 0.98% magnesium 1.6%, and small amounts of zinc, copper, cobalt and other trace elements. The reaction medium (pH) 8.64 is determined by the high content of calcium.

Aldolase (fructose-bisphosphate-(fructosodiphosphate)-aldolase) is an enzyme that catalyzes the conversion of fructose-1,6-diphosphate to dihydroxyacetone phosphate and glyceraldehyde-3-phosphate during glycolysis. The enzyme plays an important role in energy metabolism.

Alfalfa (herbaceous plant lucerne, *Medicago sativa*) is a typical species of the genus *Lucerne* family legumes (Fabaceae); fodder plant, fine honey plant. Under favorable weather conditions, honey production reaches: in the areas of irrigated agriculture 300 kg per hectare of crops, without irrigation – 25–30 kg. Nectar is colorless, contains up to 50% sugar. Liquid alfalfa honey is transparent or golden yellow.

Ambrosia (*Ambrosia artemisiifolia*, family Asteraceae) is a widespread malignant weed, readily available for collection. This plant possesses a complex of chemical substances: essential oils, glycosides, macroelements, stimulating the germination of seeds, has also antioxidant properties, is rich in vitamins B_2, B_6, C, and K, contains a lot of chlorophyll, carotene, tannic compounds, as well as the cardiac glycoside urticin. Seeds of *Ambrosia* have a hard shell.

Antioxidant activity (AOA) is the effective physicochemical parameter characterizing an ability of different substances and the cell components to protect the biological objects from the toxic effects of a number of oxygen compounds formed in the system – oxygen ions, peroxides, and free radicals.

Antioxidant enzymes (enzymes of the antioxidant defense) are several enzyme systems that catalyze reactions to neutralize free radicals and reactive oxygen species.

Antioxidants (AOs) are the molecules that inhibit free radical oxidation of other molecules. The mechanisms of actions, the most common antioxidants, consist of the termination of reaction chains: antioxidants interact with the active radicals with the formation of inactive radicals. Antioxidants give to unsaturated free radicals their electrons, while turning in the stable compounds; as a result, the continuous chain of the destruction of molecules ceases. Natural antioxidants are concentrated in generative organs of plants (pestle and stamen), in vegetative parts (leaves, roots, and shoots), in fruits, and seeds. Their amount in the product depends on the energy of photosynthesis and the climate. When a molecule donates an electron, it becomes unstable, because it's missing that crucial electron. The balance of the oxidative state of plants and animals maintain complex systems of overlapping antioxidants, such as glutathione and enzymes (e.g., catalase and superoxide dismutase) produced internally or the dietary natural antioxidants or substances possessed antioxidant properties in complex systems: ascorbic acid (vitamin C), vitamins A, E, aromatic amines, phenols, flavonoids, catalase, etc. Antioxidants have a reactive group; so they can donate an electron to stop the chain reaction because of the formation of the more reactive free radicals.

Antiradical activity is the ability of individual substances, their mixtures and lipids isolated from different biological objects to inhibit the free radical reactions. The quantitative parameters of antiradical activity allowing comparison of antioxidants of the different nature and structure are the rate constant of its interaction with peroxy radicals (k_7) and the stoichiometric coefficient of inhibition (f) which is determined by using the different model systems.

Antiviral medicals are chemical compounds with antiviral activity (antiviral drugs) used for treating viral infections. Like antibiotics for bacteria, most antiviral compounds are used for specific viral infections, but it exists a broad-spectrum of antiviral preparations against a wide range of viruses. Unlike most antibacterial antibiotics, antiviral medicines do not destroy their target pathogen; instead, they inhibit their development.

Apoptosis is one of the types of the programmed cell death that occurs in multicellular organisms. The death of cells occurs as a normal and controlled part of an organism's growth or development. Biochemical events lead to characteristic cell changes (morphology) and death.

Ascorbic acid (vitamin C) is a synergist of antioxidants as their antioxidant properties are revealed only in the presence of α-tocopherol. Ascorbic acid is contained in food, involved in tissue repair and enzymatic production of certain neurotransmitters, necessary for the functioning of several enzymes and increases the functioning of immune systems.

Biological active substances (BAS) are synthetic or natural compounds possessing high physiological activity at low concentrations with respect to certain groups of living organisms. They are also extracted from plant material at short-term ultrasound exposure with low-intensity and *may be regulators of the biological organism development, or inhibitors, or stimulators, or adaptogens.*

Biomembranes or biological membranes are the enclosing or separating membranes which are a selected barrier around cells or subcellular organelles within living biological objects. Biomembranes consist of a phospholipid bilayer with embedded integral and peripheral proteins. They play an important role in cell metabolism, specifically in the communication and transportation of chemical compounds and ions. The lipid bilayer consists of two layers which have an asymmetric organization.

Biopolymers are proteins, nucleic acids, polysaccharides, and others that are the natural polymeric molecules that are produced by living organisms. They contain monomeric units that are covalently bonded to form larger structures. Their classification is according to the monomeric units. Among them, cellulose, starch, and chitosan are the most known polysaccharides.

Bisretinoids are toxic vitamin A dimers that are amassed by retinal pigment epithelium as lipofuscin.

Bovine serum albumin, also known as *BSA*, is a serum albumin protein derived from cows. It is often used as a protein concentration standard in laboratory experiments. *BSA* is a globular protein (~66,000 Da) that is used in numerous biochemical applications due to its stability and lack of interference within biological reactions. Sigma provides a variety of Bovine serum albumins with purities ranging from 95 to 99%.

Bruch's membrane is a thin transparent plate between choriocapillaris and retina, and it consists of five layers (from inside to outside): the basement membrane of the retinal pigment epithelium; the inner collagenous zone; a central band of elastic fibers; the outer collagenous zone; and the basement membrane of the choriocapillaris. The Bruch's membrane performs a variety of important functions, primarily by selective transport of nutrients and water in the direction of the retina.

Carotenoids are lipid-soluble compounds that protect against $^1O_2$ and absorb molecular oxygen since their molecules have highly conjugated double bonds. They extinguish reactive oxygen species and can prevent lipid peroxidation in biological objects, including marine animals. Carotenoids are the natural pigments, usually red, orange or yellow. Plants, algae, several bacteria, and fungi synthesize them. Animals receive the necessary carotenoids either directly from the diet, or modify dietary carotenoid precursors through metabolic pathways.

Catalase is a ferment of the antioxidant defense system that catalyzes the decomposition of hydrogen peroxide to water, that is one of the main enzymes for the destruction of reactive oxygen species. It is a hemoprotein because of its molecule consists of four identical subunits with a molecular mass of 62,000. Catalase found in nearly all living organisms exposed to oxygen, i.e., bacteria, plants, and animals. It belongs to the class of oxidoreductases. The level of their activity differs in different tissues.

Catalyst is a substance that increases the rate of a chemical reaction.

Celsius scale is a temperature scale used by the International System of Units (SI).

Centimeter (American spelling) or centimeter (UK spelling) is a unit of length in the metric system, equal to one-hundredth of a meter; centi being the SI prefix for a factor of 1/100. The centimeter was the base unit of length in the now deprecated centimeter–gram–second (CGS) system of units.

Chemiluminescence arises as a result of the oxidation of saturated and unsaturated hydrocarbons, polymers, lipids, and proteins (so-called oxychemiluminescence). Methods of the antioxidant analysis based on the luminescent measurement of the radical reaction rates are often used in the laboratory practice. On that, some investigation methods in chemistry and biology are based. One of the main sources of chemiluminescence is the interaction of two peroxy radicals. In addition to the oxidation induces the formation of high-energy cyclic intermediates, when splitting of which form electronically excited products, relaxing to the lowest-energy state, what is accompanied by the emission of a photon.

Choriocapillaris is the layer of choroid immediately adjacent to Bruch's membrane. Consist of thin endothelial cells with special permeable transport structures such as fenestra and transepithelial canals.

Choroid (choroidea, choroid coat) is the vascular layer of the eye, lying between the retina and the sclera – the outermost layer of the ocular bulb. The outermost layer of the choroid consisting of larger diameter blood vessels, below a layer of medium diameter blood vessels and lower layer of capillaries (choriocapillaris). The choroid provides oxygen and nourishment to the retina.

Clover incarnate is species *Trifolium incarnatum* L. of herbaceous plants from the family Fabaceae. Another name is clover meat-red because of the crimson coloring of flowers—annual plant. It is suitable also for harvesting hay, but only single hay mowing of that clover. During the phase of regrowth up to the period of budding — the beginning of the flowering of the incarnate clover, an intensive growth of the vegetative mass occurs, simultaneously the nodules on the roots actively develop. Therefore, the best period for the mowing of leguminous grasses is the phase of budding — the beginning of flowering.

Clover meadow (clover red, species *Trifolium pratense*) is a perennial, less often-biennial herbaceous plant from the genus *Trifolium*, the family of legumes (Fabaceae), subfamilies of the Moths (Faboideae), used in medicine. For medicinal purposes, inflorescences are collected with the top leaves. Do this throughout the summer.

Coenzyme Q_{10} (ubiquinone) is one of the substances from the homologous series of quinones predominantly contained in mitochondria of biological objects. It is lipophilic antioxidant capable of being synthesized by de novo. Almost every cell of a living organism contains Q_{10}. It plays an important role in energy metabolism, immunological competence, and antioxidant protection.

Concentration of hydrogen ions characterizes the solution acidity. It is quantitatively presented as pH being the negative base 10 logarithmic value of the H^+ concentration expressed in moles on one liter.

Cones are photoreceptors that are concentrated in the macula of the retina are responsible for central vision and color vision and perform best in medium and bright light.

Cytochrome *c* is a small heme protein. It is a one-electron transporter, loosely bound to the inner membrane of the mitochondria and a necessary component of the mitochondria respiratory chain. Under certain conditions, it can detach from the membrane, go into the cytoplasm, and activate apoptosis.

Dalton (symbols: u, or Da) is the unified atomic mass unit is a standard unit of mass that quantifies mass on an atomic or molecular scale (atomic mass). *One* unified atomic mass unit is approximately the mass of *one* nucleon (either a single proton or neutron) and is numerically equivalent to *1* g/mol.

Decarbonylation is the type of organic reaction that involves the loss of carbon monoxide (CO). It is often an undesirable reaction since it represents degradation. In the chemistry of metal carbonyls, decarbonylation describes a substitution process, whereby a CO ligand is replaced by another ligand.

Differential scanning calorimetry (DSC) is a thermoanalytical technique in which the difference in the amount of heat required to increase the temperature of a sample and reference is measured as a function of temperature. Both the sample and reference are maintained at nearly the same temperature throughout the experiment. The temperature program for a DSC analysis is designed such that the sample holder temperature increases linearly as a function of time. The reference sample should have a well-defined heat capacity over the range of temperatures to be scanned.

Dimyristoylphosphatidylcholine *(1,2-Dimyristoyl-sn-glycero-3-phosphocholine, DMPC)* is a synthetic phospholipid used for the formation of liposomes and lipid bilayers.

DPPH is a common abbreviation for the organic stable free radical diphenyl-1-picrylhydrazyl. It accepts an electron or hydrogen radical and becomes a stable diamagnetic molecule. The reduction capability of antioxidant can be determined on the decrease in its absorbance at 517 nm.

Duff reaction is the formylation reaction used in organic chemistry for the synthesis of benzaldehydes with (hexamethylenetetramine) hexamine as the formyl carbon source. It is named after James Cooper Duff, who was a chemist at the College of Technology, Birmingham, around 1920–1950. The electrophilic species in this electrophilic aromatic substitution reaction is the iminium ion CH_2+NR_2. The initial reaction product is an iminium, which is hydrolyzed to the aldehyde. The reaction requires electron strongly donating substituents on the aromatic ring such as in a phenol. Formylation occurs *ortho* to the electron donating substituent preferentially, unless the *ortho* positions are blocked, in which case the formylation occurs *para*-substituent.

Electron paramagnetic resonance (EPR) in spectroscopy is a method for studying materials with unpaired electrons. EPR spectroscopy is particularly useful for studying metal complexes or organic radicals.

Ellipsoid is the part of the inner segment of a photoreceptor that is closer to the outer segment. It is very rich in mitochondria, also contains intermediate filaments and microtubules. The ellipsoid continues into the myoid, the lower part of an inner segment which is rich in Golgi-apparatuses and endoplasmic reticulum.

Endoplasmic reticulum is a network of membranous tubules within the cytoplasm of a eukaryotic cell, continuous with the nuclear membrane. It usually has ribosomes attached and is involved in protein and lipid synthesis.

Erythrocytes (RBCs) are the red cells in the blood of the vertebrate organisms; they contain the hemoglobin which takes part in the transport of oxygen and carbon dioxide to and from the tissues.

Erythrocyte's ghosts are the envelopes of red cells, which are exempt from hemoglobin by the method of hypo-osmotic shock.

Fahrenheit is a unit of temperature, degree of Fahrenheit (designation: °F).

Fatty acids are a class of organic compounds that enter the body with natural fats of animal and vegetable origin. They usually contain an unbranched chain with an even number of carbon atoms (from 4 to 24, including carboxyl), are mono-basic carboxylic acids with a long aliphatic chain, either saturated or unsaturated. Fatty acids are contained as esters in fats, oils, and waxes of the vegetable and animal origin.

Fenestrae are specific structures of endothelial cells in the form of pores penetrating the endothelial wall. These pores are closed by a thin diaphragm. In choroid, they are as sites of transendothelial exchange between the blood plasma and the retina.

Fluorescence is the emission of light by a substance that has absorbed light or other electromagnetic radiation. It is a form of luminescence and occurs in gas, liquid or solid chemical systems. Fluorescence is brought about by absorption of photons in the singlet ground state promoted to a singlet-excited state. In most cases, the emitted lights have a longer wavelength, and therefore lower energy, than the absorbed radiation. λex and λem are the maximal optimums of light wavelength for excitation and emission of the experimental sample.

Foxtail millet (Green foxtal, Siberian millet, Setaria italic, "chemise" in Russian) is black rice, fodder and grain crop, annual plant of the family Cereals, or Poaceae. It is an annual plant of the genus bristle family Cereals, has straight, almost cylindrical form, from 25 cm to 2 m in height. Leaves broadly lanceolate, pointed, length 25–65 cm. The inflorescence is spicule panicle, propagated by seeds. The plant is thermophilic, and drought-resistant, used for grain objectives as well as in the form of green fodder or hay for livestock is. Chumise seeds are a treat for parrots.

Free radical chain oxidation is a complex multi-stage process, the mechanism of which usually includes much intermediate elementary reactions. Among the main stages in this kinetic consequence, there are the initiation, propagation, branching, and termination. In the dependence on the oxygen in the system, the oxidation reaction can proceed both in the kinetic and diffusional conditions. The initiation oxidation is produced by means of an initiator (the initiated oxidation) or without initiator (the autooxidation). In autooxidation conditions, the oxidation reaction is an auto-initiated process with the positive feedback, which is realized through hydroperoxide.

Free radicals having an uncoupled electron characterize the high reactivity. They are able to initiate a rapid chain reaction of oxidation of various substrates, leading to a modification of organic molecules and the degradation of supramolecular cellular structures.

Galega (*Galega officinalis*) is family Legumes (Fabaceae): goat medicinal, or goat pharmacy, or Kozya Ruta, perennial rhizome herbaceous plant; has medicinal properties, is widely used in non-traditional medicine. It has antioxidant activity and is capable of satisfying up to 60–90% of their nitrogen demand thanks to biological nitrogen fixation.

Gamma radiation (symbol γ) is one of the forms of the ionizing radiation due to the radioactive decay of atomic nuclei. In the laboratory investigations, $6°Co$ and ^{37}Cs radionuclides are used as sources of gamma radiation.

Gamma-resonance spectroscopy (Mössbauer spectroscopy) is a spectroscopic technique based on the Mössbauer effect. This effect and consists of the nearly recoil-free, resonant absorption and emission of gamma rays in solids.

Glutathione is a water-soluble tripeptide. It is capable of preventing damage to important cellular components caused by reactive oxygen species such as free radicals, peroxides, lipid peroxides, and heavy metals. Glutathione exists in both reduced (GSH) and oxidized states. GSH is known both as a substrate in conjugation reactions and the component of glutathione-dependent enzymes, namely glutathione transferase, glutathione reductase, and glutathione peroxidase.

Glutathione peroxidase (GPx) *is the general name of an enzyme family with peroxidase activity whose main biological role is to protect the organism from oxidative damage. The biochemical function of glutathione peroxidase is to reduce lipid hydroperoxides to their corresponding alcohols and to reduce free radicals.*

Golgi-apparatus is a complex of vesicles and folded membranes within the cytoplasm of most eukaryotic cells, involved in secretion and intracellular transport.

Hematho-retinal barrier is the part of eye barrier system providing homeostasis of the outer part of the retina. This is complex of structures, which include choriocapillaris of the choroid, Bruch's membrane and pigment epithelium of the retina.

Hemoglobin is an oxygen-transport of metalloprotein. Along with the protein part (globin), it contains the iron-porphyrin group (heme) and has a quaternary structure that is characteristic of many multi-subunit globular proteins. Hemoglobin is also involved in the transport of other gases, including respiratory carbon dioxide. It is in the red blood cells (erythrocytes) of almost at all vertebrates.

HEPES (4-(2-hydroxyethyl)-1-piperazineethanesulfonic acid) is a zwitterionic organic chemical buffering agent; it is one of the 20 Good's buffers. *HEPES is* widely used in cell culture, largely because it is better at maintaining physiological pH.

Hindered phenol is phenol, containing bulky (for example, *tert*-butyl) substituent at both *ortho*-positions.

Hormesis is any process in a cell or organism that exhibits a biphasic response to exposure to increasing amounts of a substance or condition. In practice, it is a phenomenon of dose-response relationships in which damaging factors (heavy metals, ionizing radiation, or chemical toxic agents) that produce harmful biological effects when they act at the dose range from moderate to high concentrations can produce beneficial effects at the low doses.

Hybrid macromolecular antioxidants are the synthetic hybrid macromolecular antioxidant, and it is the group of macromolecular conjugates of hydrophilic bio- and synthetic polymers with antioxidants; they were created for uniform distributions of antioxidants at biomembranes surfaces; they are preserve the bio-object from large local concentrations of antioxidants. Thus the antioxidant stress may be is prevented.

Hydrophilic derivative is the certain substance, which was synthesized by the modification of tested substance (phenosan acid at our case) with the main aim to receive the water-soluble substance. The antiradical activities of hybrid macromolecules antioxidants were assessed in reactions with 2,2-diphenyl-1-picrylhydrazyl and the corresponding sodium sulfonate in various solvents. The mechanism that explains the substantially enhanced activities of hybrid macromolecules antioxidants in water was proposed.

Hydrophobicity is the physical property of a molecule (known as a hydrophobe) that is seemingly repelled from a mass of water (strictly speaking, there is no repulsive force involved it is an absence of attraction). Surfaces can also possess hydrophobic properties.

Hydroxyquinone (hydroxybenzoquinone) is the organic compound $C_6H_4O_3$, which can be viewed as a derivative of a benzoquinone through replacement of one hydrogen atom by a hydroxyl group. The terms usually mean specifically the compound 2-hydroxy-1,4-benzoquinone, derived from 1,4-benzoquinoneor *para*-benzoquinone, which often called just "quinone."

In situ is a Latin phrase that means "on site." Authors used this phrase as "in the reaction mixture."

Induction period (symbol τ) in chemical kinetics is a time of initial slow stage in an oxidative reaction. After the ending of the induction period, the reaction accelerates. The duration of the induction period, it is determined by the alteration of values of the different parameters in the dependence on the used model system.

Inhibitors of radical reactions (InH in the kinetic scheme of the oxidation) are compounds which inhibit the radical reaction at low doses due to their interaction with free radicals or peroxides. Among inhibitors, such compounds as phenols, amines, flavonoids and other synthetic and natural substances are presented. As a rule, the efficiency of inhibitors essentially depends on many factors, including their concentration, the condition of oxidation, and the properties of the medium.

Intensity (symbol I) is characteristic of the intensity of the EPR-signal; the value of intensity is proportional of the concentration of free radicals and is measured in relative units (spin/g).

Interferons are a group of signaling proteins releasing by cells in response to the presence of several viruses. They are immunostimulators or immunomodulators characterizing the different biological activity including an antivirus and antitumor activities. Interferon-α is one of the investigated medications being the substance that allows the organism to protect oneself from many pathologies.

Intermediate products are the molecular compounds formed from the reactants or preceding intermediate products and react further to give rise to the directly observed products of a chemical reaction. As known, most chemical reactions are complex multi-stage processes. An intermediate product is the reaction product of each of these stages, except for the last one, which forms the final product. The live time of intermediate products is often short than causes their very seldom isolation and absence among the final product of the reaction.

Intrinsic fluorescence is in dependence from the presence of any fluorescent groups or composite parts at these molecules. Such fluorescence occurs when excitation of this substance without the using of exogenous fluorescent probes or fluorescent dyes.

Kelvin scale is a *Celsius* scale in thermodynamics, and absolute zero is the temperature at which entropy reaches *minimum* value. Absolute zero or absolute 0 K (0 degrees on the *Kelvin* scale, which is typically used for absolute values) equals −273.15° on the Celsius scale.

Lactadehydrogenase (LDH or LD) is an enzyme found almost in all living cells. As known, dehydrogenase is an enzyme that transfers hydride from one molecule to another. LDH catalyzes the conversion of lactate into pyruvic acid and back, because it converts NAD + to NADH and back. LDH exists in four different forms. In is predominantly contained in the cytoplasm of cells (soluble forms). However, LDH is reversibly associated with the membranes of the subcellular structures with the formation of complex, which is one of the ways of regulating its activity in a cell.

Lectins are carbohydrate-binding proteins, highly specific for sugar moieties. Lectins play numerous roles in biological recognition phenomena on the cellular and molecular levels and used for experimental recognition of the cells, carbohydrates, and proteins.

Legumes (*Legumes,* family Fabaceae) are dicotyledonous plants. Some representatives are important food products. Legumes contain components having an antioxidant activity. Grassy representatives of the family are able to fix atmospheric nitrogen and are the main plants used for land reclamation. This task is carried out with the help of nitrogen-fixing legumes plants. On the roots of legumes, crops are formed thickenings, called nodules. They contain soil bacteria of the genus *Rhizobium* (family Rhizobiaceae) and perform symbiotic nitrogen fixation. Nodule bacteria together with roots from the rhizosphere of leguminous plants, for example, alfalfa, red clover or Galega. They are capable of satisfying up to 60–90% of their nitrogen demand thanks to biological nitrogen fixation.

Lipid peroxidation (LPO) is a normal physiological process of the oxidative conversation of lipids. This is a complex multi-stage chain process. A main of the lipid substrate of oxidation is the polyunsaturated fatty acids of the membrane phospholipids that contain multiple double bonds. Peroxide oxidation involves the stages of the interaction of lipids with free-radical compounds that result in the formation of free radicals of the lipid nature. Among the chemical products of lipid peroxidation lipid peroxides are one of the first oxidation products. In norm stationery of LPO process is provided by the physicochemical system of regulation, functioning of which is experimentally established on the membrane, cellular and organ levels.

Lipofuscin is the fluorescent pigment that accumulates with age in the lysosomal compartment of postmitotic cells in several tissues. In the eyes, large quantities of lipofuscin accumulate in granules of the retinal pigment epithelium. The components of retinal lipofuscin are produced in the membranes of photoreceptor outer segments from non-enzymatic reactions of vitamin A aldehyde. This fluorescent material is transferred to retinal pigment epithelium within phagocytosed outer segment disks. The formation of lipofuscin may be responsible for the progressive deterioration of retinal pigment epithelium function that could lead to retinal degeneration.

Lipophilicity is the ability of a chemical compound to dissolve in fats, oils, lipids, and non-polar solvents such as hexane or toluene. These non-polar solvents are themselves lipophilic (translated as "fat-loving" or "fat-liking") is the axiom that "like dissolves like" generally holds true. Thus lipophilic substances tend to dissolve in other lipophilic substances, while hydrophilic ("water-loving") substances tend to dissolve in water and other hydrophilic substances.

Macula ("yellow spot") is a small central area of the retina. At the center of the macula is a small depression called the fovea, it contains only cone photoreceptors and is the point in the retina responsible for maximum visual acuity and color vision.

Magnesium chloride (MgCl$_2$) is the white crystal substance, which is soluble at water.

Melanin (from Greek: μέλας *melas*, "black, dark") is a broad term from a group of natural pigments found in the most organisms. Melanin is produced by the oxidation of amino acid tyrosine, followed by polymerization. There are three basic types of melanin: eumelanin, pheomelanin, and neuromelanin. The most common type is eumelanin, from which there are two types: brown eumelanin and black eumelanin.

Melanolipofuscin is a complex granule, exhibiting properties of both melanosomes and lipofuscin granules, which accumulates in retinal pigment epithelial cells and may contribute to the etiology of age-related macular degeneration.

Melanosome is an organelle found in animal cells and is the site for synthesis, storage, and transport of melanin, the most common light-absorbing pigment found in the organism of animals. Melanosomes are responsible for color and photoprotection in animal cells and tissues. They contain black or brown pigment granules.

Metallothioneins are a family of cysteine-rich, low-molecular-weight proteins (molecular mass from 500 to 14,000 Da). They are localized on the membrane of the Golgi apparatus. Metallothioneins have the ability to bind both heavy metals (zinc, copper, selenium) and xenobiotic (such as cadmium, mercury, silver, arsenic) through thiol groups of cysteine residues, which represent nearly 30% of its amino-acid moieties.

Methyl oleate oxidative model is one of the model systems for the estimation an ability both the individual substances and their mixture or cell components to inhibit the oxidation reaction. The thermal methyl oleate oxidation proceeds in the autooxidation condition at the low temperature (not higher 60°C).

Microsomes are vesicular small granules (up to 100 nm in size), obtained by fractional centrifugation of cellular homogenates and re-formed from pieces of the endoplasmic reticulum. They contain ribosomes, fragments of endoplasmic reticulum, RNA, and fragments of the protein synthesis enzymes. Microsomes are used to evaluate microsomal (monooxygenase) oxidation.

Milliliter (ml) is a one-thousandth quota of a liter.

Mitochondria are a double-membrane-bound spherical or ellipsoidal organelle. They are found in all eukaryotic organisms, have a diameter range on the size from 0.75 to 3 μm. The main functions of mitochondria in the cell are the oxidation of organic compounds and the use of energy released during their decay to generate electrical potential, the synthesis of ATP, and thermogenesis. These three processes are carried out due to the motion of electrons along the electron transport chain of proteins of the inner membrane. The number of mitochondria in the cells of different organisms varies considerably. Mitochondria are involved in other tasks, such as signaling, cell differentiation, and cell death, as well as in maintaining cell cycle control and cell growth.

Mitochondrion is the plural form of mitochondria.

Molar concentration, which is also called molarity, amount concentration or substance concentration is a measure of the concentration of a substance at solution as the amount of substance in moles per unit of volume (liter).

Myeloid bodies are the specific inclusions in retinal pigment epithelium cells of birds, reptiles, and amphibians. They are specific structural forms of the smooth endoplasmic reticulum and look like an assembly of numerous membranous lamellae that were formed by a flattened saccule of the paired membranes.

Myoid is the lower part of an inner segment, which is rich in Golgi-apparatuses and endoplasmic reticulum.

Nanometer (American spelling) or *nanometre* is a unit of length in the metric system, equal to 1 billionth of a meter (10^{-9} m). The name combines the SI prefix nano with the parent unit name meter.

Nitrogen-fixing bacteria are bacteria that fix nitrogen from the air. Among the nitrogen-fixing bacteria are isolate free-living in the soil and in the nodule on the roots of leguminous plants. Under the legumes, there is a microbiological accumulation of nitrogen available for plants. The activity of nodule bacteria that infect the cells of the roots of legumes is significantly more effective. Under the area of 1 hectare, sown by clover, can be accumulated 100 times more nitrogen than by the free-living fixatives of nitrogen.

Nuclear magnetic resonance spectroscopy is the most commonly known as NMR spectroscopy, is the research technique that exploits the magnetic properties of certain atomic nuclei. This type of spectroscopy determines the physical and chemical properties of atoms or the molecules in which they are contained. It relies on the phenomenon of nuclear magnetic resonance and can provide detailed information about the structure, dynamics, reaction state, and chemical environment of molecules. The intramolecular magnetic field around an atom in a molecule changes the resonance frequency, thus giving access to details of the electronic structure of a molecule and its individual functional groups.

Numerical density is the number of organelles per 100 μm of the cytoplasm.

"Scavengers" of free radicals are the substances interacting with hydroxyl free radicals with formation of the secondary radicals, which are not participant in continuation of a chain of free radical transformations.

Oil droplets are the drops of oil that are found at some animals in photoreceptor cells of the eye. They are especially common in the eyes of birds, where are found in cone cells, suggesting a role in color vision. They are located on the cone inner segment, where they intercept and filter light before it can pass through to the cone outer segment where the visual pigment is. Some oil droplets are colored, while others appear colorless. The content of droplets is neutral lipids and carotenoid pigments.

Oil-bearing crops are plants used to produce fatty oils. They are used in the food industry. This includes the winter false flax (*Camelina silvestris*). The most common oil crops of Russia are rape, mustard, soybean, castor oil, sunflower and other, for example, perilla, lallemantia, sesame, safflower, peanuts, and many others. Processing of oilseeds is one of the main directions of the raw materials industry.

Ommatidium is a cluster of photoreceptor cells surrounded by support cells and pigment cells.

Ommochrome (or visual pigment) represents several pigments that occur in the eyes of crustaceans and insects. The eye color is determined by the ommochromes.

Oxidative equivalents are vacancies for acceptance of electrons in a molecule which is in the oxidized form.

Oxidative stress is a process of the cell damage as a result to oxidation. It reflects an imbalance between the intensification of the reactive oxygen species action and an ability of the biological system to readily detoxify the reactive intermediates or to repair the resulting damage.

Oxycarotenoids are yellow pigments that occur widely in nature and form one of two major groups of carotenoids. They are extremely efficient quenchers of singlet oxygen and often react with free radicals.

***Para*-aminobenzoic acid** (*p*-aminobenzoic acid – PABA) is 4-aminino-2-hydro-oxybezoic acid that has molecular mass 137.1; it is an organic compound with the formula $H_2NC_6H_4CO_2H$; a white-gray crystalline substance, is only slightly soluble in water, soluble in hot water (80–90°C), well soluble in benzene, ethanol, and acetic acid; it is classified as non-toxic vitamin-like compound of group B, also known as vitamin H_1 or vitamin B_{10}; microorganisms use it as a precursor in the folic acid synthesis. It takes part in the synthesis of purines and pyrimidines, ultimately in the synthesis of DNA and RNA. PABA as a chemical compound is known since 1863, but its high biological activity is established only at 20th century. In dependence of concentrations, PABA has of the antimicrobial, promoter and antioxidant properties, and also can decrease an action of harmful mutagens.

Permafrost is the upper part of the earth's crust, frozen deposits of minerals, the temperature of which does not rise above 0°C for a long time (from 2–3 years to millennia).

Peroxidases are a group of oxidation-reduction enzymes of the class of oxidoreductases, which use hydrogen peroxide (H_2O_2) as an electron acceptor. Peroxidase reacts similar to catalase, but the reaction catalyzed is the oxidation of a wide variety of organic and inorganic substrates by hydrogen peroxide. They are subdivided into two superfamilies: plant peroxidases and animal peroxidases. For many of these enzymes, the optimal substrate is hydrogen peroxide, but others are more active with organic hydroperoxides such as lipid peroxides. Peroxidases can contain a heme as a cofactor in their active sites.

Peroxisomes are intracellular organelles containing enzyme peroxidase at high local concentration.

Phagosome is a vacuole formed in the process of phagocytosis, inside which there are substrates to be digested. Retinal pigment epithelium cells' phagosomes contain shedding ends of photoreceptor outer segments.

Phase transition is the most commonly used term to describe a transition between the different physical states of matter as well as the structural arrangements of biomembranes when a temperature is changed. For example, it may be transition from gel phase to liquid crystal phase in the lipid component of biomembranes or under the denaturation of proteins which are associated with biomembranes.

Phenosan acid (3,5-ditretbutyl-4-oxiphenyl propionic acid) is a highly effective biologically active substances with a wide range of therapeutic effects. It is also used as base for the synthesis of other chemical and biologically active substances.

Phospholipids are a class of lipids which are one of a major component of all cellular and subcellular membranes. The structure of phospholipid molecule consists of two hydrophobic fatty acid "tails" and a hydrophilic "head" consisting of a phosphate group. Two components are joined together by a glycerol and sphingosine molecules. Consequently, they can form bilayer because of their amphiphilic properties. The phosphate group can be modified with simple organic groups such as choline, ethanolamine, inositol, or serine. Phospholipids take an active part in a regulation of the cell metabolism. They are predominantly synthesized in liver and small intestine.

Photodissociation photolysis or photodecomposition is the chemical reaction in which a chemical compound is broken down by photons. It is defined as the interaction of one or more photons with one target molecule. Photodissociation is not limited to visible light. Any photon with sufficient energy can affect the chemical bonds of a chemical compound. Since a photon's energy is inversely proportional to its wavelength, electromagnetic waves with the energy of visible light or higher, such as ultraviolet light, X-rays and gamma rays are usually involved in such reactions.

Photoreceptor cell consists of the outer segment with photosensitive membrane discs, the inner segment with organelles and the cell's nucleus and synaptic terminal.

Photoreceptor outer segments (POS) are exposed to constant photo-oxidative stress, and as a result, they are destroyed. They are constantly renewed by shedding the ends and then retinal pigment epithelium phagocytes and digest these segments.

Phytohormones (plant hormones) are represented by five classes: auxins, cytokinins, gibberellins, abscisic acid. In agronomy, they are used to enhance the growth of seedlings or roots, and for advance along the stages of plant ontogeny.

Pigments of the eye are the part of intracellular organelles melanosomes carrying out the function of optical protection of photosensitive cells from excess light by its absorption. These pigments possess antioxidant activity.

Polymers synthetic are polyvinylpyrrolidone, polyacrylic acid, polyvinyl alcohol, and others.

Potassium chloride (KCl) is the white crystal substance, which is soluble at the water.

Potassium dihydroortophosphate (KH_2PO_4) is the white crystal substance, which is soluble at the water.

Potassium salt of phenosan (potassium phenosan) (1-β-4-hydroxy-3,5-di-*tert*-butylphenyl-1-propionate potassium) synthesized as an antioxidant for polymer stabilization in Emanuel Institute of Chemical Physics of Academy of Sciences is a hindered phenol. Its molecular formula $C_{23}H_{25}NO_4$, and molecular weight 379.453.

It is not well solubilized at aqua solutions but very well solubilized at lipids or another organic material. It is white crystalline powder with the melting temperature 85°C and dissolved in acetone, benzene, toluene, diethyl ether, and heptane. Potassium phenosan possesses a wide spectrum of biological activities and also an ability to inhibit the lipid peroxidation processes and to induce the structural rearrangements in biomembranes.

Pro-oxidants are substances accelerating the oxidation reaction. They can induce oxidative stress, either by generating reactive oxygen species or by inhibiting antioxidant systems. Some substances can be both antioxidants and pro-oxidants, depending on conditions of oxidation and their concentration.

Psychrophilic organisms are extremophilic organisms that are capable of growth and reproduction in low temperatures, ranging from −20°C to +10°C. They are found in regions that are permanently cold, such as the Polar Regions and the deep sea.

Quasi-reductive state is dynamic redox state of natural water at which dissolved oxygen is present in water (in the thermodynamic sense, water is in the oxidative state); however, instead of hydrogen peroxide, as a carrier of reactive oxidative equivalents in water, water-soluble substances of reduced nature are found. They effectively interact with H_2O_2, but inert to the oxidative action of O_2.

Quaternization is the conversion of a *tertiary* amine into a quaternary ammonium salt by reaction with an alkyl halide or acid.

Quinone is the class of organic compounds that are formally "derived from aromatic compounds such as benzene or naphthalene by conversion of an even number of –CH= groups into –C(=O)– groups with any necessary rearrangement of double bonds," resulting to "a fully conjugated cyclic dione structure." The class includes some heterocyclic compounds.

Radioresistance is a level of ionizing radiation which organisms are able to withstand.

Reaction inhibitor is the substance that decreases substantially the reaction rate or prevents completely a chemical reaction.

Reactive oxygen species (ROS) are oxygen-containing particles and molecules accelerating the oxidation reactions. Among ROS, there are superoxide anion-radical and hydroxyl radicals, singlet oxygen, peroxides of both inorganic and organic origin. At present, ROS are considered by initiators of oxidative stress.

Redox reaction is a chemical reaction in which the oxidation states of atoms are changed, involves both a reduction process and a complementary oxidation process, involved with electron transfer processes. Redox reactions involve the transfer of electrons between chemical species. The chemical species from which the electron is stripped is said to have been oxidized, while the chemical species to which the electron is added is said to have been reduced. Oxidation is the loss of electrons or an increase in oxidation state by a molecule, atom, or ion. Reduction is the gain of electrons or a decrease in oxidation state by a molecule, atom, or ion.

Reducing agent (also called a reductant or reducer) is the element (such as calcium) or compound that loses (or "donates") an electron to another chemical species in a redox chemical reaction. Since the reducing agent is losing electrons, it is said to have been oxidized.

Regioselectivity is the preference of one direction of chemical bond making or breaking over all other possible directions. It can often apply to which of many possible positions a reagent will affect, such as which proton a strong base will abstract from an organic molecule, or were on a substituted benzene ring a further substituent will add.

Relative volume is the volume occupied by the organelles per unit volume of the cytoplasm.

Retina is the inner coat of the eye and is composed of several layers of neurons interconnected by synapses, including one lay that contains specialized light-sensitive cells called photoreceptors.

Retinal pigment epithelium (RPE) is the pigmented cell layer of retina, just outside the neurosensory layer that nourishes retinal visual cells, and is firmly attached to the underlying choroid and overlying retinal visual cells. The RPE has several functions, namely, light absorption by melanin granules, epithelial transport, spatial ion buffering, visual cycle, phagocytosis of photoreceptor outer segments, secretion, and immune modulation.

Rods are photoreceptors detecting motion, providing black-and-white vision and functioning well in low light. These are located throughout the retina. These structures directly related to the process of photoreceptor outer segment shedding.

Selen is the chemical element of the 16th group, the 4th period in the periodic system of Mendeleyev has an atomic number of 34, is denoted by the symbol Se (Latin Selenium), a brittle shining black, at the fracture (stable allotropic shape, unstable form – cinnabar-red) is no metal. Refers to chalcogens.

Semiquinone is the free radical resulting from the removal of one hydrogen atom with its electron during the process of dehydrogenation of a hydroquinone, such as hydroquinone itself or catechol, to a quinone or alternatively the addition of a single H atom to a quinone. It is highly unstable.

Sodium chloride (NaCl) is the white crystal substance, which is soluble at water.

Sodium selenate (Na_2SeO_4) is an inorganic compound, an alkali metal salt of sodium and selenic acid, colorless crystals, soluble in water. In the latest inventions is used for the activation of seeds of cereals and oilseeds. In the latest inventions is used for stimulating the growth of plants and seeds of cereals and oilseeds. Sodium selenate has antioxidant qualities and is used in plant growing, veterinary medicine and medicine as a mineral additive. In the latest inventions is used for stimulating the growth of plants and seeds of cereals and oilseeds.

Sofora (*Sophora japonica*) is a large deciduous tree of the legume family, reaching 25–30 meters in height. He has a powerful root system, branched stems and a wide crown. It contains a large number of alkaloids, fatty oil, rutin, the roots contain phenolics. Kaempferol, quercetin, flavonoids, organic acids, and vitamin C are distinguished from different parts of the plant. It is used in medicine.

Solid-phase synthesis is the method in which molecules are bound on a bead and synthesized step-by-step in a reactant solution; compared with normal synthesis in a liquid state, and it is easier to remove excess reactant or byproduct from the product. In this method, building blocks are protected at all reactive functional groups. The two functional groups that are able to participate in the desired reaction between building blocks in the solution and on the bead can be controlled by the order of deprotection. This method is used for the synthesis of peptides, deoxyribonucleic acid, and other molecules that need to be synthesized in a certain alignment. More recently, this method has also been used in combinatorial chemistry and other synthetic applications.

Solubility is the solubility of a substance fundamentally depends on the physical and chemical properties of the solute and solvent as well as on temperature, pressure, and the pH of the solution.

Solvation is the interaction of solvent with molecules or ions in a solution. Ions, and to some cases, molecules, interact strongly with solvent. Solvation is the process of reorganizing solvent and solute molecules into solvation complexes. Solvation involves bond formation, hydrogen bonding, and van der Waals forces.

Sphingomyelin (SM) is a type of sphingolipid of membranes in the myelin sheath of axons. Sphingomyelin is one of the more poorly oxidizable phospholipids because of high saturation of its fatty acids. It plays an important role the regulation of the cell metabolism.

Stinging nettle (*Urtíca dioica* L.) from family Urticaceae. It is a perennial herbaceous plant 60–170 cm high, densely covered with burning hairs. Other names are: a burning spot, a bonfire, a coffin, a jalousie, a stalking, a flyover. Seed processing by this method significantly increased the number of sprouted seeds, energy germination of seeding, having initially low sowing qualities. It is spread everywhere like a weed.

Super-oxidizing states are the state of an aquatic environment with an abnormally high oxidizing ability. Rather long-living carriers of reactive oxidizing equivalents in natural waters are metastable microcolloid particles of manganese with the mixed valency Mn (III, IV).

Superoxide anion radical (hyperoxide) is one of the reactive oxygen species which occurs widely in nature. It is particularly important as the product of the one-electron reduction of dioxygen O_2.

Superoxide dismutase belongs to the group of the antioxidant defense enzymes. It alternately catalyzes the dismutation (or partitioning) of the superoxide anion radicals (O_2^-) in either ordinary molecular oxygen (O_2) or hydrogen peroxide (H_2O_2).

Synaptosome contain synaptic vesicles, presynaptic membranes, synaptic complexes detached as a result of fractional centrifugation from axons, and nerve endings, are used for studies by the subcellular fraction of the nervous system. With osmotic shock the cytoplasm and intracellular organelles are separated from them, thus obtaining preparations enriched with presynaptic and/or postsynaptic membranes can be obtained.

Thin-layer chromatography is performed on a sheet of glass, plastic, or aluminum foil, which is coated with a thin layer of adsorbent material. This layer of adsorbent is known as the stationary phase. After the sample has been applied on the plate, a solvent or solvent mixture (known as the mobile phase) is drawn up the plate via capillary action. Different substances in mixture ascend the TLC plate at different rates. The mobile phase is different from the stationary phase.

Tmax is the main characteristic of DSC melting. The maximal microdomains at experimental sample change of its phase station at this temperature. Tmax is measured at peak top on the melting curve. It is a general assumption that if a *temperature maximum* is observed in the melting region, then the compound melts congruently.

Tocopherols are a group of organic chemical compounds, many of which have vitamin E activity. Tocopherols, especially α-tocopherol (vitamin E), are natural lipid-soluble antioxidants which are located within the bilayers of cell membranes and protect them against reactive oxygen species. The term vitamin E includes two families of compounds, tocopherols, and tocotrienols, which have similar structure and differ from each other only in their saturation.

Transendothelial channels are present in the wall of fenestrated type endothelial cells also in choroid, they are patent pores spanning the endothelial cell body from lumen to albumen and are provided with two-stomach diaphragm (one luminal and one abluminal). Along with fenestrae, transendothelial channels occur in the attenuated part of the endothelial cell near Bruch's membrane.

Transesterification is the process of exchanging the organic group R″ of an ester with the organic group R′ of an alcohol.

Tryptophan emission is the one type of intrinsic fluorescence, when the tryptophan residues are at protein composition, and then the emission was occurred.

Two-electronic innerspheric charge transfer is the process proceeding in the coordination sphere of an ion of metal of variable valency, which is followed by transfer of a couple of electrons between coordinate molecules of the donor and an acceptor.

Ubiquinone: see Coenzyme Q_{10}.

Vitamin A is a group of unsaturated organic compounds that includes retinol, retinal, retinoic acid, and several provitamin A carotenoids (most notably beta-carotene). Vitamin A is needed for the retina of the eye. Vitamin A also functions in the role as retinoic acid (an irreversibly oxidized form of retinol), which is an important hormone-like growth factor for epithelial and other cells.

Vitamin K (phylloquinone) is a group of structurally similar, lipid-soluble vitamins. It can protect the organism against reactive oxygen species. It participates in the blood coagulation and in controlling binding of calcium in bones and other tissues.

Winter false flax is a plant of the genus *Camelina*, species *Camelina silvestris* (L.) (Granz, 1762). It is cultivated as oil-bearing crops, and it is a valuable Shrovetide culture of winter sowing. It is characterized by cold resistance (seeds are able to germinate at +1°C, sprouts can withstand frosts to −10°C), has satisfactory drought resistance. A short vegetation period (225–230 days) of the redhead allows the effective use of spring moisture reserves. Potential productivity reaches 3.0 t/ha with an oil content of 40–42% in seeds. Oil cake, obtained after extraction from the seeds of oil, contains a high protein food. The oil is used for food and for technical purposes.

Index